Advanced Technology of Horticulture

The Editors

Dr. Manoj Kumar Singh (1972) is working as Associate Professor (Horticulture), College of Horticulture, Sardar Vallabhbhai Patel University of Agriculture and Technology, Meerut (UP). He did his Ph.D. in 2002 from Chaudhary Charan Singh University, Meerut (UP). He has more than 25 years of teaching and research experience. He has published 70 Research Papers, 90 Popular Articles, 12 Bulletin and 100 Abstract. He has guided 03 Doctoral and 19 Master students Dr. Singh was awarded Young Scientist Award (2006), Excellence in Teaching Award (2018) and HTHS Gold Medal Award (2019) by various Professional Societies in field of Horticulture. He has handled 02 Projects as PI.

Mr. Khursheed Alam is pursuing his Ph.D. in Vegetable Science from Sardar Vallabhbhai Patel University of Agriculture and Technology, Meerut (U.P.). He has qualified ICAR-ASRB NET in Vegetable Science during 2021. He has completed his B.Sc. (Hons.) Horticulture in 2018 from CSA University of Agriculture and Technology, Kanpur. He did his M.Sc. (Ag) Horticulture in 2020 from SVPUA and T, Meerut. He has published 25 Research Paper, 02 Review Paper, 30 Popular Articles, 05 Book Chapters, 02 Book and 35 Abstract. Mr. Alam was awarded Young Achiever Award, Best Research Scholar Award and Best Master Thesis Award by Various Professional Society.

Mr. Vibhu Pandey is pursuing his Ph.D. in Fruit Science from Sardar Vallabhbhai Patel University of Agriculture and Technology, Meerut. He did his B.Sc. (Horticulture) from Narendra Dev University of Agriculture and Technology, Kumarganj, Ayodhya during 2013-2017. He has completed his M.Sc. (Ag.) Horticulture from Sardar Vallabhbhai Patel University of Agriculture and Technology, Meerut in 2020. He has attended several National and International conferences. He has published 1 Book, 1 Practical Manual, 10 research papers, 5 Book Chapter and 15 Popular articles.

Ms. Shalini Singh is presently pursuing her Ph.D. in Fruit Science from College of Horticulture, Sardar Vallabhbhai Patel University of Agriculture and Technology, Meerut (U.P.). She did her Graduation in Agriculture and Post Graduation in Agriculture with a major in Horticulture from Institute of Agricultural Sciences, Banaras Hindu University. Shalini Singh has also qualified ASRB NET and UGC NET. She has 7 publications (1 research paper, 2 review paper, 1 book chapter and 3 popular articles) and has attended various International and National conferences/seminars.

Dr. Mukesh Kumar is working as a Professor and HOD in the Department of Floriculture and Landscaping, College of Horticulture, Sardar Vallabhbhai Patel University of Agriculture and Technology, Meerut, UP, India. Dr. Kumar has published more than 240 research papers in National as well as International repute Journals, three books, eight book chapters, 40 popular articles and 80 abstracts. He has received an international grant (World bank) under National Agriculture Technology Project and worked as a visiting scholar at Michigan State University, East Lansing, Michigan, USA during the year 2010.

Advanced Technology of Horticulture

— Editors —

Dr. Manoj Kumar Singh

Mr. Khursheed Alam

Mr. Vibhu Pandey

Ms. Shalini Singh

Dr. Mukesh Kumar

2023

Daya Publishing House®

A Division of

Astral International Pvt. Ltd.

New Delhi – 110 002

ISBN: 9789359191980

Publisher's Note:

Every possible effort has been made to ensure that the information contained in this book is accurate at the time of going to press and the publisher and author cannot accept responsibility for any errors or omissions, however caused. No responsibility for loss or damage occasioned to any person acting, or refraining from action, as a result of the material in this publication can be accepted by the editor, the publisher or the author. The Publisher is not associated with any product or vendor mentioned in the book. The contents of this work are intended to further general scientific research, understanding and discussion only. Readers should consult with a specialist where appropriate.

Every effort has been made to trace the owners of copyright material used in this book, if any. The author and the publisher will be grateful for any omission brought to their notice for acknowledgement in the future editions of the book.

Published by : **Daya Publishing House®**
A Division of
Astral International Pvt. Ltd.
– ISO 9001:2015 Certified Company –
4736/23, Ansari Road, Darya Ganj
New Delhi-110 002
Ph. 011-43549197, 8130496929
E-mail: info@astralint.com
Website: www.astralint.com

Printed at : **Replika Press Pvt Ltd**

सरदार वल्लभभाई पटेल कृषि एवं प्रौद्योगिक विश्वविद्यालय, मेरठ 250110
SARDAR VALLABHBHAI PATEL UNIVERSITY OF AGRI. & TECH.,MEERUT 250110

डा० के० के० सिंह
कुलपति
Dr. K.K. Singh
Vice Chancellor

Phone : 0121-2888522 (O)
0121-2888566 (R)
Web : svbpmeerut.ac.in
E-mail : vc@svpuat.edu.in

FOREWORD

Horticulture has been studied and practiced for a long time. Horticulture aided in the transition of human populations from nomadic to sedentary or semi-sedentary lifestyles. Horticulture is divided into several categories which focus on the newly adopted cultivation practices, post-harvest management and value addition of different horticultural produces for specific purposes. In order to conserve the science of horticulture, multiple organizations worldwide educate, encourage and promote the advancement of horticulture through hi-tech horticultural techniques. Horticulture encompasses 8.5% of arable land accounting about 27.48 million hectares with an annual production 334.60 million tones during 2020-21. Horticulture provides a dynamic tool for enhancing economic returns, attaining nutritional security, generation of employment and ensuring ecological sustainability. The sector covers a broad spectrum of activities like production, post-harvest management, marketing, processing and export of fruits, vegetables, flowers, medicinal and aromatic plants, plantation crops, spices etc.

India is the one of the largest producer of fruits and vegetables next to China. Despite the above positive indicators, performance of the sector in value addition, processing and export are not commensurate with its progress on the production front. In spite of having a 12% share of the global production of both fruits and vegetables, just 2.2% is processed. The per capita availability of fruits and vegetables is also very low leading to nutritional insecurity, ill health, and loss in working hours. Now-a-days, the demand quality planting materials is much needed to obtain high yield with high quality produce. On the other hand, increase water use efficiency in horticultural crops, living and productive soils, use of advance technologies like high-tech nursery, precision horticulture, protected cultivation of vegetable, fruit and commercially important ornamental crops etc.

The present book "Advanced Technology of Horticulture" by the authors is a welcome effort. It is an excellent up-to-date compilation of all major and relevant technologies in field of horticulture. This book would impart relevant knowledge and will be of immense value for students, teachers, scientists, progressive farmers and all those who are concerned with promotion of horticulture.

I personally congratulate and appreciate efforts of the authors to put together information on "Advanced Technology of Horticulture", planning the contents and bringing out this book. I am sure that this publication will be very useful to all, who are concerned with horticulture.

(K. K. Singh)

Preface

Horticulture is said to be the backbone of agriculture. Due to the excessively broad scope of this modern technology, the students were required to read numerous books, absorb their information, and write down pertinent notes. This activity is frequently difficult, and when there aren't enough books or time, it's actually impossible. In light of these circumstances, we made an effort to produce a book that would be appropriate for exam preparation and revision while upholding academic rigour and faculty accuracy. According to their respective technology and levels of importance, each chapter in the book had been appropriately covered. This book satisfies the demand for a quality reference-cum edited book among students.

I would like to extend my sincere gratitude to all the authors for their contributions to the material selection, compilation, and editing. Every effort has been taken to assure the accuracy of the data gathered from a variety of sources, including people, organisations, institutions, and publications. We expect that this book will be a helpful resource for educators, professionals, and learners seeking employment in ICAR Institutes, State Agricultural Universities, State Agriculture/Horticulture Departments, as well as for competitions like State Public Services/UPSC/ARS/JRF/SRF/NET along with entrance examinations for B.Sc., M.Sc., and Ph.D. Degree in Horticulture and also very useful for progressive farmers, horticultural extension workers, and boost farmer's incomes.

We believe that the readers will benefit from this effort in their endeavours. Any ideas for making the book more educational are always appreciated.

Editors

Contents

Chapter 1

Advance Horticulture: Present Status and Future Prospectus

Manuj Awasthi[1] * *and Jaidev Chauhan*[2]

[1]*Teaching Associate, Department of Fruit Science, Chandra Shekhar Azad University of Agriculture and Technology, Kanpur – 208 002, Uttar Pradesh*

[2]*Research Scholar, Department of Plant Physiology (HAPPRC), Hemvati Nandan Bahuguna Garhwal University (A Central University), Srinagar Garhwal – 246 174 Uttarakhand*

**e-mail: manujawasthi24@gmail.com*

INTRODUCTION

Horticulture is a branch of agriculture which deals with the cultivation of fruits, vegetables, ornamentals and flower crops, spices and condiments, plantation crops, medicinal and aromatic plants *etc.* Horticulture can be defined as the branch of agriculture concerned with intensively cultured plants directly used as sources of food and also for medicinal and aesthetic purposes. The horticulture sector is considered one of the leading sectors of agriculture in various parts of the world. Guaranteed plant production and world trade make more and more demands, not only quantitatively and qualitatively, but also legal quarantine requirements and respect for the environment. The fast and world-wide spread of plant identities, owing to modern commercial management, also introduces new geo-phyto pathological interactions, and requires strong phytosanitary control. Horticulture is the main industry to improve food security for rapidly increasing population. China and India are the world largest producer of fruits and vegetable. More than 70 per cent of the world cut flowers are grown by Netherlands, Columbia and Kenya.

Horticulture Status in India

India is rich in diversity of horticultural crops, large number of fruits, vegetable, spices, floricultural and ornamentals, medicinal and aromatic plants are distributed in various phyto-geographical regions of India. The Horticulture Statistics Division in the Department of Agriculture, Cooperation and Farmers' Welfare has taken various initiatives to improve the database of horticulture crops. The total horticulture production is estimated to be 342.33 million tons in the year 2021-22, showing an increase of about 7.73 million tons (an increase of 2.3 per cent) over the year 2020-21. The area under horticulture crops has increased to 28.08 million hectares in 2021-22 from 27.47 million hectares in 2020-21 (Anonymous 2022. MoA and FW. horticulturedatabase.2022). India is the second largest producer of fruits and vegetables in the world with first rank in the production of Banana, Mango, Lime and Lemon, Papaya and Okra.

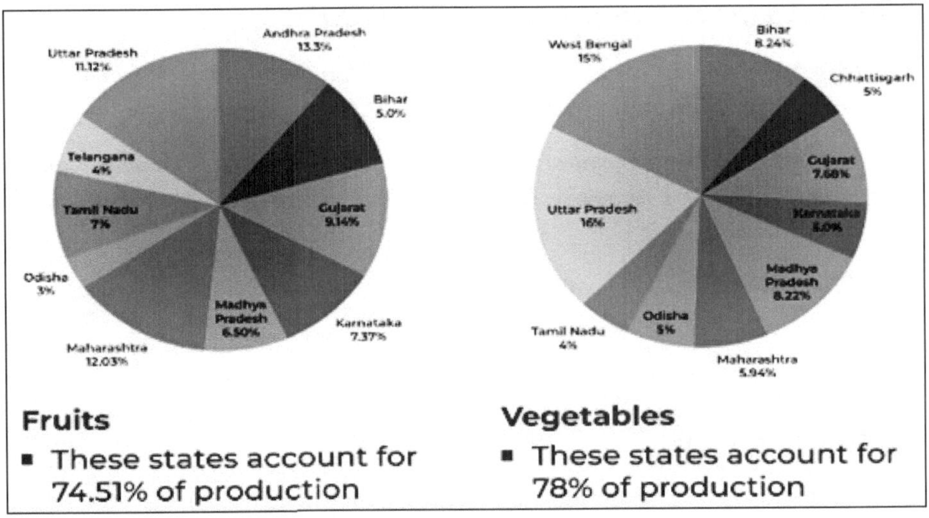

Figure 1.1
(*Source*: DoA and CFW horticulturedatabase.2022).

Mission for Integrated Development of Horticulture (MIDH) is being implemented by adopting an end-to-end approach for increasing production of horticulture crops and reducing postharvest losses. As a part of the initiatives taken by the government, the Mission for Integrated Development of Horticulture (MIDH) was implemented in 2014. It is a centrally sponsored scheme with a focus on devising plans and processes to facilitate horticulture production efficiently and cost-effectively.

MIDH is the Centrally Sponsored Scheme for the holistic growth of the horticulture sector covering fruits, vegetables and other areas. Under MIDH, Government of India contributes 60 per cent of the total outlay for developmental programs in all the states (except North Eastern and Himalayan states where GOI contributes 90 per cent) and 40 per cent is contributed by State governments.

It has five major schemes on horticulture:

- ☆ National Horticulture Mission (NHM)
- ☆ Horticulture Mission for North East and Himalayan States (HMNEH)
- ☆ National Horticulture Board (NHB)
- ☆ Coconut Development Board (CDB)
- ☆ Central Institute of Horticulture (CIH), Nagaland

The main objectives of this scheme are to ensure the overall growth of the horticulture sector through region-specific strategies and action plans. The various activities involve research, technology adaptation and application, harvest management, transportation, storage, processing, and marketing. It takes into account the diverse agro-climatic features of different regions and focuses on improving productivity with the use of quality planting materials and advanced irrigation facilities. The scheme encourages skill development and participation of local youths in horticulture production that generates employment opportunities. The government of India sponsors 60 per cent of the total developmental expenditure incurred by different states and 40 per cent is contributed by state governments. For North-Indian states, the government contribution is 90 per cent. The farmers involved in horticulture production are aggregated into groups to achieve efficiency in terms of volume and cost.

Table 1.1: Area and Production of Horticultural Crops (First dip in output since 2014-15)

Year	*Area (000' Mha)*	*Production (000' MT)*
2014-15	23.4	281
2015-16	24.5	286
2016-17	24.9	300
2017-18	25.4	310
2018-19	25.7	311
2019-20	2.65	320
2020-21	27.5	334
2021-22	27.6	333
2022 (up to 3rd advance estimate)	28.08	342

Source: DoA and CFW horticulture database. 2022.

Government Contribution to Horticulture Development in The Different States (as per MIDH):

- ☆ Union Territories – 100 per cent

- North Indian States – 90 per cent
- Other States -60 per cent

The production of fruits and vegetables has overcome the production of food grains in the country.

Modern Day Horticulture

1. Protected Horticulture

Protected horticulture refers to the cultivation of crops under controlled conditions. Commercial protected horticulture appeared first in Northern Europe in the early decades of the last century and developed broadly after World War II. After the advent of plastics in early 1960's, green house crops started to move to mild winter regions such as Mediterranean Basin countries (in particular Italy, Spain, Morocco). The rise in oil prices in the 1970's which increased the heating costs further enhanced the diffusion of crops in Southern countries. More recently, protected crops expanded to some Asian countries such as India, Korea and especially China (Jiang *et al.*, 2004).

Some objectives and advantages are described below:

(a) **Higher yields:** Protected cultivation, in its various forms, efficiently uses natural resources of land, water, sunlight, and atmospheric carbon dioxide. Simultaneously, yields may be increased several folds over those in open field conditions.

(b) **Early season crop:** Fruit crops like strawberry can be obtained early in season by providing desirable temperature, sunlight intensity and carbon dioxide required for growth and development through protected cultivation.

(c) **Round the year propagation of planting material:** Plant material can be developed throughout the year as the environmental conditions are controlled.

(d) **Protection from frost and wind:** Plants are grown under various structures like low tunnels, high tunnels, green houses, *etc.* These protect crops from high wind velocities and frost.

(e) **Minimized cases of insects and diseases:** Protected cultivation ensures minimum incidence of insects and diseases as compared to open field conditions.

(f) **Minimized cases of insects and diseases:** Protected cultivation ensures minimum incidence of insects and diseases as compared to open field conditions.

(g) **Highly useful in biotechnology:** Structures in protected cultivation are highly employed for hardening of tissue cultured plants.

(h) **Stabilizing market for high-quality horticultural crops:** Protected cultivation circumvents many of the hazards of climate and weather and greatly increases the chances of availability of high value horticultural crops like fruits, vegetables and flowers in early season and off season and thus stabilizing the market for these crops

2. Precision Horticulture

Precision horticulture is a science whereby growers use technology to fine-tune crop production practices. It helps growers determine exactly when and what to plant and what kind of inputs to use, how to control the environment, lighting, irrigation and nutrition for optimum yield and monitor how crops are growing, by greenhouse bay, zone and bench. The equipment and systems growers use are all part of the grand scheme of Precision Horticulture, helping growers ultimately do their jobs as effectively and efficiently as possible. From pot fillers, seeders and robotic transplanters to automated irrigation, lighting technology and environmental control systems, all the way to the software that provides inventory control and space management capabilities, 3D printers, RFID benches, cart readers and more, using Precision Horticulture, growers can optimize production processes from start to finish, to reduce costs and improve profits.

3. Hydroponics and Aeroponic Produce

In hydroponics, plants may be suspended in water full-time or fed by an intermittent flow of water. Aeroponic plants are never placed into water, instead being given nutrients from a mist that's sprayed onto their roots. This key difference is what gives each method their unique name. The term 'Aeroponics' stems from the Greek word *aero*, meaning air, and *ponos*, meaning work. Hydroponics on the other hand, uses the Greek term *hydro*, meaning water, and is translated to the term "working water'! Another major distinction between Aeroponics and Hydroponics lies in the design and methodology of the farming setup. Aeroponic farms are completely controlled and enclosed environments, with crops growing suspended in air. The plants are contained in a humid atmosphere where frequent misting delivers a nutrient rich solution right to the roots, keeping the crop from drying out. Because the entire process is enclosed, the mist is able to remain around the plants for longer, helping them grow much quicker than in a traditional outdoor farm. Hydroponic farming also utilizes nutrient-rich water, but the process of delivering these nutrients is different. Instead of spraying the solution on to the roots, the plants are either grown in containers of nutrient solution, or the solution is circulated past the roots.

4. Organic Horticulture

Organic horticulture is the science and art of growing fruits, vegetables, flowers, or ornamental plants by following the essential principles of organic agriculture in soil building and conservation, pest management, and heirloom variety preservation. Organic horticulture is a system which avoids or largely excludes the use of synthetic inputs (such as fertilizers, pesticides, hormones, feed additives *etc.*) and to the maximum extent feasible rely upon crop rotations, crop residues, animal manures, off-farm organic waste, mineral grade rock additives and biological system of nutrient mobilization and plant protection

5. Peri-urban Horticulture

The world urban population is expected to double in 30 years leading to a growing number of urban poor. The urban population expansion is more pronounced in developing countries as result of the immigration from rural areas, as people flock to the cities in search of food, employment and security. The trend is accelerating, and by the year 2030, it is expected that about 60 per cent of the world's population will be living in cities. Such rapid urbanization and the harsh reality of urban poverty require ad hoc strategies to ensure adequate food supply and distribution systems to address escalating levels of urban food insecurity. Besides the growing demand for food, there is a rapid increase of poverty, unemployment, hunger and malnutrition in the urban and peri-urban environment around the world - issues that are of great concern to central and municipal authorities. Within this reality, urban and peri-urban agriculture/horticulture (UPA/UPH) has become a key component of the survival strategies of poorer sections of the population while also providing a significant contribution to the urban fresh food supply chain. UPA/UPH is also a source of employment and income, and has the potential to improve the nutrition of disadvantaged urban residents. A key challenge is developing policy, strategies and technical support mechanisms ad hoc for the sustainable management of urban and peri-urban agricultural systems, addressing production issues and marketing needs within a broader framework of environmental planning and management, water supply and utilization schemes, and food safety assurance.

6. Vegetable Grafting

For many, when we hear about grafted plants we may think of trees. Grafting has been used in tree fruit and nut production for many years, with great success. Though the practice of grafting is a relatively new technique in the vegetable world, it has been utilized for centuries (Kubota, *et al.*, 2017). Originating in Asia, grafting was done on gourds to increase the root system and size of gourd fruit (Lee and Oda, 2003). This fact sheet will provide a brief overview of the benefits of using grafted vegetables along with the grafting procedure. Vegetable production with grafted seedlings was originated in Japan and Korea to avoid the serious crop loss caused by infection of soil-borne diseases aggravated by successive cropping. This practice is now rapidly spreading and expanding over the world. Vegetable grafting has been safely adapted for the production of organic as well as environmentally friendly produce and minimize uptake of undesirable agrochemical residues (Lee *et al.*, 2010). Grafting of watermelon onto bottle gourd enhances its tolerance to soil-borne diseases (Heidari *et al.*, 2010) and reduces Fusarium wilt (Rivard and Louws, 2008). The major vegetable crops being grafted are: tomato, cucumber, eggplant, melon, pepper and watermelon (Nichols, 2007). Grafting is also very much useful for increasing yield for example as much as 106 per cent with the use of certain rootstocks for watermelon production in Australia. Some rootstock varieties have been bred specifically to be used as rootstocks, such as the Maxifort rootstock used in greenhouse tomato production systems. Use of vigorous rootstock varieties can increase water and nutrient uptake in grafted plants. Many growers worldwide are utilizing these rootstocks to increase fruit yields, even where little disease pressure is evident.

The main objective of grafting is to avoid soil borne diseases such as Fusarium wilt in Cucurbitaceae (cucumber, melon *etc.*) and bacterial wilt in Solanaceae (tomato, eggplant and pepper *etc.*). Expanding the use of resistant rootstocks in grafting in combination with Integrated Pest Management (IPM) practice, may help to reduce the need for soil fumigation with methyl bromide for many crops. This may prove boon in organic farming of vegetables. Further, continuous cropping is in vogue in greenhouse which results reduced yield and quality of the produce. An estimated loss of 68 per cent in vegetable yield caused by soil borne diseases under continuous cropping was reported by Takahashi 1984. In brinjal yield is very low due to numerous diseases and parasites, in particular to *Ralstonia solanacearum*, *Fusarium* wilt, *Verticellium* wilt and Bacterial wilt, nematodes and several insect pests (Collonnier *et al.*, 2001). Among all, bacterial wilt is extreme disease expressed rapid wilting by yellowing of foliage followed by collapse of entire plant. Soil treatment with chemicals and resistant sources were used so far but due to residual nature of chemicals only alternative left is resistant sources. The number of wild relatives of Solanum species was resistant and graft compatible to eggplant. Grafting of vegetable crops is a simple method of propagation in which preferred rootstocks are used to improve vigour, precocity, enhanced yield and quality, better survival under abiotic and biotic stress conditions (Pandey and Rai, 2003). Kumar *et al.* (2017) reported that among all rootstocks used in grafting, *Solanum torvum* was found best rootstock followed by *Solanum khasianum* and promising for resistance towards bacterial wilt, whereas *Solanum surathense* and Solanum *xanthocarpum* showed maximum susceptible reaction against bacterial wilt infection among all grafted plants. The non-grafted control plants showed highly susceptible compare to grafted ones. The results recommended that eggplant could be grafted on *Solanum torvum* and *Solanum khasianum* for graft compatibility controlling bacterial wilt in north eastern region of the country.

Advantages of Vegetable Grafting

The use and demand for grafted vegetables have been growing worldwide over the last few decades. Advantages of using grafted vegetables include better resistance to pathogens, drought and other environmental stresses, more vigorous growth, and higher yield. These advantages also allow for fewer inputs such as pesticide applications and an extended harvest season. The need for grafted plants may also add a revenue source for growers wanting to sell.

How it Works

A grafted plant consists of two parts: a scion and a rootstock with each being a different plant. The scion provides the shoot and the rootstock provides the root system. The point of attachment is called the union, which is the point where the vascular bundles of the two plants fuse together and become one plant. A scion may be chosen for its fruit quality and the rootstock for its vigor, resistance to pathogens, and environmental stresses.

How to Begin

Once a scion and rootstock are chosen for grafting, they must be close to the same stem diameter when grafted. This allows for the cambium of the plants to be well aligned and thus join successfully. Considering this, a grower must know the rate of germination and development of the scion and rootstock to allow for stem diameters to match. Another factor in choosing plants to graft is whether to use the same species or the same family of plants. An example of the same species would be grafting a tomato scion onto a tomato rootstock. For the same family of plants, a watermelon scion may be grafted onto a squash rootstock.

Before grafting, both scion and rootstock should be watered up to twenty-four hours in advance, twelve hours minimum. Watering immediately prior to grafting should be avoided if possible. Cuts should be made with a clean, sharp blade. A new double edge razor blade split in half is generally used for making cuts, though there are many tools available to use for making the cuts. Plants are ready to graft when they have two true leaves, make sure to cut the rootstock below the cotyledons to discourage new shoot growth. Persons performing the grafting should make sure their grafting environment as well as themselves are as clean as possible. Make sure to wash hands thoroughly or use surgical-style gloves.

7. Biotechnological Interventions

Use of microorganisms like bacteria, viruses, fungi, yeast, animal for specific uses.

Importance of Horticultural Crops

1. Genetic Transformation

Li-feasible has developed a powerful plant transformation platform to generate genetically modified plants (also known as transgenic plants). We provide both transient and genetically stable transformation services in a wide range of plant species, including model plants, major food crops, economic plants, and new-era bio-energy feedstock plants (please see our plant species list for details). Plant transformation refers to the transfer and incorporation of engineered genes or plasmids into the plant genome. It is not only an essential biotechnological approach for in-depth research of plant growth, but also a promising tool for modern agriculture, specifically for the introduction and selection of desired crop traits. Some of the most popular plant transformation methods include gene gun delivery, thermal/electric shock, microinjection, virus/*Agrobacterium*/Polyethylene Glycol (PEG)-mediated transformation, protoplast fusion, as well as silicon carbide fibers mediated transformation. Which method to choose is dependent on the target species and specific study purposes. Comparing to conventional plant breeding methods, the plant molecular engineering approach is more efficient, as it allows targeted modifications of plant growth. Therefore, the application of genetic modification techniques in crop plants empowers the production of crops with desired features, such as increased productivity, higher nutritional value, drought resistance, disease resistance, and so on.

Experimental Process

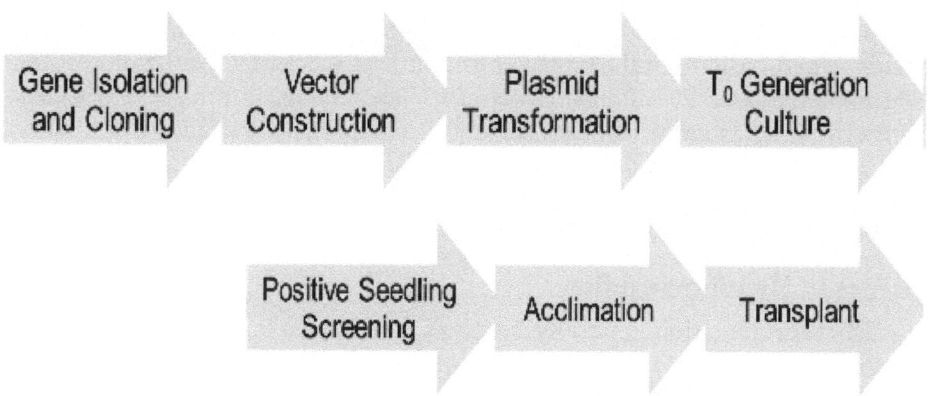

Li-feasible has established a specialized molecular biology laboratory with cutting-edge equipment's, excellent experts, as well as close collaboration with renowned institutions. We proudly offer worldwide customers with high quality and short turnaround services in plant transformation. We devote ourselves to be the leader in plant engineering and crop improvement to ensure that each project is given the attention it needs.

Materials Provided by Clients

Objective gene sequences or readily made vectors, names of target species.

Service Process

- ☆ Communicate with clients.
- ☆ Design experiments.
- ☆ Sign contracts and confidentiality agreements with clients.

- ☆ Update project status.
- ☆ Final reports with organized data and figures.

2.. Micropropagation

Multiplication of genetically identical copies of a cultivar by asexual reproduction is called clonal propagation. In nature, clonal propagation occurs by apomixis (seed development without meiosis and fertilization) and/or vegetative propagation (regeneration of new plants from vegetative parts). Tissue culture has become popular method for vegetative propagation of plants. Aseptic method of clonal propagation is called as Micropropagation and it offer the advantage of large number of true-to-type plantlets can be produced with relatively short time and space from a single individual. It is the fact that micropropagation is the only commercially viable method of clonal propagation of most of the horticultural crops. *e.g.*, Orchids.

Explants Used in Micropropagation

Different kinds of explants were used in micropropagation. For example, in case of orchids, shoot tip (*Anacamptis pyramidalis, Aranthera, Calanthe, Dendrobium*), axillary bud (*Aranda, Brassocattleya, Cattleya, Laelia*), inflorescence segment (*Aranda, Ascofinetia, Neostylis, Vascostylis*), lateral bud (*Cattleya, Rhynocostylis gigantean*), leaf base (*Cattleya*), leaf tip (*Cattleya, Epidendrum*), shoot tip (*Cymbidium, Dendrobium, Odontioda, Odontonia*), nodal segment (*Dendrobium*), flower stalk segment (*Dendrobium, Phalaenopsis*) and root tips (*Neottia, Vanilla*) are being used in micropropagation.

Stages in Micropropagation

Micropropagation generally involves five stages. Each stage has its own requirements.

Stage 0: Preparative Stage

This stage involves the preparation of mother plants to provide quality explants for better establishment of aseptic cultures in stage I. To reduce the contamination problem in the subsequent stages, mother plant should be grown in a glasshouse and watered so as to avoid overhead irrigation. This will also reduce the need for a harsh sterilization treatment. Stage 0 also includes exposing the stock plants to suitable light, temperature and growth regulator treatments to improve the quality of explants. In the case of photosensitive plants, it may be possible to obtain suitable explants throughout the year by controlling photoperiod in the glasshouse. For example, red-light treated plants of Petunia provided leaf explants which produced up to three times as many shoots as did the explants from untreated plants.

Stage I. Initiation of Culture

1. **Explant:** The nature of explant to be used for in vitro propagation is

governed by the method of shoot multiplication. For enhanced axillary branching, only the explants which carry a pre-formed vegetative bud are suitable. When the objective is to produce virus-free plants from an infected individual it becomes necessary to start with sub-millimeter shoot tips. If the stock is virus-tested or virus eradication is not necessary, then the most suitable explant is nodal cuttings. Small shoot-tip explants have a low survival rate and show slow initial growth. Meristem tip culture may also result in the loss of certain horticultural characteristics which are controlled by the presence of virus, such as the clear-vein character of the Geranium cv. Crocodile. Generally, the clear vein character is transmitted in petiole-segment culture but not in shoot-tip culture.

2. **Sterilization:** Special precautions need to be taken when explants are derived from field-grown materials, which is often necessary in cloning an elite tree. In such cases an ideal approach would be to take cuttings from the selected plant and grow them in greenhouse. Discarding the surface tissues from plant materials while preparing the explants also minimizes the loss of cultures due to microbial contamination.
3. **Browning of medium:** A serious problem with the culture of some plant species is the oxidation of phenolic compounds leached out from the cut surface of the explant. It turns the medium dark brown and is often toxic to the tissues. This problem is common with the adult tissues from woody species.

Stage II. Multiplication

This is the most crucial stage since it is the point at which most of failures in micropropagation occur. Broadly three approaches have been followed to achieve in vitro multiplication.

1. **Through callusing:** The potentiality of plant cells to multiply indefinitely in cultures and their totipotent nature permit a very rapid multiplication of several plant types. Differentiation of plants from cultured cells may occur via shoot-root formation (organogenesis) or somatic embryogenesis. Somatic embryogenesis is most appealing from a commercial angle. A somatic embryogenesis system once established lends itself to better control than organogenesis. Since somatic embryos are bipolar structures, with defined root and shoot meristems, the rooting stage required for micro shoots gets eliminated. Above all, somatic embryos being small, uniform and bipolar are more amenable to automation at the multiplication stage and for field planting as synthetic seeds, offering cost advantages from labour savings, can also be stored through cold storage, cryopreservation or desiccation for prolonged periods. These characteristics make somatic embryogenesis potentially a less expensive and flexible system for

micropropagation. The most serious objection against the use of callus cultures for shoot multiplication is the genetic instability of their cells.

2. **Adventitious bud formation:** Buds arising from any place other than leaf axil or the shoot apex are termed adventitious buds. The shoots differentiated from Calli should also be treated as adventitious buds. In many crops, vegetative propagation through adventitious bud formation from root (blackberry, raspberry) and leaf (Begonia, Crassula) cuttings is standard horticultural practice. In such cases the rate of adventitious bud development can be considerably enhanced under culture conditions. For most bulbous plants (*e.g.*, Lilley) adventitious bud formation is the most important mode of multiplication and the best explants are obtained from bulb scales. A serious problem may arise when this method of propagation is applied to varieties which are genetic chimeras. Adventitious bud formation involves the risk of splitting the chimeras leading to pure type plants. For example, in variegated geranium cv. Mme *Salleron*, the chimera is perpetuated in meristem culture but broken down in petiole culture.
3. **Enhanced axillary branching:** In cultures the rate of shoot multiplication by enhanced axillary branching can be substantially enhanced by growing shoots in a medium containing a suitable cytokinin at an appropriate concentration with or without auxin. Due to continuous availability of cytokinin, the shoots formed by the bud, present on the explant, develops axillary buds which may grow directly into shoots. This process may be repeated several times and the initial explant transformed into a mass of branches.

Stage III. Rooting of Shoots

Somatic embryos carry a pre-formed radical and may develop directly into plantlet. However, these embryos often show very poor conversion into plantlets, especially under in vitro conditions. They require an additional step of maturation to acquire the capability for normal germination. Adventitious and axillary shoots developed in cultures in the presence of a cytokinin generally lack roots. To obtain full plants the shoots must be transferred to a rooting medium which is different from the shoot multiplication medium, particularly in its hormonal and salt compositions. For rooting, individual shoots measuring 2.0 cm in length are excised and transferred to the rooting medium.

Stage IV. Hardening of Plants

The ultimate success of commercial propagation depends on the ability to transfer plants out of culture on a large scale, at low cost and with high survival rates. The plants multiplied in vitro are exposed to a unique set of growth conditions (high levels of inorganic and organic nutrients, growth regulators, sucrose as carbon source, high humidity, low light, poor gaseous exchange) which may support rapid growth and multiplication but also induce structural and physiological abnormalities in the plants, rendering them unfit for survival under in vivo conditions. The two main deficiencies of in vitro grown plants are; poor control of water loss and heterotrophic mode of nutrition. Therefore, gradual acclimatization is necessary for these plants to survive transition from culture to the greenhouse or field. During acclimatization the invitro formed leaves do not recover but the plant develops normal leaves and functional roots. While transferring out shoots/roots their lower part is gently washed to remove the medium sticking to them. The individual shoots or plantlets are then transferred to potting mix and irrigated with low concentration of inorganic nutrients. This probably recommissions the photosynthetic machinery of plants, enabling them to withstand the subsequent reduction in the ambient relative humidity and survive under field conditions. A variety of potting mixtures such as peat, perlite, polystyrene beads, vermiculate, fine bark, coarse sand *etc.* or their mixtures in different combinations are used for transplantation. For initial 10-15 days, it is essential to maintain high humidity (90-100 per cent) around the plants, to which they got adapted during culture. The humidity is gradually reduced to ambient level over a period of 2-4 weeks.

Advantages of Micropropagation

1. Clonal mass propagation - extremely large numbers of plants can be produced. Rather than getting 10,000 plants per year from an initial cutting in vegetative propagation, one can obtain more than 10,00,000 plants per year from one initial explant through micropropagation.
2. Culture is initialized from small parts of plants: - so no need of much space: from 1.0 m^2 space in culture room, 20,000 – 10,0000 plants can be produced per year.
3. Production of disease and virus free plantlets: This leads to simplification of international exchange of plants.
4. Micropropagation enables growers to increase the production of plants that normally propagate very slowly such as Narcissus and other bulbous crops.
5. Introduction of disease-free new cultivars is possible through micropropagation.
6. Vegetative propagation of sterile hybrids can be used as parent plants for seed production. *e.g.* Cabbage.
7. One of the rapid methods for cloning of disease-free trees.
8. In vitro cultures can be stored for long time through cryopreservation.
9. Breeding cycle can be shortened.

Disadvantages of Micropropagation

1. Expensive laboratory equipment and service.

2. No possibility of using mechanization.
3. Plants are not autotrophic.
4. Poor Acclimatization to the field is a common problem (hyperhydricity).
5. Risk of genetic changes if 'de novo' regeneration is used.
6. Mass propagation cannot be done with all crops to date. In cereals much less success is achieved.
7. Regeneration is often not possible, especially with adult woody plant material.
8. More problems in inducing rooting.
9. May not get uniform growth of original plant from tissue culture. Each explant has different in vitro growth rates and maturation.

Thus, can't be used for floriculture crop production where uniformity is critical.

Horticultural uses for plant tissue culture:

1. **Clonal mass propagation:** The important point here is that extremely large numbers of plants can be produced. Rather than getting 10,000 plants per year from an initial cutting, one can obtain upwards of 10,00,000 plants per year from one initial explant.
2. **Difficult or slow to propagate plants:** Micropropagation enables growers to increase the production of plants that normally propagate very slowly such as narcissus and other bulbous crops.
3. **Introduction of new cultivars** *e.g.* **Dutch iris:** Get 5 daughter bulbs annually. Takes 10 years for commercial quantities of new cultivars to be built up. Can get 100-1000 bulbs per stem section.
4. **Vegetative propagation of sterile hybrids used as parent plants for seed production**: *e.g.*, cabbage.
5. **Pathology:** Eliminate viruses, bacteria, fungi *etc.* Use heat treatment and meristem culture. Used routinely for potatoes, carnation, mum, geranium, garlic, gypsophila
6. **Storage of germplasm**: Generally, the only successful method to date is keeping them in refrigerator. Slows down, but does not eliminate, alterations in genotype.

3. *In vitro* Conservation of Germplasm

- ☆ In vitro method is an advanced technology of *Ex-situ* conservation for the preservation of genetic materials.
- ☆ In vitro methods employing shoots, meristems and embryos are ideally suitable for the preservation of germplasm of vegetatively propagated plants.

- ☆ This approach can also preserve plants with recalcitrant seeds and genetically engineered materials along with orthodox plants.
- ☆ The conservation implies preservation of cells, calluses or tissues of selected plant species in sealed test tubes in in vitro method.
- ☆ This conservation depends on principle that plant cells are totipotent and plant materials can be kept alive for infinite period of time as in vitro cultures.

Advantages

- ☆ Requires less area for preservation of large quantities of materials.
- ☆ The germplasms are preserved in pathogen-free environment.
- ☆ Genetic materials are protected against the nature's hazards.
- ☆ Large number of plants can be obtained from the germplasm stock whenever required.
- ☆ Since the germplasm is kept under aseptic conditions, it can be easily transported.

Disadvantage

- ☆ It requires constant electricity, skilled manpower and high technology.

Techniques for the *In vitro* Conservation of Germplasm

- **I.** Cryopreservation (freeze-preservation),
- **II.** Cold storage, and
- **III.** Low-pressure and low-oxygen storage.

I. Cryopreservation

- ☆ Cryopreservation (Greek, krayos-frost) actually means preservation in the frozen state.
- ☆ Cryopreservation is based on the principle where the metabolism or division of plant cells and tissue cultures are completely halted by reducing temperatures with the involvement of cryoprotectants.
- ☆ This approach is highly applicable for the conservation of plant species in danger of extinction.

  Cryopreservation broadly refers to the storage of germplasm at very low temperatures:

  - **i.** Over solid CO_2 (at -79°C)
  - **ii.** Low temperature deep freezers (at -80°C)
  - **iii.** In vapour phase nitrogen (at -150°C)
  - **iv.** In liquid N_2 (at -196°C)

- Out of these, the most commonly used cryopreservation is use of liquid nitrogen.
- At the temperature of liquid nitrogen (-196°C), the cells remain in totally inactive state and thus can be preserved for long time.
- Cryopreservation has been applied for germplasm conservation of several horticultural species *e.g.*, fruits and tuber crop species.
- Various plants can be regenerated from cells, meristems and embryos stored in cryopreservation.

II. Cold Storage

- Germplasm conservation at a low and non-freezing temperatures (1-9°C) is involved in this stage.
- In contrast to complete stoppage of cryopreservation, the growth of the plant material is slowed down in cold storage.
- Thus, cold storage is considered as a slow growth germplasm conservation method.
- This approach avoids the cryogenic injuries of plant material.
- This step is simple, economical and results germplasm with good survival rate.
- This approach stores many in vitro developed shoots/plants of fruit tree species for *e.g.*, grape plants, strawberry plants.
- With the addition of a few drops of medium periodically (once in 2-3 months), virus- free strawberry plants could be conserved at 10°C for about 6 years.

III. Low-Pressure and Low-Oxygen Storage

- Low-pressure storage (LPS) and low-oxygen storage (LOS) are other alternatives for the cryopreservation and cold storage which have been developed for germplasm conservation.

Low-Pressure Storage (LPS)

- The atmospheric pressure surrounding the plant material is decreased in low pressure storage.
- This yields in a partial reduction of the pressure exerted by the gases around the germplasm.
- The lowering of partial pressure decreases the in vitro growth of plants (of organized or unorganized tissues).
- Low-pressure storage systems are essential for both short and long-term storage of plant materials.

- ☆ The short-term storage is specifically useful to enhance the shelf life of many plant materials *e.g.*, fruits, vegetables, cut flowers, plant cuttings.
- ☆ The storage of germplasm grown in cultures can be done for long term under low pressure.
- ☆ Besides germplasm preservation, LPS decreases the activity of pathogenic organisms and prevents spore germination in the plant culture systems.

Low-Oxygen Storage (LOS)

- ☆ In the low-oxygen storage, the oxygen concentration is decreased, but the atmospheric pressure (260 mm Hg) is maintained by the addition of inert gases (specifically nitrogen).
- ☆ There is reduction in plant tissue growth if the partial pressure of oxygen is below 50 mm Hg.
- ☆ It is because, with less availability of O_2, the production of CO_2 is low.
- ☆ As a result, the photosynthetic activity is decreased, thereby halting the plant tissue growth and dimension.

Limitations

- ☆ The long-term conservation of plant materials by low-oxygen storage may halt the plant growth after certain dimensions.

8. Synthetic Seed Technology

Synthetic seed can be defined as the artificial encapsulation of somatic embryo, shoot tips or any tissues which has the ability to form a plant in *in-vitro* or *ex-vitro* condition. Synthetic seeds can be stored for a long time in appropriate condition. Synthetic seed production and used technology is rapidly growing branch of seed biotechnology. Synthetic seed or Encapsulation or artificial seed are used to provide protection to the artificially produced propagules. Encapsulation techniques are used to produce artificial or synthetic seeds of species belonging to angiosperm and gymnosperm families.

Synthetic seed production is requiring the large-scale production of viable plant material in laboratory condition using *in vitro* culture system.

Production of Synthetic Seed

- ☆ Explant is selected from choice of plant.
- ☆ In a laboratory using tissue culture techniques callus is induced in the explants.
- ☆ Using tissue-culture techniques somatic embryo is induced in the callus.
- ☆ Somatic embryos are proliferated.

- ✰ Histodifferentiation and maturation of somatic embryo.
- ✰ *In vitro* germination of transported to field for germination.

Based on the techniques two types of synthetic seeds are produced such as desiccated and hydrated.

Desiccated Synthetic Seeds

- ✰ Desiccated synthetic seeds are produced naked or poly-oxyethylene glycol encapsulated somatic embryo.
- ✰ This type of synthetic seeds is produced on desiccation tolerance species of plant.

Hydrated Synthetic Seeds

- ✰ Hydrated synthetic seeds are produced by encapsulating the somatic embryos in hydrogels like sodium alginate, potassium alginate, carrageenan sodium pectate or sodium alginate with gelatin.

Application of Synthetic Seeds

- ✰ Hybrid plants can be easily propagated using synthetic seed technology.
- ✰ Generally modified plant can be propagated using synthetic seed technology.
- ✰ Endangered species can be propagated using this technology.
- ✰ Elite genotypes can be preserved and propagated using artificial seed technology.
- ✰ Synthetic seed production is cost effective when compared to traditional method.
- ✰ Synthetic seeds can be directly used in fields.
- ✰ Synthetic seeds can be transported from one country to another without obligations from quarantine department.
- ✰ Fruits and medicinal plants can be studied anywhere in the world using synthetic seeds.
- ✰ Synthetic seeds transportation is easy as these do not contain any disease-causing agents. As synthetic seeds are produced using sterile plant materials produced using plant tissue culture techniques.
- ✰ Synthetic seeds are small therefore they are easy to handle.
- ✰ Synthetic seed encapsulation provides aseptic condition to the plant materials or explant, which is present inside the capsule.
- ✰ While producing the synthetic seeds encapsulation herbicides can be added to the formulation. This herbicide will provide extra protection to the explants against pests and diseases.

- ☆ Synthetic seeds plantation can be done by using the sowing farm machinery.
- ☆ Synthetic seed crops are easy to maintain because of uniform genetic constituent.
- ☆ This technology improves the food production and also produces environment friendly plantation.

9. Viruses Cleaning

Production of virus-free plants is necessary to control viral diseases, import novel cultivars from other countries, exchange breeding materials between countries or regions and preserve plant germplasm. In vitro techniques represent the most successful approaches for virus eradication. In vitro thermotherapy-based methods, including combining thermotherapy with shoot tip culture, chemotherapy, micrografting or shoot tip cryotherapy, have been successfully established for efficient eradication of various viruses from almost all of the most economically important crops. Virus diseases cause great losses of crop yield and have long been a constraint for sustainable developments of agricultural production. For example, potato leafroll virus (PLRV), potato virus-S (PVS), potato virus-X(PVX) and potato virus-Y (PVY) are among the most serious viruses attacking potato. Single infection caused yield losses of 40-60 per cent by PLRV, 10-20 per cent by PVS, 10-50 per cent by PVX and 20-50 per cent by PVY. Mixed infection with two viruses resulted in a much larger loss of yield than the single infection. Plum pox virus (PPV), one of the most serious viral

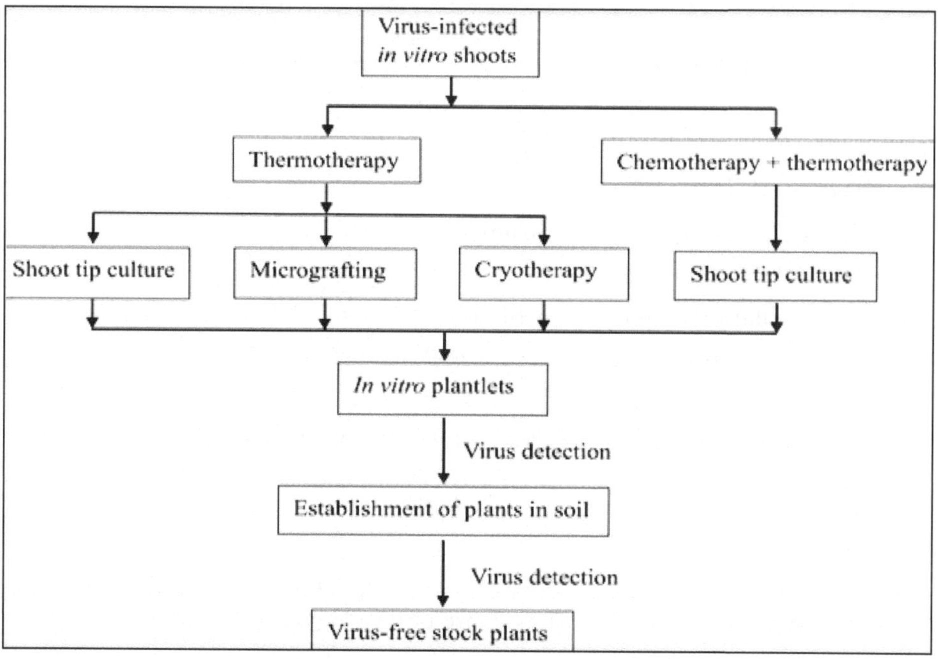

diseases attacking *Prunus* fruit trees, widely occurred in almost all stone fruit producing countries. *In-vitro* culture techniques represent the most successful strategies for production of virus-free plants. So far, various methods have been established for eradication of plant viruses, including shoot tip culture (also called meristem culture), micrografting, chemotherapy, thermotherapy and shoot tip cryotherapy.

10. Biofertilizers and Biopesticides

The large population of a specific or a group of beneficial microorganisms for enhancing the productivity of soil. Either by fixing atmosphere nitrogen or by solubilizing soil phosphorus is called biofertilizers.

Types of Biofertilizers

1. ***For Nitrogen***

   Rhizobium for legume crops

   Azotobactor/Azospirillum for non-legume crops

2. ***For Phosphorus***

   Rhizobium, Azotobactor, Azospirillum and *Acetobacter*

Bio-compost

- ☆ Eco-friendly organic fertilizer.
- ☆ Prepared from the sugar industry waste material.
- ☆ Consists of nitrogen, phosphate solubilizing bacteria and various useful fungi like decomposing fungi, bacteria, *Trichoderma virida* which protects the plants from various soil borne disease.
- ☆ Increase soil fertility which results to a good quality product to the farmers.

Advantages of Biofertilizers

- ☆ Renewable source of nutrients
- ☆ Sustain soil health
- ☆ Supplement chemical fertilizers
- ☆ Increase the grain yields by 10-40 per cent
- ☆ Decompose plant residues, and stabilize C: N ratio of soil
- ☆ Improve texture, structure and water holding capacity of soil.
- ☆ No advance effect on plant growth and soil fertility.

Biopesticides

- ☆ Biopesticides are biochemical pesticides that are naturally occurring substances that control pests by nontoxic mechanisms.

- ☆ Biopesticides are sustain types of pesticides derived from such materials as animals, plants, bacteria and certain minerals.
- ☆ All the living organisms which are cultivated in the laboratory on large scale and are used and exploited experimentally for the control of harmful organism.

Three Major Classes of Pesticides

1. **Microbial pesticides:** consist of a microorganism (*e.g.*, a bacterium, fungus or protozoa) as the active ingredient. Microbial pesticides can control many different kinds of pests, although each separate active ingredient is relatively specific for its target pest. For example, there are fungi that control certain weeds, and other fungi that kill specific insects.
2. **Plant pesticides:** are pesticidal substances that plants produce from genetic material that has been added to the plant. For example, scientists can take the gene for the Bt. pesticidal protein and introduce the gene into the plant's own genetic material.
3. **Biochemical pesticides:** are naturally occurring substances that control pests by non-toxic mechanisms. Conventional pesticides, by contrast, are generally synthetic materials that directly kill or inactivate the pest. Biochemical pesticides include substances, such as insect sex pheromones, that interfere with mating, as well as various scented plant extracts that attract insect pests to traps.

The Advantages of Using Bio-pesticides

- ☆ Biopesticides are usually inherently less toxic than conventional pesticides.
- ☆ Cheap, renewable can be handled safely.
- ☆ Difficult for insects to develop resistance to these pesticides.

11. High-health Nursery

Nursery is defined as an area where plants are raised for eventual planting out. It comprises of nursery beds, paths irrigated channels *etc.* Nursery bed is defined as a prepared area in a nursery where seed is sown or into which seedlings or cuttings are raised. On the bases of kind of plants growing in them nursery beds are classified into seedling beds and transplant beds, seedlings, beds are those nursery beds in which seedlings are raised either for, transplanting in other beds or for planting out. A nursery which has only seedling beds *i.e.*, in which seedlings are only raised for transplanting is called seedlings nursery. Transplant beds are those nursery beds in which seedlings raised in seedling beds are transplanted before planting out in forest. A nursery that has only transplant beds *i.e.* in which seedlings are transplanted in preparation for forest planting is called transplant nursery. In India separate seedling and transplant nurseries are seldom made in the same nursery. Generally, whatever is grown in nursery for planting out is called nursery stock. The aim of good nursery management is to provide planting material of the highest possible quality for new development areas and replanting. This aim is of the greatest important as the areas planted are likely to have a productive life span of 25 years or more. Poor planting materials will lead to low yield and unnecessary thinning cost top rid of runts in planted field. So, the selection of good planting materials and strict culling in nursery are the important step. The importance of the best quality planting material as an initial investment is a well realized factor for persons engaged in horticulture field. So, nurseries have great demand for the production of plants, bulbs, rhizomes, suckers and grafts. But in general, good quality and assured planting material at reasonable price is not available. So, persons having a skill of propagation of plants can go for this avenue as an agro-business of future. Seedling production is a major expense of afforestation and every effort should be made to produce good quality seedlings at a reasonable cost. To this end mastering the techniques of nursery operations is essential means high tech nursery management is very essential. (Mbora *et al.*, 2008).

The main suppliers of perennial tree seedlings are the departmental/ government and industrial nurseries. They are producing seedlings and vegetative propagules to meet their own seedling demand and also supply them to public to meet their raw material demand. Mostly the vegetable and ornamental seedlings are produced by the farmers themselves, due to the market availability of improved seed and requirement of minimum inputs to establish them. Since the price of ornamental seedlings mainly depends on the buyer's interest, size of planting material, the small private nurseries mostly concentrate on the ornamental seedling production to fetch more profit. The industrial nurseries are well equipped with infrastructure, manpower, automation and target to produce seedling/propagules of short rotation tree species to meet their factory raw material demand such as pulp and paper, plywood, small timber for furniture, juice, jam and pickle making. Hence, different kind of nurseries targets various end products. But nursery is pre requisite for meeting the quality seedlings demand and nursery management is a potential tool to execute the activity in successful way (Krishnan *et al.*, 2014).

There are four main tools for planning nursery operations. A nursery calendar to help plan necessary actions and purchases of seed, supplies and equipment. A plant development register for collecting species-specific information about seed treatment, germination requirements and duration, plant development, special requirements for potting substrate, watering, shading or disease control. A nursery inventory to keep track of the species and numbers of seedlings in different stages of development.

Tools for High-Tech Nursery Management

1. Nursery calendars
2. Plant development registers
3. Nursery inventories
4. Records of nursery experiments.

These are needed for production management as well as for research.

Management of Nursery

Nursery plants require due care and attention after having either emerged from the seeds or have been raised from other sources like rootstock or through tissue culture technique. Generally, they are grown in the open field under the protection of Mother Nature where, they should be able to face the local environment. It is the duty and main objective of a commercial nursery grower to supply the nursery plants with suitable conditions necessary for their development and growth. This is the major work of management in the nursery which includes all such operations right from the emergence of young plantlet till them are fully grown-up or are ready for uprooting and transplanting in the main fields.

12. Containerized Nursery

In horticulture various types of containers are used for sowing of seeds of fruits, vegetables, ornamental and flower crops. A flower pot or plant pot is a container in which flowers and other plants are cultivated. Containers for plant propagation are available in various forms, sizes and in different materials. There are two types of nursery production: *i.e.* Field and container. Field stock is either direct seeded or transplanted from seedlings and then lifted as bareroot stock for use as nursery liners, fruit trees seedlings for Christmas trees, wind breaks and conservation plantings. Field stock is also grown for balled and bur lapped landscape or shade trees. Container stock, which is propagated from seed, rooted cuttings and field grown seedlings, is common in both forestry and landscape nursery production.

There are several factors to keep in mind when deciding what containers to use. The factors include:

- ☆ Cost
- ☆ Appearance
- ☆ Design (Shape and size)
- ☆ Features that control root growth
- ☆ Durability
- ☆ Availability
- ☆ Shipping capacity

- ☆ How affects the growing medium moisture content and temperature.
- ☆ How suits the particular needs of the nursery.

Round black pots are the industry standard, but can cause root constriction that leads to plants with poorly developed root systems. There are other kinds if containers that promote better root systems. A wide selection of ornamentals is produced in containers. Home owners usually prefer to buy containerized plants because the plants are easier to transport and transplant than balled and bur lapped plants.

Advantages of Containerized Production

- ☆ Achieving high plant densities - Using land unsuited for production
- ☆ Planting at times independent of the weather
- ☆ Eliminating some operations like root pruning
- ☆ Lowering transportation costs because of light weight media
- ☆ Experiencing less root loss and greater chance of survival than with field grown trees

Disadvantages

- ☆ Small containers need frequent watering
- ☆ Nutrients deplete rapidly
- ☆ Plants require winter protection
- ☆ Plants easily become root bound
- ☆ Wind can knock over trees
- ☆ Containers are costly
- ☆ Labour cost to pot up the plants is high
- ☆ Temperature extremes stress roots.

Pots and containers designed for enhanced root growth are an important feature in containerized nursery production. Each pot and container offer its own advantages and disadvantages. Inadequate drainage and aeration of container media is a limiting factor in the production of quality nursery crops. In severe cases this may lead to or contribute to the pre-mature death of the plant. In addition to appropriate growing medium the bags and containers should be provided with vents and drainage holes at the bottom to drain the excess water and for aeration in root zone.

Container Depth

The only way to reduce the effects of perched water table is to increase the depth of container. The deeper the container, the lower the impact of the saturated area at the bottom of the container and the deeper the container, the greater the overall aeration and the less the pores are filled with water.

Containers

- Lat pots
- Earthern pots
- Large sized pots (30×30×30cm or bigger): - These pots are used for growing perennial ornamental plants like Areliya, Acalypha, croton, Thuja *etc.*
- Medium sized pots: (20×20×20 cm): - These pots are used for sowing seedlings small shrubs, indoor plants and rooted cuttings.
- Small sized pots: (15×15×15 cm): These pots are used for sowing seedlings of fruit plants and ornamental plants and they are easy for transporting.
- Metal pots: These pots are used for raising perennial ornamental indoor plants.
- Cement pots: These are pots for raising perennial plants for public garden and big bungalows garden.
- Plastic pots: These pots are light in weight and easy to transport and used for indoor plants.
- Ceramic pots: These pots are used for indoor gardening and as they have to keep under shade. Different shape and size of ceramic pot plants are used for Bonsai.
- Baskets:
  - Hanging basket: These pots are used for hanging plants and are hanged in balcony or verandah. Bamboo basket.
  - Plastic basket:
  - Thali:
  - Earthen seed pan: These shallow pots are used for raising seedling from seeds.
  - Iron thali:
  - Plastic tray:
  - Polythene bags: Small, medium and big sized bags are available, which are widely used in place of pots because plastic preserved moisture for longer time and cheaper and easy to transport.

Containers for Plant Propagation

Comes in various forms, sizes and different materials – Polystyrene, polyethylene, fiber or paper.

Polybags

Containers most commonly used in tree nurseries in developing countries. It is usually made of black polyethylene and have several drainage holes at the bottom.

Poly Sleeves

Made from same material and are cut from a continuous roll and have no bottom Above both types are often locally made and cheap and readily obtained in most countries. They come in various gauges and volumes between 0.3 to 15 liters. Bags or sleeves are not very durable and can break prematurely. For plants with a long nursery cycle, this can be a problem. They are normally used only once. The discarded polybags and poly-sleeves are a problem for nursery waste management, as they do not decay and are often burnt

Jiffy Pots

Made from compressed peat and need to be filled with a growth substrate. Roots can easily penetrate the container walls, so root curling does not occur. The seedling is planted together with the pot which is biodegradable.

Root Trainers

They are usually rigid containers with internal vertical ribs, which direct roots straight down if prevent spiral growth. The containers are set on frames or beds above the ground to allow air-pruning of roots as they emerge from the containers. Seedlings grown in root trainers have more vigorous and rapid root growth than seedlings grown in polybags.

13. Automation/Robotics in Horticulture

A robot is a mechanical, artificial agent and is usually an electromechanical system. Robot is a machine especially programmable by a computer capable of carrying out a complex series of actions automatically". Robots can be guided by an external control device or the control may be embedded within. Most existing crop robotics studies focus on horticultural or industrial crops, but grains and oilseeds may be of low preference for robotics entrepreneurs because most grain and oilseed production are already mechanized.

Robots in the Mining

Robots are now doing jobs like lying explosives, going underground after blasting to stabilize a mine roof or mining in areas where it is impossible for humans to work of even survive.

Scope of farm robots in India our farm equipment companies and researchers have developed a lot of small and heavy farm equipment for traditional farming needs but some kinds of robotic and pneumatic mechanism are required in precision farming, if the robots are being used for weed control, that will help to reduce the herbicides usage and the produces will turn into an organic, the same way robots can be used for transplanting the seedlings to avoid intensive labour. There are lot of hurdles taken in the agriculture sector in all countries but specially in India. Farmers decreasing in India a daily report from the newspaper. They are great men who provide food for us but now a day's farmers are reducing more in number. Many are leaving the farming profession by telling some repeated common sentences that it is no longer profitable and none want to get losses and it is becoming risky day by day. So many disadvantages. Also, the youngsters are not interested in that. So, they are ready to work in construction companies and not in farmland.

Most of the farm operation is done by using the tractor as a prime mower with attaching different implements on it. At present intercultural operation mainly carried out by using manually operated implements such as hoe and weeder are very popular for horticultural crops. Robots in horticultural sector are widely used in harvesting, drone spraying, field monitoring, sorting, grading, packing of final horticultural produce and in nurseries and greenhouses to some extent.

The Problems of Traditional Farming Lack of Knowledge of using Fertilizer

Adequate amount of fertilizer will benefit the farmers but surplus quantity damages the crop as well as the yield. The lack of knowledge of the usage is damaging the crops and hence giving the problems to the farmers. Even some farmers mistakenly drink the water added with fertilizer for the plants and lose their lives.

Lack of Man Power

The farm workers are paid less than the construction labour. So, to earn livelihood many of workers are choosing the straight path work in construction and different areas but not farming.

Farmers Suicides

It is a report from an agriculture region Maharashtra, Andhra Pradesh and Karnataka states of India that more than 17,000 farmers have suicide and end their lives from 2006-2007. All the above-mentioned crisis can be eradicated and Indian farming and agriculture sector can be improved for a very large extent by the use of Robotics with smart intelligence devices and making smart robots that benefit to the agriculture sector and farming and cattle grazing. All the researches are on and we need a good time for the establishment of the technology in the farming.

Need of Robotics in Horticulture

- ☆ Labour a major cost for horticulture nearly 50 per cent of production costs for fruit are for hired labour. Intensive horticultural crops require much more skilled labour than broad scale agriculture.
- ☆ Land mechanization.
- ☆ Reduce crop losses by harvesting at proper time.
- ☆ Skill and availability when it's needed.
- ☆ Cost of labour.

- Skill level and experience of labour often not available.
- Machines with sensors are objective and product can be consistently monitored.
- Uniform quality of output.

How Robots Work?

Robots can move and sense. They require multiple sensors and controls that allow them to move in an unknown environment.

Robots usually have five parts:

1. **Sensors:** Sensors send information in the form of electronic signals back to the controller. They can give robot controller information about its surroundings. Robots can be designed and programmed to get specific information that is beyond what our senses can tell us
2. **Controller:** It is also called as computer. Controller functions as the brain of the robot. Controller also allows the robot to be networked to other system, so that it may work together with other machines, processors or robots.
3. **Drivers/Actuators:** It is the engine of the robot. It is defined as "a mechanical device that produces motion.

   Types

   - Hydraulic motor.
   - Stepper motor.
   - Pneumatic motor.
   - DC motor.
   - Servo motor.
4. **Arms:** Usually, a robot arms are like human arms with a shoulder, elbow, wrist and fingers. The arm is a part of robot that positions the end-effectors and sensors to do their preprogrammed business.
5. **End effectors:** End effectors are the last link (or end) of the robot. In a wider sense, end effectors can be seen as the part of a robot that interacts with the work environment.

   Ex: Gripper, vacuum pump, tweezers, scalpel, blow torch.

Types of Robots Used in Horticulture

1. **Demeter:** Demeter is a robot that can cut crops and looks like a normal harvester, but can drive by itself without any human supervision. It has cameras on it that can detect the difference between the crop that has been cut and crop that hasn't. this information tells it where to drive, where to put its cutter head and when it has come to end of a crop row so it can turn around. This robot can also be driven by remote

control. It can be taught a path and then follow the path using its on board sensors and computer control systems.

2. **Weed controller**: A four-wheel drive weed seeking robot was developed and the task of the weed removing device is to remove and destroy the weed. Crops that are grown in rows can be weeded by running a hoe between the crop rows. An intelligent hoe uses vision system to identify the rows of crops, steer itself accurately between them, considerably reducing the need of herbicides. Weed identification is based on colour photography. The equipped robots help production of weed maps identifying plants.
3. **Fruit picking robot**: These robots need to pick up fruits without damaging the branches or leaves of the tree. The robots must be able to access all areas of the tree being harvested. The robot can distinguish between fruits and leaves by using video image capturing. The camera is mounted on the robot arm and the colours detected are compared with properties stored in the memory. If a match is obtaining, the fruit is picked. If fruit is hidden by leaves, an air jet can be used to blow leaves out the way for a clearer view. The pressure applied to the fruit is sufficient for removal from the tree, but not enough to crash the fruit. The shape of the gripper depends on the fruit being picked.
4. **Drones**: A drone is a flying robot. To get a bird's eye view of the land. Offers a quick and easy way to check on the progress of crops and determine where they may need to replant or direct pesticide application.

Applications

- ☆ Weeding
- ☆ Crop scouting
- ☆ Micro spraying
- ☆ Irrigation
- ☆ Selective harvesting
- ☆ Sorting and Packing
- ☆ Parks and golf courts

Advantages of Robots

- ☆ Robots can work 24 hours a day without rest.
- ☆ Robots are extremely accurate compared to humans.
- ☆ Robots can perform tasks more quickly than humans.
- ☆ Robots can work in very dangerous conditions.
- ☆ It can reduce farms use of pesticide.

- Maximum robots are small to move easily between plants.
- Robots are lighter in weight than equipped tractors used for inter cultural operators which avoid soil compacting.
- Robots avoids maximum problems associated with labour.
- The robots are not getting sick or tired, and the time off is not needed.
- The robots can deliver products of high quality and lower the cost of production.

Disadvantages

- It costs a lot of money to make or buy robots.
- They need maintenance to keep them running.
- The labours can lose their jobs.
- The robots can change the culture/the emotional appeal of agriculture.
- High energy cost and maintenance.
- The high cost of research and development.
- Lack of access to poor farmers.
- Robots lack decision making power.

Challenges about Horticultural Produce

- Horticulture does not enjoy a safety net like the Minimum Support Price (MSP) for food-grains.
- Lack of good cold chain storage and transport networks to extend the life of perishable products.
- Very less or limited input by machinery and equipment so it is tough to minimize the time restraints.
- Higher input costs than food-grains make it a difficult set up, especially when there is no support from the local governments to the smaller farmers.
- It gets challenging for marginal farmers to cope with the high price fluctuations.
- Limited availability of market intelligence, mainly for exports makes it a tougher option to choose.
- Insufficient availability of quality seed and planting materials.
- Imbalanced use of inputs (fertilizer, irrigation and plant protection).
- Significant losses (biotic and abiotic stresses). Inadequate storage and processing/packaging facilities).
- Poor market infrastructure and access to infrastructure.

The Future Scenario of Horticulture in India

The horticulture sector as a diversified function of the agriculture sector has become a promising source of income opportunities. It contributes towards a steady growth for the sector as well as for the economy. India has evolved as a leading producer of horticulture products and emerged as a reliable source for employment generation, income creation, and export promotion. Horticulture production supplies diverse varieties of fruits and vegetables that are a useful source of nutrients as a part of a balanced diet and enhances a healthy lifestyle. So, there is a steady increase in demand for these products as people become more concerned about their health and well-being. The increased affluence and trendy lifestyle have also led to a greater requirement for flowers for gift and decoration purposes. The production of horticulture products is economic and convenient as they are more resilient to changing weather conditions. The requirement of area, water, and other resources are much less than other agricultural products and hence reduces the chance of failure thereby giving better returns and income opportunities. Add to foreign exchange (Date palm, Citrus, Mango, Apricot and Cashew nut *etc.*)

India is likely to gain a prominent place in the world forum with the help of better resource allocation, infrastructure development, technological upgrade, and better policy implementation for the development of the horticulture sector. A conscious effort is required to implement certain initiatives that would lead to the future growth and development of the horticulture sector. Some of the future initiatives are mentioned as follows:

- ☆ Import of technologies to improve the productivity and skill development in the horticulture sector.
- ☆ Application of organic farming methods in horticulture to achieve environmental and economic benefits in the production process.
- ☆ Adoption of more standardized processes in horticulture production that help farmers to apply knowledge and skills effectively towards improving production.
- ☆ The involvement of more private firms in the horticulture sector can bring more efficiency and export opportunities.
- ☆ The government should promote organic horticulture production by encouraging farms with incentives and providing support during the conversion from conventional to organic farming.
- ☆ Encourage farmers to adopt Good Agriculture Practices (GAP) by providing certifications. These practices help in producing fruits and vegetable products with improved quality. This in turn will generate more demand from foreign countries and boost export activities.

Future Projections

- ☆ The area under horticulture is estimated to increase by 2.7 per cent in Financial Year 2021 as compared to the previous year.
- ☆ The horticulture production in Financial Year 2021 is estimated to gain an increase of 5.8 M.T. over Financial Year 2020.
- ☆ The agriculture ministry has identified 10 globally popular fruit crops and 10 indigenous fruit crops for promotion.
- ☆ There is a scope for enhancing productivity to cater to the estimated demand of 650 M.T. of fruits and vegetables in India by the year 2050.

CONCLUSION

India is among the top producers of fruits and vegetables worldwide and is considered one of the best performers in horticulture production across the globe. The horticulture sector can experience further growth and profitability through improved techniques for production, farm mechanization to increase harvesting and processing efficiency, and procurement of best quality planting materials to produce high-quality products. The diversification in the agricultural sector mainly of the horticulture sector has become a major source of positive growth for the sector itself and for the nation.

Some Suggestions about Better Horticulture Produce

- ☆ Achieve technology led development in Horticulture.
- ☆ Postharvest and value addition in horticulture crops.
- ☆ Modified atmosphere packaging for long storability and transportation of fruits and vegetables.
- ☆ India can emerge as a far bigger producer and exporter if sufficient emphasis is given to resource allocation, infrastructure development, more R and D, technological upgradation and better policy framework for horticulture sector.
- ☆ Insect pollinators for improving productivity and quality of the crops. Development of varieties for cultivation in non-traditional areas.
- ☆ Nutrient dynamics and interaction.
- ☆ Bioenergy and solid waste utilization to make horticulture more efficient and eco-friendlier.
- ☆ Plan, coordinate and monitor R and D programs at national level as well as to serve as knowledge repository in Horticulture sector.
- ☆ It has emerged as a promising source of income acceleration, employment generation, poverty alleviation and export promotion.
- ☆ Horticulture sector with strong forward and backward linkages as an organized industry can stimulate and sustain growth.

REFERENCES

Collonnier, C., Fock, I., Kashyap, V., Rotino, G. L., Daunay, M. C., Lian, Y., Mariska, I. K., Rajam, M. V., Servaes, A., Ducreux, G. and Sihachakr, D. (2001). Applications of biotechnology in eggplant. Plant Cell, Tissue and Organ Culture, 65: p.91-107.

Heidari, A. A., Kashi, A., Saffari, Z. and Kalatejari, S. (2010). Effect of different Cucurbita rootstocks on survival rate, yield and quality of greenhouse cucumber cv. Khassib. *Plant Ecophysiology,* 2: 115-120.

https: //www.fao.org/fcit/index.asp

Jiang, Y., Yung, Y. L., Sander, S. P. and Travis, L. D. (2004). Modeling of atmospheric radiative transfer with polarization and its application to the remote sensing of tropospheric ozone. *J. Quant. Spectrosc. Radiat. Transfer,* 84, 169-179.

Krishnan, P. R., Kalia, R. K., Tewari, J. C. and Roy, M. M. (2014). Plant Nursery Management: Principles and Practices, Central Arid Zone Research Institute, Jodhpur, pp. 6.

Kubota, C., Meng, C., Son, Y. J., Lewis, M., Spalholz, H. and Tronstad, R. (2017). Horticultural, systems-engineering and economic evaluations of short-term plant storage techniques as a labor management tool for vegetable grafting nurseries. *PLoS One* 12: e0170614. 10.1371/journal.pone.0170614, PMID.

Kumar, A. B., Raja, P., Pandey, A. K. and Rabindro, P. (2017). Evaluation of wilt resistance of wild Solanum species through grafting in brinjal. *Int. J. Curr. Microbiol. App. Sci.* 6(9): 3464- 3469.

Lee, J. M., Kubota, C., Tsao, S. J., Bie, Z., Echevarria, P. H. and Morra, L. (2010). Current status of vegetable grafting: diffusion, grafting techniques, automation. *Sci. Hortic.* 127, 93-105.

Lee, J. M. and Oda, M. (2003). Grafting of herbaceous vegetable and ornamental crops. *Horticultural Reviews,* 28: 61-124.

Mbora, A., Lillesø, J. P. B. and Jamnadass, R. (2008). Good Nursery Practices: A Simple Guide (ISBN: 978-92-9059-235-8). Nairobi. The World Agroforestry Centre. 36 pp.

Nichols, M (2007) Grafting. Massey University, New Zealand.

Pandey, A. K and Rai, M. (2003). Prospects of grafting in vegetables: An appraisal. *Vegetable Science.* 30(2): 101-109.

Rivard, C. L. and Louws, F. J. (2008). Grafting to manage soil borne diseases in heirloom tomato production, *Hortscience,* 43(7): 2104–2111.

Takahashi, K. (1984). Replant failure problems in vegetables. *Res. Data Natl. Res. Inst. Vegetables,* 18, 87-99.

Chapter 2

Challenges of Greenhouse Gases in Horticultural Crops

Vibhootee Garg[1], Deepak Singh[2], Raushan Kumar[3] and Vedant Gautam[4]*

[1]Ph.D. Research Scholar (Vegetable Science), Department of Hoticulture, College of Agriculture, Jawaharlal Nehru Krishi Vishwa Vidyalaya, Jabalpur, (M.P.)

[2]Ph.D. Research Scholar (Fruit Science), Department of Horticulture, College of Agriculture, Jawaharlal Nehru Krishi Vishwa Vidyalaya, Jabalpur, (M.P.)

[3]M.Sc. Fruit Science, Department of Horticulture, College of Agriculture, Jawaharlal Nehru Krishi Vishwa Vidyalaya, Jabalpur, (M.P.)

[4]Ph.D. Research Scholar, Department of Mycology and Plant Pathology, Institute of Agricultural Sciences, Banaras Hindu University, Varanasi (U.P.)

**e-mail: vibhootee007garg@gmail.com*

INTRODUCTION

A greenhouse gas (GHG or GhG), is a gas that creates the greenhouse effect by absorbing and radiating heat infrared light which are coming from sun. Those gases that trap heat in the atmosphere are called greenhouse gases.

The climate system can be affected by gas molecules that absorb thermal infrared radiation and are present in significant quantities. These particular gas molecules are referred to as greenhouse gases. -*By Michael Daley*

The main greenhouse gases' atmospheric emissions and removal are described in this section. The increase in temperature that the Earth experiences as a result of the atmosphere's gases (such as methane, carbon dioxide, nitrous oxide, and water vapor) trapping solar energy is known as the greenhouse effect.

The highest temperature increase that may be prevented without the planet reaching "hazardous" levels by 2050, according to the United Nations'

Intergovernmental Panel on Climate Change (IPCC), is 2 °C (3.6 °F). Burning fossil fuels, notably coal, petroleum (including oil), and natural gas, is the main cause of anthropogenic carbon dioxide emissions. The manufacture of fertiliser, cement, and other building materials, as well as deforestation and other changes in land use, are other sources of emissions.

Arrival of Radiation

- One of the main sources of greenhouse gas (GHG) emission in recent years is from horticulture crops. In the spectrum of light that the Earth emits, infrared light is absorbed and released by greenhouse gases. Because they act similarly to greenhouse glass, some trace gases are referred to as greenhouse gases. Some of the radiation that enters the earth is absorbed. The earth becomes heated as a result, and radiates heat in the infrared region of the spectrum.
- Some greenhouse gases (GHGs), including carbon dioxide, methane, nitrous oxide, and halogenated chemicals, are produced naturally, but most are caused by human activity. The GHGs increase the natural greenhouse effect, or global warming, by absorbing infrared radiation and trapping heat in the atmosphere.
- Infrared (IR) radiation is captured like a blanket by carbon dioxide (CO_2) and other greenhouse gases, preventing it from escaping into space. The overall result is a gradual heating of the Earth's atmosphere and surface, a phenomenon known as global warming.
- Without this natural occurrence, the low temperature would make it impossible for life to exist on earth. This natural phenomenon warms the atmosphere.

Key Properties

There are three crucial characteristics of a greenhouse gas:

- The wavelength of the energy the gas absorbs,
- The amount of energy it absorbs, and
- The amount of time the gas stays in the atmosphere

In the infrared portion of the spectrum, which we typically associate with heat, greenhouse gas molecules absorb energy. More than 90 per cent of the energy in the atmosphere is absorbed by greenhouse gases in a very small area of the energy spectrum. Each greenhouse gas has a different absorption energy, but they all work together to absorb radiation across a significant portion of the infrared spectrum. Methane has a 12-year atmospheric half-life while fluorocarbons have a 270-year half-life. A small amount of atmospheric carbon dioxide will last for thousands of years, but the majority will disappear within the first century of its release.

Abundance and Half-Life of Greenhouse Gases in the Atmosphere

Parameters	*CO_2*	*CH_4*	*N_2O*	*Chlorofluoro-carbons*
Average concentration 100 years ago (ppbV)	290,000	900	270	0
Current concentration (ppbV) (2007)	380,000	1,774	319	3-5
Projected concentration in the year 2030 (ppbV)	400,000-500,000	2,800-3,000	400-500	3-6
Atmospheric lifetime (year)	5-200	9-15	114	75
Global warming potential (100 years relative to CO_2)	1	25	298	4750-10900

Source: IPCC (2007).

Types of Greenhouse Gases

There are five main greenhouse gases on Earth:

1. The ozone (O_3)
2. Carbon dioxide (CO_2)
3. Methane (CH_4) and
4. Nitrous oxide (N_2O).
5. Water Vapour

- ☆ Carbon dioxide (0.04 per cent), nitrous oxide (0.08 per cent), methane (0.012 per cent), and ozone are the trace gases that account for around 0.1 per cent of the Earth's atmosphere and have a pronounced warming effect.
- ☆ The average surface temperature of the Earth would be closer to 18°C (0 °F) without greenhouse gases than it is right now, which is 15 °C (59 °F).

Global Emissions of Anthropogenic GHGs by Sector

Sector	*Gas Emission (Per cent)*
Forestry	17 per cent
Energy supply	26 per cent
Waste and waste water	3 per cent
Agriculture	14 per cent
Industry	19 per cent
Transport	13 per cent
Residential and commercial	8 per cent

- The majority of the gases in the Earth's atmosphere, including argon (Ar-0.9 per cent), oxygen (O_2-21 per cent), and nitrogen (N_2-78 per cent), do not vibrate in a way that alters the distribution of their electrical charges, making them not greenhouse gases. Monatomic gases, like N_2, do not have vibrational modes.

Effects of a Rise in Greenhouse Gases

High temperatures as a result of GHG buildup and global warming

- More rain than usual
- More frequent droughts.
- Repeated floods
- Repeated hot and cold waves
- Frost occurrences

Important GHGs

Carbon Dioxide (CO_2)

- CO_2 makes for around 76 per cent of all greenhouse gases in the atmosphere and accounts for roughly 65 per cent of all emissions of greenhouse gases in India.
- Although CO_2 has a lower warming potential, it contributes between 55 and 60 percent to global warming.
- The main sources of CO_2 emissions are cars, industrial power plants, and deforestation.
- It results from manmade and natural activity.
- Plants, animals, and microorganisms all engage in respiration and fermentation as natural activities.
- It is a crucial gas since only CO_2 causes both global warming and has a favourable impact on the C_3 Plant's production.
- CO_2 production Temperature rise, increased precipitation, frequent droughts, frequent floods, and frequent heat waves are all effects of climate change.

Methane (CH_4)

- Methane is also the second most harmful greenhouse gas produced by human activity after CO_2.
- Methane concentration has increased by 149 per cent since 1850, contributing 15-20 per cent to global warming with a 21–25 times higher warming potential than CO_2 and

- ☆ Accounting for 31 per cent of all greenhouse gas emissions in India, according to the Telegraph newspaper.

Source

- ☆ About 78 per cent of India's total methane emissions come from agriculture.
- ☆ In India, the major contributors to methane emission are ruminant animals like cattle and buffaloes (59 per cent) and rice paddy fields (23 per cent).
- ☆ Rice plants' aerenchyma is responsible for 80–90 per cent of the methane emission that occurs through soil through anaerobic decomposition or fermentation of organic materials in paddy fields.
- ☆ Rice fields in wetlands have lately been discovered to be a significant source of atmospheric methane. The first thorough measurements of methane fluxes in rice fields were only reported in the early 1980s, despite the fact that the potential for methane emission from rice fields has long been observed.
- ☆ Diffusion: A direct diffusion mechanism results in the emission of 5–10 per cent of the methane produced in rice paddy fields.
- ☆ Ebullition: This process results in the small-scale release of 5–10 per cent of the CH_4 produced in rice fields into the atmosphere. Under muddy or marshy rice fields or mangrove ecosystems, this procedure is very prevalent.

Nitrous Oxide (N_2O)

- ☆ It has 140 times more warming potential than CO_2, contributing 4 per cent of all GHG emissions with a 5 per cent impact to global warming.
- ☆ Nitrous oxide generated in crop fields is released into the atmosphere immediately.
- ☆ Both nitrification under anaerobic conditions and denitrification under aerobic conditions cause nitrous oxide to be produced in the soil.
- ☆ Organic matter in the soil decomposed.

Organic Matter Ammonification (CO_2 + H_2O + NH_2)

1. Nitrification Process (Aerobic Oxidation)
2. It contributes just 4 per cent to greenhouse gas emissions, 5 per cent to global warming, and has 140 times greater warming potential than CO_2
3. Agriculture accounts for 84 per cent of all N_2O emissions in India.

Effect

1. NO is a factor in the ground-level troposphere's ozone production (plant detrimental effect).
2. Acid rain is a factor in water pollution (water quality deterioration)
3. Toxic chemicals, global warming, and reduced visibility due to climate change as an air contaminant

Sulphur Dioxide (SO_2)

- Sulfur compounds are present in coal and petroleum, which when burned produce sulphur dioxide.
- Through an intramolecular interaction with the hydroxyl radical, SO_2 is oxidised in the gas phase.

Uses

- reducing substance that serves as a precursor to sulfuric acid, which is used to preserve wine.
- Uses in Biochemistry and Biomedicine
- Being A Refrigerant

Unhealthy Effect

- Influence the habitat's fitness for both human life and plant communities.
- Atmosphere particles and an acid rain precursor
- Causes oxidative stress in plants, which causes cell damage
- Global warming's causes

Water Vapour

- The most powerful greenhouse gas in the Earth's atmosphere is water vapour, yet it behaves very differently from the other greenhouse gases.
- As climate feedback, or a reaction within the climate system that influences the system's continuous activity, water vapour plays the principal function in the climate system rather than as a direct agent of radiative forcing.
- This discrepancy results from the fact that air temperatures, rather than human behaviour, generally cannot directly modify the amount of water vapour in the atmosphere. The rate at which water evaporates from a surface increase with surface temperature.

- Since water vapour can absorb infrared radiation and radiate it back to the surface, increased evaporation results in a higher concentration of water vapour in the lower atmosphere.

Sources of Greenhouse Gas

Emissions

Challenges of Greenhouse Gases

General Effects

- The melting of the Himalayan ice cap has a chilly impact.
- Horticultural crops' physiological disorders will be more pronounced, such as the mango's spongey tissue and the fruit cracking of the litchi, apricot, and cherry
- A shorter growing season, less water availability, and inadequate vernalization (FAO, 2004)
- Fruit and blossom dropping, as well as floral abortions, were common.
- Crops will mature earlier and develop more quickly.
- Production dropped from 10.8 to 5.8 t/ha.

A. Vegetables

Lettuce

Increasing the average temperature from 16.3°C to 21.1°C increased early-season growth and decreased crop duration, according to studies. Thus, the total yield was effectively lowered by 17 per cent.

- Tip burn can happen in warm, humid conditions. This condition is brought on by accelerated growth rates and is linked to localised calcium insufficiency. A few days of 32°C heat can result in enough visual damage to wipe out an entire crop. The issue is not solved by increasing the calcium availability of the soil. Application of foliar calcium may be partially effective, but it is not a workable option.
- Tip burn can also occur at somewhat high nighttime temperatures (15–16°C), especially when combined with high RH.
- Bolting, poor colour development, and bitterness are all risks that are exacerbated by the high daytime temperatures (32°C).
- According to some research, nighttime temperatures play a greater role in the initiation of bolting than do daytime temperatures. If nighttime temperatures are warm (21°C) as opposed to cold (16°C), iceberg lettuce blooms 21 days earlier.

- Lettuce seed will not normally germinate at over 25°C. Utilizing seedlings or priming seed in water or 1 per cent potassium phosphate will solve this problem.
- Lettuce can survive in cold temperatures, even briefly below zero. Low temperatures, however, can cause anthracnose, russetting, and external cracking and can extend the growing season by up to 150 days.

Capsicum

- Pollen viability decreases at high temps (32°C), which results in poorer fruit set.
- High temperatures, especially in extreme circumstances, also increase the risk of sunburn.
- Pollen generation and release are impacted by basic physiological processes, making breeding for high temperature tolerance challenging. The Asian Vegetable Research and Development Center is currently developing varieties that can withstand temperatures as high as 32°C.
- Breeding for low temperature development is possible, with varieties perhaps available that resist night temperatures of 8–10°C.
- Fruit fail to develop normally in size and shape below the lower threshold for capsicum growth (8–10°C).

Broccoli

- Broccoli is a crop that is extremely sensitive to temperature
- It can withstand temperatures of 30 to 32°C as long as the nighttime low is at least 15°C. However, issues including blindness, brown heads, bracing in the head, catseye, unequal floret size, uneven cluster development, and loose clusters may cause a yield reduction of up to 45 per cent. The best varieties are frequently not those that can thrive in hot climates.
- Low temperatures (4°C) can lead to the growth of wet patches, purple hues in the head, and early fading of colour. 25 per cent of the yield could be lost.

Beans

- Pollen viability decreases above 28 °C, which causes flower buds to fall off and limited pod development. Additionally, pods are more likely to be fibrous and flavourless. Low temperatures stunt the growth of the root system and limit photosynthesis, perhaps reducing the overall production.
- Bean breeding strategies over the long run might increase climatic tolerance:

- ☆ It is possible to breed for superior root systems and increased drought resistance.
- ☆ The USDA has previously done some breeding for high and low temperature tolerance.
- ☆ In the tropics, snake beans (*Vigna unguiculata*), which are more resilient to high temperatures, can replace French beans (*Phaseolus vulgaris*).

Cauliflower

- ☆ The ideal temperature range for growing cauliflower is 15–18°C, when yields can reach 23–30 t/ha.
- ☆ Many cauliflowers have exceptional cold tolerance, allowing them to be grown over winter for spring harvest, despite the fact that cold temperatures, even down to zero, can diminish output by about 15 per cent.
- ☆ A bigger issue is high temperatures because anything above 32°C can lead to bolting, curd yellowing, and uneven floret development.

Beetroot

- ☆ Temperatures between 18 and 25°C are ideal for beetroot growth. It is sensitive to heat; 27°C has the potential to cause bolting, poor colour, and irregular root shape. Particularly, high temperatures might result in the root developing light and dark rings (zoning), which is not desired for processing. In addition, "damping off" diseases become more problematic when soil temperatures are high.
- ☆ Exacerbation of nutrient diseases such boron shortage and "hollow heart" caused by calcium and boron deficit are the main problems with cold temperatures (5°C or less).

B. Fruit

Blueberries

- ☆ CO_2 reduced deterioration in blueberries, although concentrations above 15 kPa softened and discoloured the flesh.
- ☆ Fruit maintained acceptable quality for more than six weeks when kept at 0 °C and 15 kPa O_2. This was discovered to be equivalent to 10 kPa CO_2.

Citrus

- ☆ They noted that at enrichment CO_2 levels, citrus leaf photosynthetic rate more than doubled and citrus leaf respiration rate decreased by a third.

- ✰ Citrus trees may respond proportionally more strongly to CO_2 enrichment than annual plant species since they sequestered more than three times as much above ground carbon as ambient controls.
- ✰ Baker and Allen observed that the CO_2 enrichment significantly improved CO_2 uptake by citrus trees and, in most cases, reduced water loss, albeit to a smaller extent. This was confirmed by newly expanded flushes of citrus leaves.

Sl.No.	*Fruit Crops*	*Physiological or Quality Parameter*	*Effect of High CO_2*
1	Blueberries Tomatoes Pears	Respiration	Decreased
2	Lemons Cucumber Mango Apples	-	Increased
3	Strawberry Raspberry	Firmness	Increased
4	Grape	Color Intensity	Increased
5	Strawberry Orange	Ascorbic acid	Increased
6	Grape Strawberry	Total Phenolics	Increased
7	Grape	Flavonoids	Increased
8	Grape Strawberry	Anthocyanins	Increased
9	Strawberry	Antioxidant Capacity	Decreased

Sl.No.	*Fruit Crops*	*Physiological or Quality Parameter*	*Effect of High O_3*
1	Blueberry	Respiration	Increased
2	Strawberry	Photosynthesis	Decreased

Sl.No.	*Fruit Crop*	*Injury Symptom*
1.	Apple	Skin discoloration Pigment breakdown and Water-soaked areas
2	Avocado	Skin and flesh browning,increase decay susceptibility
3	Lime	Juice Vesicle Rupture, Formation of brown spots on fruit surface
4	Pineapple	Flesh with Scattered Water- soaked areas and Translucent fruit flesh

Effect of CO_2 in Various Plant Processes

1. Respiration

- Various crops showed an inhibitory effect of CO_2 consumption. Pears' intake of O_2 was found to be inhibited by CO_2. The rate of respiration was slowed by elevated CO_2 partial pressures.
- 'Bartlett' and 'Bose' pears retained higher contents of organic acids despite elevated CO_2 levels (up to 3 per cent) that decreased ethylene production rates. However, one of the main issues with storing pear fruits in high CO_2 environments is the potential for CO_2 injury.

2. Colour

- Different impacts on the colour change of bananas and plantains were brought about by elevated CO_2. Plantains entirely turned yellow after 6 days of ripening in 20 per cent CO_2, whereas bananas maintained a high chlorophyll content and remained green.
- The patterns of softening, starch degradation, and soluble sugar accumulation in the pulp of 20 per cent CO_2 treated bananas and plantains, in contrast to the differentiated colour changing patterns, exhibited similarly to the patterns in the fruits ripening in regular air, showing that the pulp ripening was not inhibited by 20 per cent CO_2, and the abnormal ripening of bananas in 20 per cent CO_2 can be considered as green ripening.

3. Crop Growth

- Most vegetable crops tend to produce more or grow faster when atmospheric CO_2 levels rise. The glasshouse business, where CO_2 enrichment has been used for crops like tomatoes, cucumbers, capsicums, and leafy vegetables for many years, is well aware of this effect.
- The amount of CO_2 in the atmosphere has lately increased to 400 ppm from 320 ppm in 1960 and 350 ppm in 1985. By 2035, it is anticipated that atmospheric CO_2 levels will be around 435 ppm, barring considerable worldwide mitigation efforts.
- High CO_2 levels can enhance growth and increase photosynthesis, especially if soil water is scarce. If atmospheric CO_2 levels hit 500 ppm, the effect might be severe, with growth rates rising by up to 20–50 per cent. For instance, the growth of capsicum increased by 46 per cent at 450 ppm CO_2 compared to 360 ppm, and other crops like eggplant and tomatoes also showed small growth increases of 24 per cent and 31 per cent, respectively.

- However, yield increases are frequently less than anticipated in practise. The interaction between the variety and its environment has an impact on yield, which is complex. It has been demonstrated that perennial crops, such as fruit trees and forest trees, can adapt to elevated CO_2, reducing its effects over time.

4. Water Use Efficiency

- Most plants consume less water or transpire less when there is more CO_2. This occurs as a result of the stomata, or leaf pores, partially closing to minimise water loss. The result might have a big impact. For instance, a free air CO_2 enrichment (FACE) research on wheat revealed that under enhanced CO_2, water use efficiency was 19 per cent greater.
- If dry spells lengthen and rainfall declines as expected, this might be especially crucial.
- For the vegetable business, a significant source of greenhouse gas emissions is electricity used to pump water. It may be feasible to minimise the amount of water needed for crops by improving agricultural water management and detecting soil moisture levels. This would allow irrigation pumps to run for less time, which would cut emissions and energy costs associated with production.
- The occurrence of physiological problems like tip burn in lettuce and brassica crops and blossom-end rot in tomatoes, capsicums, and watermelons may be reduced under elevated CO_2 levels because these disorders are occasionally linked to excessive transpiration.

Effects of O_3 on Carbon Assimilation and Allocation in Plants

- Both CO_2 and O_3 enter through the stomata, and both may affect the conductance of the stomata and, consequently, the flux of the two gases into the leaf.
- Ozone may reduce the rate of CO_2 assimilation, and the availability of carbohydrates (CH_2O), which is influenced by CO_2 and O_3 concentrations,
- Above threshold flux ozone may cause injury
- About 10,000 times less UV radiation reaches the surface of the Earth thanks to the ozone layer.
- Impact of UV radiation from increasing UV-B radiation due to stratospheric ozone loss
- Variations in the thickness of the epidermal, palisade, and mesophyll layers led to changes in leaf anatomy.

- ☆ A reduction (10–70 per cent) in the chlorophyll concentration
- ☆ Higher UV absorbing substances (10–300 per cent)
- ☆ 3–90 per cent less photosynthesis, especially at higher UV dosages
  - ❖ Express (effect on photosystem)
  - ❖ Deceptive (decrease in pigments and leaf area)
- ☆ Through the stomata, ozone enters plant tissues and damages cells directly, especially palisade cells.
- ☆ Leafy vegetables' leaves will become more prone to yellowing as ozone concentrations rise.

Effects of Methane (CH_4) in Plants Processes

- ☆ Black carbon (from fossil and biomass fuels), halocarbons (used in refrigeration and air conditioning), and methane (mostly from natural gas production and cattle) are all climate change contributors to varied degrees.
- ☆ Black carbon and gases used in refrigeration do the most damage to crops per tonne in the immediate future–the first decade after emissions are discharged. Because methane emissions create surface ozone, which damages plants by yellowing the leaves, methane emissions are also very bad for the environment.
- ☆ Overall, the study predicts that non-carbon dioxide emissions, particularly methane, will be responsible for 93 per cent of crop losses over the course of the rest of this century. The non-carbon dioxide damage is still nine times greater than that brought on by carbon dioxide, even if the impact of surface ozone is eliminated from the calculations.

Remedies

Utilisation of Sustainable Fuels

Using traditional and genetically modified (GMO) crops like sugarcane, oilseed, rapeseed, and jatropha to produce biofuels will help to mitigate the negative effects of transportation-related CO_2 emissions. As a result, energy-efficient farming will employ machinery that runs on bioethanol and biodiesel rather than traditional fossil fuels.

Green energy initiatives involving perennial plants that produce non-edible oil-seeds will aid in air purification and the production of biodiesel for use directly in the energy sector. They will also help reduce the use of fossil fuels to some extent by blending biofuels with them in specific ratios.

Less Fuel Consumption

Composting and mulching techniques in organic farming reduce the need for spraying weed and herbicides because there is less ploughing involved. Reduced fuel consumption and decreased irrigation would both lower the amount of CO_2 released into the atmosphere. Using modern biotechnology such as GMOs and other related technologies facilitate less fuel usage by decreasing necessity and frequency of spraying and reducing tillage or excluding the tillage practice. For example, insect resistant GM crops reduce fuel usage and CO_2 production by reducing insecticides application.

Biotechnology's use of reduced fuel usage resulted in savings of roughly 962 million kg of CO_2 emitted in 2005, whereas the use of reduced tillage or no tillage techniques resulted in reductions of CO2 emissions due to less fuel usage of 40.43 kg/ha or 89.44 kg/ha, respectively.

Carbon Sequestration

Carbon sequestration is the term used to describe the capture or ingestion of carbon-containing compounds, particularly carbon dioxide (CO_2). It is frequently used to refer to any rise in soil organic carbon content brought on by a change in land use, with the implication that the increased soil carbon storage reduces climate change. Soil carbon sequestration is a crucial approach to reduce the rise in atmospheric CO_2 levels. Enhancing carbon sequestration can be done, for example, by reducing conventional tillage.

No-till agriculture decreases CO_2 loss from agricultural systems by leaving at least 30 per cent of residue on the soil surface. It may also help boost soil stability, decrease water loss through evaporation, and create a cooler soil microclimate. Conservation measures that reduce soil erosion may simultaneously increase methane (CH_4) consumption and sequester soil carbon.

The benefits of increasing soil organic carbon on climate change from improved crop growth (such as when using industrial fertilisers) must be weighed against greenhouse gas emissions from the production and application of such fertilisers. Up to 95 per cent of no-till land in the United States of America (USA) and Argentina has been planted with genetically engineered Round up Ready TM (herbicide resistant) soybeans, which has resulted in the storage of 63,859 million tonnes of carbon dioxide (CO_2). The modified crops enable farmers to switch to "no till" agricultural methods by reducing the requirement for tillage or ploughing. This approach improves soil quality and stores more carbon in the soil, both of which are beneficial for reducing climate change.

Less Use of Artificial Fertilizer

Dependence on agricultural chemicals to maintain productivity in sparsely populated areas has resulted in widespread environmental contamination with toxins that alter the path of biogeochemical cycles. Utilizing less fertilisers also results in less nitrogen contaminating ground and surface waters. When they interact with common soil microorganisms, artificial inorganic nitrogenous fertilisers like ammonium sulphate, ammonium chloride, ammonium phosphates, sodium nitrate, and calcium nitrate cause the creation and release of greenhouse gases (especially N_2O) from the soil to the atmosphere. The use of environmentally friendly biotechnology-based fertilisers is being promoted in order to lessen the adverse impacts of artificial fertilisers.

Biofertilizers

Some of the traditional biotechnology methods for minimising artificial fertiliser use include organic farming techniques using bio-based fertilisers (composted humus and animal manure), crop rotation, and intercropping with leguminous plants with nitrogen-fixing capacities. Rhizobium inoculants have been modified by the use of mutation or genetic engineering techniques in contemporary biotechnology, leading to strains with increased nitrogen-fixing properties. A promising biotechnological development could enable non-leguminous plants to fix nitrogen in the soil: the induction of nodular structures on the roots of cereal crops like rice and wheat.

Growing GM crops that utilize nitrogen more effectively is an additional choice. The nitrogen-efficient GM canola is an illustration of such a crop. It not only minimizes the amount of nitrogen fertilizer that is wasted to the atmosphere or leached into soil and waterways, but it also has a positive effect on farmers' economies by increasing profitability. Managing soil nitrogen in accordance with crop requirements can lower N_2O emissions and prevent negative effects on water quality. Additionally, changing the animal diet and managing the manure can lower the CH_4 and N_2O emissions from animal husbandry.

Reduction N_2O Emission

First, as CO_2 concentrations rise, less nitrogen is present in plant tissue. This is because plants growing in high CO_2 environments utilise nitrogen more effectively, which means that less nitrogen is needed to achieve maximum yields than would be necessary at CO_2 levels from 1990. For instance, N concentrations in the leaves of cucumbers, spinach, and fenugreek were roughly 16 per cent lower at high CO_2 than they were for plants grown at ambient CO_2 levels. This has the implication that as CO_2 levels rise, vegetable crops will become more effective at using nitrogen, requiring less nitrogen to be applied.

This may work well with methods used as part of the Carbon Farming Initiative to reduce N_2O emissions based on better nitrogen management. The stimulation of growth and yield that can be anticipated at increased CO_2 levels is nitrogen's second major effect. Since plants require enough nitrogen to benefit from rising CO_2, this reaction is largely dependent on the availability of nitrogen to the plants.

REFERENCES

Bhattacharyya, P., Munda, S. and Dash, P. K. (2020). Climate Change and Greenhouse Gas Emission. New India Publishing Agency, 150p.

Cassia, R., Nocioni, M., Aragunde, N. C. and Lamattina, L. (2018). Climate Change and the Impact of Greenhouse Gasses: CO_2 and NO, Friends and Foes of Plant Oxidative Stress. Plant Abiotic Stress. Frontier Plant Science,

Change, C. (2007). IPCC fourth assessment report. The physical science basis, 2: 580-595.

Food And Agriculture Organization of The United Nations. (2011) Rome

Ritchie, H., Roser, M., Rosado, P. (2020). CO and greenhouse gas emissions. *Our world in data.*

Tuckett, R. (2019). Greenhouse gases. In Encyclopedia of Analytical Science Elsevier (pp. 362-372).

Chapter 3

Challenges of Sustainable Fruit Production

Khushboo Sharma[1], *Vibhu Pandey*[2], *Khursheed Alam*[3] *and Devendra Pal*[2]

[1]*Ph.D. Research Scholar, Division of Vegetable Science, Sher-e-Kashmir University of Agricultural Sciences and Technology, Jammu and Kashmir*

[2]*Ph.D. Research Scholar, Department of Fruit Science*

[3]*Ph.D. Research Scholar, Department of Vegetable Science*

Sardar Vallabhbhai Patel University of Agriculture and Technology, Meerut – 250 110, Uttar Pradesh

**e-mail: tks33557799@gmail.com*

INTRODUCTION

The idea of "sustainable fruit production" has been formed, with sustainability including both evolving farming mindsets and raising environmental awareness. Improved irrigation efficiency, avoidance of nutrient accumulation and leaching concerns, and prevention of soil erosion and root suffocation are all possible with sustainable and creative soil management techniques that place a special emphasis on water management. The main factors affecting fruit tree growth and productivity include biotic and abiotic stress, which also triggers a variety of reactions at the molecular, cellular, physiological, and developmental levels. Application, optimization, and invention of environmentally friendly, sustainable agricultural methods allow fruit agro-ecosystems to regain or improve their typical levels of total fertility while sustaining high yields of high-quality fruit.

Sustainable Fruit Production

Today's fruit production faces a wide range of economic, environmental, and social sustainability challenges. Pesticides, energy, biodiversity, water consumption, and air quality issues are all related to sustainability, but they frequently have a longer time horizon than economic issues. The scale, the distribution methods, and the geographical setting will all affect how sustainably fruit is produced. A large-scale, export-focused apple producer will encounter distinct difficulties than a small-scale, direct market apple grower.

Sustainable Advantages of Fruit Crops

- Permanent and perennial
- Good soil protection and cover in many cases
- Add leaf litter back to soil
- Good for soil conservation on sloped lands
- Some of the most advanced IPM and pest management tools available for crops
- Relate to local cultural heritage

Sustainability Issues for Fruit Crops

Economic

- Increasing resource and input production costs
- Declining wholesale value relative to production costs
- Increasing market product competition
- Increasing imported fruit

Environmental

- Small number of fruit breeding programs with public access
  - Lack of development of locally adapted cultivars
  - Few programs working on insect resistance
- Reliance on synthetic pesticides and fertilizers
- High energy requirement for production, harvest, transport and storage of products

Social

- High labor requirement but now wages
- Worker safety with significant equipment and chemical use
- Loss of local fruit farms, related businesses, and dissolution of communities with lost heritage

Challenges with Some Fruit Crops

Challenges with pest management, particularly the use of insecticides that left a persistent legacy (*e.g.*, lead arsenate), led to new pest outbreaks, provided the impetus for considering sustainable fruit production. Similar to how sustainable agriculture emphasises internal biological resources, integrated control and integrated pest management emerged in the middle of the 20th century with a focus on limiting unplanned pesticide problems and maximizing natural control forces **(Glass, 1975).**

- ✰ Insecticides, fungicides, bactericides, herbicides
- ✰ Fruits, especially apple, have public attention for pesticide use although some pesticides such as herbicides are used much less in fruit crops than other crops
  - ❖ Moderate amount of fertilizers but low nutrient uptake efficiency
  - ❖ More hours human labor/crop yield than most other crops
- ✰ Especially pruning, training, and hand harvest
  - ❖ Have very high capital investments (Rs/acre), especially for tree fruit, grape, guava and strawberry production systems
  - ❖ High grading standards for fruit and intense market competition (at wholesale)
- ✰ Intense competition for the Indian diet as many fruits are regarded as "snack foods"

Fruit Systems moving Towards Sustainability

Integrated Fruit Production

In the 1970s, an Integrated Production (IP) framework was developed by the International Organization for Biological and Integrated Control of Noxious Animals and Plants (IOBC) as an outgrowth of its pioneering work on biological control and integrated pest management (IPM). Eleven principles of IP were identified, with guidelines for crops, nutrient management, soils, biological diversity and landscape, pest control, product quality, and animal production **(IOBC, 1993).** The initial systems of interest were perennial fruit. IOBC then produced technical guidelines under the IP umbrella for pome fruits, stone fruits (*Prunus spp.*), and vines (*Vitis vinifera*), specific to Europe (IOBC, 1994). The guidelines for pome fruit detail the following elements for integrated fruit production:

- ✰ Professionally trained, environmentally and safety conscious growers
- ✰ Should be conserving the orchard environment
- ✰ Site, rootstocks, cultivar and planting system for new orchards
- ✰ Proper soil management and tree nutrition

- Weed-free strip and Irrigation
- Training and management
- Integrated plant protection and management
- Efficient and safe spray application methods
- Harvesting, storage and fruit quality and Postharvest chemical treatments
- Mode of application, controls, certification, and labeling.

Planning for Sustainable Production

- Crop and Cultivar Selection
  - Select appropriate rootstocks for region and for sustainable production systems (tree fruits and grapes)
  - Crop must be adapted to the growing region
  - Crop and cultivar must match a market
  - Crop and cultivar should have multiple pest resistances
- Site and Soil
- Management affects pest problems
  - Crop susceptibility and resistances
  - Farm scope affects pest prevalence

Orchard Biodiversity

- Dealing with the Monoculture
  - Diversify crops (multiple fruit crops; add vegetables)
  - Diversify cultivars
- Spreads both environmental and market risk
- Have managed vegetation belts; Farmscape
  - Pesticide and runoff catchments
  - Reservoir of beneficials
  - Trap crops
- Have diverse ground cover vegetation
  - Minimize use of herbicides and cultivation,
- Encourage mammal and bird predators
  - Raptor boxes, perch poles

Ground Cover and Vegetation Management

(Note: Young fruit plants are very sensitive to vegetative competition)

- ☆ Cultivation
  - ❖ Although seasonal clean cultivation is often used as an orchard practice, it is not sustainable
  - ❖ Limited, under crop/vine/tree cultivation may be useful
- ☆ Use targeted herbicides
  - ❖ Minimize/eliminate soil-active residual herbicides
  - ❖ Targeted contact herbicides in conjunction with other practices (*e.g.* mowing, cultivation)
- ☆ Allow some non-competitive vegetation
  - ❖ Either temporally, spatially, or genetically

Alternatives for Ground Cover Management

- ☆ Mulches
  - ❖ Organic mulches:
- ☆ Sawdust, Wood chips: Blueberries, blackberries, tree fruits, grapes
- ☆ Straw mulch: all fruits
- ☆ Plastic and geo textile mulch: strawberries, grapes, tree fruits
- ☆ Cultivation: grapes, tree fruits
- ☆ Flaming; very energy consuming

Between Row Vegetation

- ☆ Cover Crops: Fruit crops do not compete well directly with cover crops. However, cover crops can be grown between fruit rows to provide various services to the farm system and ecosystem
- ☆ Plant vegetation that provides benefits
  - ❖ Mechanical support of equipment
  - ❖ Habitat for beneficial predators and parasites
  - ❖ Helps dry soils during wet periods
- ☆ Avoid vegetation that may host pests
  - ❖ Alternate hosts for insect pests and disease
  - ❖ Habitat and food stuffs for mice
  - ❖ Does not compete with fruit crop for water or nutrients
  - ❖ Does not spread into the crop row

Nutrition Management

- ☆ Develop healthy soil
  - ❖ It is a challenge in a perennial crop sensitive to competitive vegetation

- ☆ Maintain soil pH
- ☆ Apply nutrients based upon soil and foliar analysis
- ☆ Apply nutrients at a time when they will be absorbed by the plant
- ☆ Multiple applications may be more efficient than single application
- ☆ Apply nutrients to the crop with banding or target applications
    - ❖ But, be sure to provide nutrients to drive-row cover crops
- ☆ Apply foliar nutrients as needed (more efficient)
- ☆ Use of organic nutrients is somewhat untested

Reducing Energy Use

The majority of fruit production energy use is in chemicals and application, tractor operations, storage and transport

- ☆ Resistant Cultivars require less spray
- ☆ Small, high density and diversified orchards reduce less chemical inputs
- ☆ Alternative ground covers reduce mowing, herbicides
- ☆ Develop multiple operation equipment
    - ❖ Try to do multiple operations with one orchard tractor pass
- ☆ Develop a market strategy that includes regional sales
- ☆ Capture your local market ("build a wall around the state")
- ☆ Farm Energy Sources for Buildings and Storage facilities
    - ❖ Wind turbines, solar panels
    - ❖ Belowground storage

Vertebrate Pests

- ☆ Significant Vertebrate pests of fruit crops
    - ❖ Birds
    - ❖ Deer
    - ❖ Mice
- ☆ Best controls are
    - ❖ Exclusion; fencing and netting
    - ❖ Scare tactics; visual and sound
- ☆ Include dogs in the management plan
    - ❖ Predatory birds - include habitat or perches for raptors (hawks, owls, *etc.*)

Trends

- ☆ Organic tree fruit consumption – significantly up
  - ❖ News reports stimulate: "It pays to purchase."
  - ❖ Report on costs and profitability of organic tree fruit production stimulating industry in west
- ☆ 15-20 per cent of WA apple crop now organic

Organic Fruit Crop Production

Organic Fruit Production Systems

Growing quickly and driven by the market, organic farming is one strategy for improving sustainability in the agricultural sector. The roots of organic farming can be found in a focus on enhancing the organic matter in the soil to grow robust plants that can fend off pests and illnesses and provide the best possible health to people and animals that eat them. Certification systems became important as organic farming grew in the 1980s in order to assure consumers that the product they were buying–and typically paying more for–was truly produced as intended. As opposed to confirming that a farm is "operating with natural systems," organic certification programmes may more quickly ascertain whether a grower has used or has not used a specific item. Sustainable agriculture generally does not share this focus, although IFP programs often do exclude or restrict the use of certain more toxic or disruptive pesticides. Since tree fruit and vine systems are perennial and typically involve little tillage after planting, they can be very conducive to increasing soil organic matter. On the other hand, tree fruit and vine crops typically require a high level of pest management to produce marketable crops. Key challenges for organic fruit production include nutrient management, weed control, and control of replant diseases. Resistance management is another challenge, given the fewer effective tools for pest control. When a new tool comes along, such as Spinosad, there is a tendency to overuse it and thus increase the likelihood of inducing pest resistance.

This conflict between the potential for improved environmental sustainability of organic systems and their challenge in maintaining economic sustainability has limited expansion of organic fruit production in certain regions. Use of scab-resistant (*Venturia inaequalis*) apple varieties have expanded the potential for organic apple production in humid regions where this disease is difficult to control.

Challenges for Organic Fruit Production

- ☆ Crop Adaptability
- ☆ Weed Control
- ☆ Pest Management

- ☆ Crop Nutrition
- ☆ Annual Cropping
- ☆ Maintaining product quality

Organic Fruit Production – Still Developing

- ☆ Limited research, on organic production systems
- ☆ Reports in other regions have indicated significant reductions in yield, especially pack-out, declining production over time, *etc.*, due to problems with competitive vegetation, sufficient nutrient supply and adequate pest management
- ☆ More work is needed

Organic Fruit

- ☆ Blueberries – a real opportunity
  - ❖ Many locally adapted types and cultivars
  - ❖ Strong local consumer demand
  - ❖ Few insect pests and diseases – spotted wing drosophila
  - ❖ Very sensitive to weeds; require high and well drained soils; very drought sensitive
  - ❖ Birds are a problem
- ☆ Blackberries – good opportunity
  - ❖ Adapted to the region
  - ❖ Strong local seasonal markets
  - ❖ Few insect and disease pest
  - ❖ Cane borers are the big challenge for organics
  - ❖ Japanese beetles are a problem

  - ❖ Spotted wing drosophila
  - ❖ Not competitive with many grasses
  - ❖ Pest of blueberry, strawberry, blackberry, raspberry
  - ❖ Monitor populations with traps
  - ❖ Spray to manage
- ☆ **Strawberries**
  - ❖ Locally adapted
  - ❖ Frost protection is essential
  - ❖ Floating row covers or low tunnels
  - ❖ Strong seasonal demand
  - ❖ High value crop
  - ❖ Sanitation and crop rotation critical
  - ❖ Diseases can be managed by cultivar selection, cultural practices and with organic fungicides
- ☆ **Apples**
  - ❖ Several insects and disease pests
- ☆ Plum curculio has risen as a primary pest with unreliable organic controls
  - ❖ Other primary insect pests controlled with mating disruption and targeted biological and parasitic sprays
  - ❖ Secondary pests controlled by introduction of predaceous insects and parasites
  - ❖ Trees are sensitive to competitive vegetation
  - ❖ Organic nutrition at appropriate times is difficult
- ☆ **Peaches**
  - ❖ Most insect pests can be controlled culturally, with mating disruption and trapping, and targeted biological, botanical, bio-rational sprays

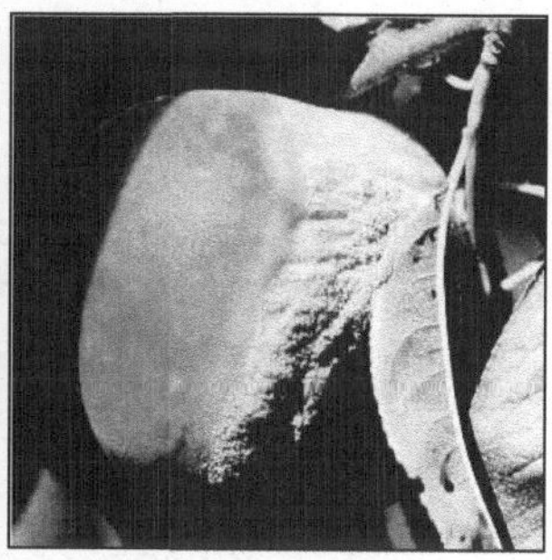

- **BROWN ROT – the limitation**
- No satisfactory organic controls or technology at this time
- Grapes
    - Most insect pests can be controlled culturally, with mating disruption and trapping, and targeted biological, botanical, bio-rational sprays
    - Phylloxera root louse is a problem
- Controlled with rootstocks
- BLACK ROT– the limitation
- No satisfactory organic controls or technology at this time

Observations

- Slow response with organic mulches and organic, natural fertilizer sources
    - Young trees and vines are very sensitive to vegetation competition
    - However, can maintain tree nutrient status, cropping and quality with organic nutrient sources
- Some "tweaking" needed
    - Increased soil biological activity, significant increase in soil, pH maintenance
    - Increased water infiltration, retention

Pest Management

- Select multiple resistant cultivars
    - Generally available disease resistance but not insect resistance
- Strong cultural prevention program required
    - Sanitation, exclusion
- Implement IPM program
    - Scouting, monitoring, trapping
    - Employment of models to predict insect and disease outbreaks
    - Treat prior to outbreak; prevent problems rather than correct problems
    - Use mating disruption, trapping, predators and parasites for insect control; supplement with targeted bio-rational, biological and botanical insecticides
    - Use cultural controls to reduce potential for disease
    - Maintain a proactive fungi stat and bacteristatic spray program in advance of anticipated disease problems

Annual Cropping

- ☆ Annual cropping and avoiding biennial bearing is a significant issue in tree fruits, grapes, and blueberries

  Flowers are formed the year prior to bloom and are influenced by crop health, crop load and environment that year

  - ❖ Must have strong, consistent annual pruning
  - ❖ Must practice crop load management
- ☆ Few reliable growth regulators; must rely on hand labor
- ☆ The goal is to optimize crop load, thereby maximizing crop size and quality, and return bloom
  - ❖ Over cropped plants may have high yield of small, poor quality fruit; become biennial

Opportunities

- ☆ Tremendous local and regional markets for fresh fruit
  - ❖ Must still meet quality/grade standards to be acceptable
- ☆ Potential for local labeled fruit, eco-labeled fruit, and organically produced/labeled fruit

Summary Crop Comments

Natives

- ☆ Limited experience with commercial production
- ☆ Pest complex is not well understood
- ☆ Limited number of cultivars
- ☆ Marketing challenges

Grapes

- ☆ Sustainability metrics being adapted faster than many other specialty crops because of wine production and retail pressures
  - ❖ Moving beyond IPM in vineyards
- ☆ Requires annual pruning; high labor requirement for pruning
- ☆ Mechanization and pesticides may drive LCA (life cycle assessment-true costs) upward
- ☆ Difficult to produce organically in MO because of insect pests and particularly black rot of clusters

Blueberries

- ✰ Wide range of adapted species, hybrids, and cultivars for the region
  - ❖ Cultivars can be used to spread harvest and extend season
- ✰ Special care needed for soil management, weed control, and irrigation
- ✰ Few insect and disease pests; most can be managed
- ✰ Highly adaptable to sustainable production and readily adaptable to organic production
- ✰ Proper nutrition, pruning, irrigation are biggest management issues

Blackberries

- ✰ Widely adapted; strong local/regional market that is expanding
- ✰ Select disease resistant cultivars
- ✰ Readily adapted to sustainable and organic production
  - ❖ Limitations; weed control
  - ❖ Cane borers (can be difficult in organic planting)

Strawberries

- ✰ Now grown as an annual crop in plasticulture
  - ❖ Planted in Sept/Oct; overwintered; harvest in spring (March–May) until daily temps consistently exceed about 90°F
  - ❖ Require plastic mulch
  - ❖ Generally, require a row cover or organic straw mulch in the winter for winter and frost protection
  - ❖ High nutritional requirements
  - ❖ Require strong crop rotation schedule; probably a 3-5 year rotation on the same ground in the absence of soil fumigation or pasteurization
  - ❖ Some disease problems need to be controlled; apply fungi static sprays for organics
  - ❖ Few insect pests; can use row covers to protect from insects.

CONCLUSION

Helpful to achieve holistic approach of farming system through appropriate planning and applying best management practices Due to over use of chemical fertilizers other chemical based products in fruit crop production, fruits are losing their quality *viz.*, chemical compound, color, fragrance, shape and size. We have to move towards sustainable fruit crop production and this may possible by using organic farming, organic fertilizers, natural farming system *etc.* These farming systems benefited to fruit production as Protects biological flora, reduce soil erosion, Include ecofriendly management system. It comprises integrated pest and disease management system.

REFERENCES

Ahirwar, K. S. and Dwarka (2021). Sustainable Fruit Production: Principle and Practices. Agriculture and Environment. 2(20). pp. 28-30.

Altieri, M. A., Nicholls, C. I. and Fritz, M. (2005). Manage Insects on your Farm. A Guide to Ecological Strategies. Sustainable Agriculture Network. Beltsville, MD.

Gliessman, S.R., Swezey, S. L. and Rosemeyer, M. (2009). The Conversion to Sustainable Agriculture: Principles, Processes, and Practices. CRC Press.

IOBC (1994). Guidelines for Integrated production of pome fruits in Europe. IOBC/WPRS Bulletin 17(9): 1-8.

Liang, W. and Huang, M. (1994). Influence of citrus orchard ground cover plants on arthropod communities in China: A review. Agriculture, Ecosystems, Environment 50: 29-37

Chapter 4

Fertigation Concept and Practices for Horticultural Crops

Imran Ali[1], Prashant Gautam[2], Rahul Kumar[3] and Vijay Pratap Yadav[4]*

[1]M.Sc. Fruit Science, [2]Ph.D. Research Scholar, Department of Fruit Science,

[3]M.Sc. Vegetable Science, Department of Vegetable Science,

Sardar Vallabhbhai Patel University of Agriculture and Technology, Modipuram, Meerut, Uttar Pradesh

[4]M.Sc. Ag. (Horticulture), NPG College, Barahalganj, Gorakhpur (DDU University Gorakhpur), Uttar Pradesh

**e-mail: imranaligkp09@gmail.com*

INTRODUCTION

Fertigation is one procedure of applying fertilizers through water system frameworks, primarily in trickle promotion and other miniature water system frameworks; the alludes to the utilization of fertilizers to an area of soil, where most of the dynamic roots are available and furthermore matches the planning of wholesome interest of the tree. Numerous creators have contended that a profoundly precise application, both for water and fertilizers, can be accomplished by the synchronous organization of fertigation. Consequently, we acquire the benefit of synchronous stockpile of mineral components as per the tree needs (Neilsen *et al.*, 2001; Weinbaum *et al.*, 1992). The foliar utilization of compost related with ferti-water systems, safeguards in incredible measure the natural product trees needs for provisions, the dirt treatment is reducing and the composts utilization too (Marangoni *et al.*, 2001). Then again, assuming is applied continually, the restricted water system gives into the plants a supported sap circuit, with the ability to take-up the minerals particularly by the roots created in the dirt volume provided with the water-by-water system. Fertigation disposes of change of dampness in soil, which brings about upgraded plant advancement what's more, better returns (Bar-Yosef, 1999; Tojnko and Cmelik, 2004; Kipp, 2005).

Water and fertilizer are the two significant contributions for crops further more; connect with one another in impacting the development, yield and nature of harvests. As the vast majority of the green crops are water-and supplement comprehensive in nature, proficient administration of these valuable sources of info is a must for accomplishing higher efficiency while keeping up with natural quality. India is honored with assorted agro-climatic circumstances with particular seasons, making it conceivable to develop wide cluster of plant crops. Also, supported better returns with high-yielding assortments rely completely upon the feasible utilization of the restricted water and energy assets, explicitly in water-scant bone-dry and semi-dry locales. There is a pressing need to decrease the utilization of water system water by embracing compelling water-saving innovations and techniques.

Figure 4.1

Definition

Fertigation is a process in which fertilizer is dissolved and distributed along with water in your drip or spray irrigation system (Figure 4.1). There is abundant research available that supports the superiority of fustigations as compared to traditional fertilizing techniques. Most fertigation systems for residential use require the use of liquid fertilizers since they are unable to solubilise (dissolve) higher-concentration, water-soluble fertilizers. Most companies that produce fertilizers for fertigation produce liquid fertilizers since that is the only option that their dispensing units can handle.

The practice of supplying crops with fertilizers via the irrigation water is called fertigation (Bar Yosef, 1992). Fertigation is a modern technique and provides an excellent opportunity to maximize yield and minimize environmental pollution (Hagin *et al.*, 2002). This is provided by increasing fertilizer use efficiency, minimizing fertilizer application and increasing return on the fertilizer invested.

In a fertigation system, the timing, quantity, concentration and ratio of the nutrients can be easily controlled. Nutrients can be applied during the growing season based on crop need, hence saving in amount of fertilizer applied, due to reduction in leaching losses and increased water use efficiency. It is safer application method, as it minimizes the chances of damage to the roots due to higher dose. Due to this improved control, optimum yield and higher quality of produce are achieved than those produced by traditional fertilizer application. The use of fertigation system to apply fertilizer reduces the need of mechanical operations or sometimes completely eliminates them. Smaller amounts of fertilizer can be applied quickly to solve any deficiency issues and highly mobile nutrients such as nitrogen can be carefully managed to ensure uptake by the crop at faster rate. Fertigation has become a striking method of fertilizer application in modern intensive agriculture system.

Meaning of Fertigation

Fertigation is the application of soluble fertilizers (plant nutrients) through an irrigation system. Reduce fertilizer needed by 70-90 per cent which virtually eliminates fertilizer runoff. Save from 20-50 per cent on your water us age Improve the health and vitality of your landscape.

Fertigation is a technique of fertilizer application in which fertilizer is incorporated within the irrigation water by the drip system where timing, amounts and concentrations of fertilizers applied are easily controlled. Fertigation ensures saving in fertilizer (30-50 per cent), due to "better fertilizer use efficiency" and "reduction in leaching" (Kumar and Singh, 2002).

Fertigation is a recent innovative cultural method, by which fertilizers are applied along with irrigation water through irrigation systems like drip to get higher fertilizer use efficiency as well as increasing crop yields. Nutrient solutions are injected into the irrigation water using an appropriate injection device. Fertigation provides essential elements directly to the active root zone, thus minimizing losses of expensive nutrients, which ultimately helps in improving productivity and quality of farm produce and reduces the risk of environmental pollution. Fertigation involves specific equipment according to the crop type and irrigation system and precise selection of suitable fertilizer and its combination.

History of Fertigation

The first reported example of fertigation dates back to ancient Athens (400

B.C.) where city sewage was used for the irrigation of tree groves. The practice of fertigation started commercially in the mid - 20th century. Liquid ammonia was probably the first commercially produced liquid fertilizer but in modern fertigation the use of ammonia as the nitrogen source is negligible. Since the early 1950s, in The Netherlands, the fertilizers were applied with the irrigation water in the glasshouses. In mid 1950s, in the United States, mixing of fertilizers with irrigation water was used on a limited scale, in surface, flood and furrow irrigations. The fertilizers used were gaseous ammonia, aqua ammonia and ammonium nitrate. In the early 1960s, in Israel, the development of fertigation technology was parallel to the development and introduction of micro-irrigation. Electrical pumps and mixing tanks were developed for the precise application of nutrients. Initially, nutrient distribution by fertigation was relatively uneven when fertilizer tanks were used. Later, a more uniform distribution was achieved when venturi suction pumps and fertilizer injectors were used. Further considerable improvements were achieved by fully computerized fertigation units.

Significance of Fertigation in Horticulture

In India, most of the horticultural crops are cultivated on raised beds, with soil texture ranging from loamy clay to sandy loam, where downward percolation of water is rapid and often leads to leaching of nutrients beyond root zone under the conventional irrigation methods. Under such soil conditions, frequent occurrences of the deficiencies of macro and micro nutrients become a major production constraint exacerbated by the low moisture and nutrient holding capacities of these soils. Under conventional methods of irrigation, rapid hydraulic conductivity, faster infiltration rate cause accelerated leaching of basic cations to the lower layers (below root zone). Net result is deterioration of soil fertility. This condition may load the ground water with pollutants like leached down fertiliser salts, other toxic pesticides, herbicides *etc.* Hence, the horticultural crops in such soils under conventional irrigation system require large quantities of nutrients to support its growth and yield for compensating the leaching losses of nutrients. Considering the soil and crop constraints, nutrient should be applied in synchrony with crop demand in smaller quantities during the growing season. The right combination of water and nutrients is a prerequisite for higher yield and good quality of produce. Method of fertilizer application is also important in improving the use efficiency of nutrients. Fertigation is the only irrigation strategy, which helps in adequate supplies of water and nutrients with precise timing and uniform distribution to meet the nutrient demand of horticultural crops. Besides, fertigation ensures substantial saving in fertilizer usage and reduces leaching losses (Mmolawa and Or, 2000).

Need of Fertigation

When pressurized irrigation systems are used, fertigation is not optional but absolutely necessary. In drip irrigation, without fertilization only ~30 per cent of the soil is wetted by the drippers so fertilizer efficiency decreases because the nutrients will not be dissolved in the dry zones where the soil is not wetted. Thus, the benefits of irrigation and fertilizer will not be expressed. Therefore, fertigation is the most suitable method to apply fertilizers to micro irrigated crops (Figure 4.2).

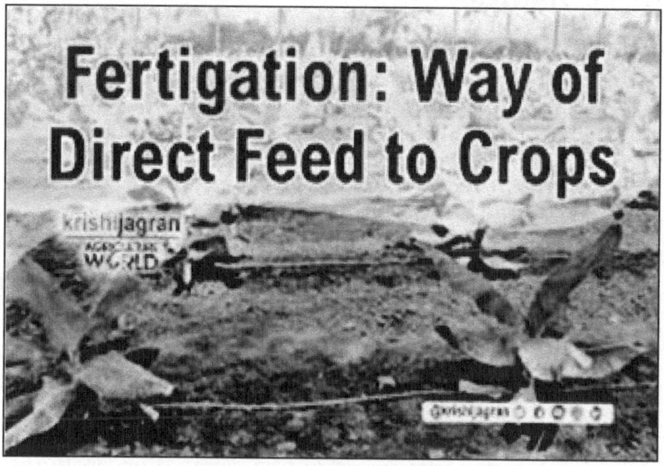

Figure 4.2

It is the most effective and convenient means of maintaining optimum fertility level and water supply according to the specific requirement (Shirgure, *et al.*, 2000).

1. Fertigation permits application of a nutrient directly at the site of a high concentration of active roots and as needed by the crop. Scheduling fertilizer applications on the basis of need offers the possibility of reducing nutrient element losses associated with conventional application. Methods that depend on the soil as a reservoir of nutrients thereby increasing nutrient use efficiency. Fertilizer savings through fertigation can be to the tune of 25-50 per cent (Haynes, 1985).
2. The concept of fertigation is new but the rapid increase in area under micro irrigation, fertigation is getting popular. Fertigation is the technique to apply water soluble solids or liquid fertilizers through the drip irrigation on weekly or monthly basis so as to reach each and every plant regularly and uniformly.
3. Fertilizers and pesticides applied through a drip irrigation system can improve efficiency, save labour and increase flexibility in scheduling of applications to fit crop needs. However, all chemicals must meet the following criteria for the successful maintenance of the drip irrigation system (Bucks and Nakayama, 1980).

However, increasing water scarcity and value crops and green houses to ensure higher escalating fertilizer prices may lead to greater efficiency of the two most critical inputs in crop adoption of the technology especially in high production.

Benefits of Fertigation

1. **Higher nutrient use efficiency**: Nutrient use efficiency by crops is greater under fertigation compared that under conventional application of fertilizers to the soil.
2. **Less water pollution:** Intensification of agriculture led by the use of irrigation water and indiscriminate use of fertilizers has led to the pollution of surface and ground waters by chemical nutrients. Fertigation helps lessen pollution of water bodies through the leaching of nutrients such as N and potassium (K) out of agricultural fields.
3. **Higher resource conservation:** Fertigation helps in saving of water, nutrients, energy, labor and time. Nutrients can be applied to the soil when crop or soil conditions would otherwise prohibit entry into the field with conventional equipment.
4. **Efficient delivery of micronutrients:** Fertigation provides opportunity for efficient use of compound and ready-mix nutrient solutions containing small concentrations of micronutrients, which are otherwise very difficult to apply accurately to the soil when applied alone.
5. **Healthy crop growth:** When fertigation is applied through the drip irrigation system, crop foliage can be kept dry thus avoiding leaf burn and delaying the development of plant pathogens.
6. **Helps in effective weed management:** Fertigation helps to reduce weed menace particularly between the crop rows. Use of plastic mulch along with fertigation through drip system allows effective weed control in widely spaced crops. However, to get maximum benefit of fertigation, care must be taken while selecting the fertilizer and injection equipment and the management and maintenance of the system.

Advantages of Fertigation

1. The availability of the nutrient is high due to maintenance of soil moisture near root zone under drip irrigation.
2. Chances of loss of nutrients through leaching are minimum and are around 10 per cent as compared to 40-55 per cent in the traditional system, which further help in minimizing environmental pollution. It also reduces soil erosion.
3. There is about 25-50 per cent reduction in the quantity of fertilizer that leads to higher fertilizer use efficiency.

4. Timely as well as precise application of fertilizers directly to the roots zone improves fertilizer use efficiency and reduces nutrient leaching below the root zone.
5. Fertigation reduces the cost of labour and saves energy in the application of fertilizers.
6. It is a comparatively safer application method, as it eliminates the danger affecting roots due to higher dose.
7. Ensures a uniform and regular flow of water as well as nutrients, resulting in increased growth rates for higher yields and quality.
8. Offers greater versatility in the timing of the nutrient application to meet specific crop demands. In most cases, fertilizers can be applied regardless of weather or field conditions.
9. Improves availability of nutrients and their uptake by the roots.
10. Safer application method, as it eliminates the danger of burning the plant root system.
11. Timely applications of small but precise amounts of fertilizer directly to the root zone allows growers to effectively manage fertilizer programs, this improves fertilizer use efficiency and reduces nutrients leaching below the root zone.

Disadvantages

- ☆ High initial system cost.
- ☆ Power costs.
- ☆ Difficulties with emitter uniformity.
- ☆ Careful system design is essential.
- ☆ Soil salinity issues must be addressed as well as the effects of excess calcium carbonate dissolved in the irrigation waters.
- ☆ Emitter clogging will affect distribution uniformity.
- ☆ Algae growth and scale build-up ($CaCO_3$) must be controlled.
- ☆ Provisions must be made for utilising the flush water, same as with all systems that use filters.
- ☆ Water must be available on a regular basis.
- ☆ Problems with deficit irrigation strategies during the early stages of cultivation may result in limited bulb size and root entry in the dripper.
- ☆ The depth of drip line installation limits soil tillage.
- ☆ Uneven nutrient distribution when the irrigation system is faulty.
- ☆ Chemical reactions of fertilizer with calcium and bicarbonate in water, which can lead to clogging of drip fertigation system.

- ☆ Non-target chemical application can occur as a result of malfunctioning equipment.
- ☆ Beneficial only when the drip irrigation system is adequately designed, fully functional and properly managed.

Limitations of Fertigation

- ☆ Uneven nutrient distribution when the irrigation system is faulty.
- ☆ Chemical reactions of fertilizer with Ca and Mg may leads to chemical clogging.
- ☆ Phosphatic fertilizers and some micro nutrients may precipitate in micro irrigation systems.
- ☆ Corrosion resistant fertigation equipment's are needed.
- ☆ Potential chemical backflow in to water supply source.
- ☆ The major factors limiting its large-scale adoption are high initial cost and lack of information on various aspects such as crop water requirement, scheduling of irrigation and fertigation.

Fertigation System

Fertigation technology distributes fertilizers to the crops/plant root zone along with water through drip or sprinkler irrigation at frequent intervals in controlled manner so as to allow for steady uptake of nutrients by plants and to effect savings in the costly inputs of both water and fertilizer (Figure 4.3).

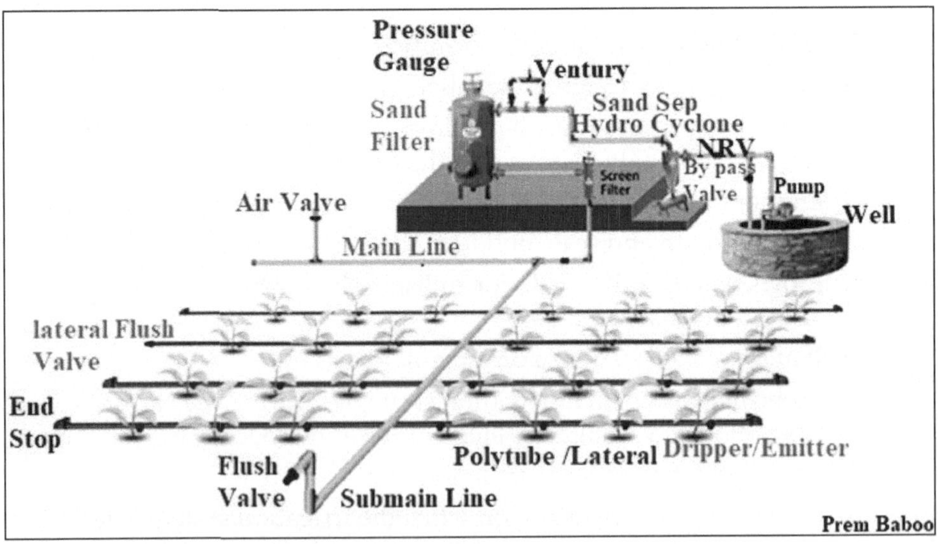

Figure 4.3: Fertigation Through Drip Irrigation System.

Fertilizer Injection Equipment

The selection of correct injection equipment is as important as the selection of correct nutrients. The pressure of injected fertilizer solution has to be greater than that of the internal pressure. A filter should be there to prevent the dripper clogging by any solid particles of fertilizer solution reaching the dripper. A back-flow preventing valve is also needed to prevent the chemical backflow to the water source. Fertilizer injectors are the most important part of the fertigation unit. There are three types of injectors:

1. Pressure Differential (By-pass tank)

This system works on the principle of a pressure differential created by valve and pressure regulation. The pressure difference forces the water to enter through a by-pass pipe into a pressure tank which contains the fertilizer and to go out again, carrying a varying amount of dissolved fertilizer. The application of nutrients is quantitative and inaccurate, therefore is adapted for perennial crops like citrus and fruit trees (Figure 4.4).

Figure 4.4: Fertilizer Tank.

Advantage

- ☆ Very simple to operate, the stock solution does not have to be pre-mixed.
- ☆ Easy to install and requires very little maintenance.
- ☆ Easy to change fertilizers.
- ☆ Ideal for dry formulations.
- ☆ No electricity or fuel is needed.

Disadvantages

- ☆ Tank must be able to withstand irrigation line pressure.
- ☆ Concentration of solution decreases as fertilizer dissolves.
- ☆ Accuracy of application is limited.
- ☆ Pressure loss in main irrigation line.
- ☆ Proportional fertigation is not possible.
- ☆ Limited capacity.
- ☆ Not adapted for automation.

2. Vacuum Injection (Venturi)

This is based on the principle of venturi tube (Figure 4.5). A pressure difference is needed between the inlet and the outlet of the injector which cause a reduced pressure (vacuum) that sucks the fertilizer solution into the line.

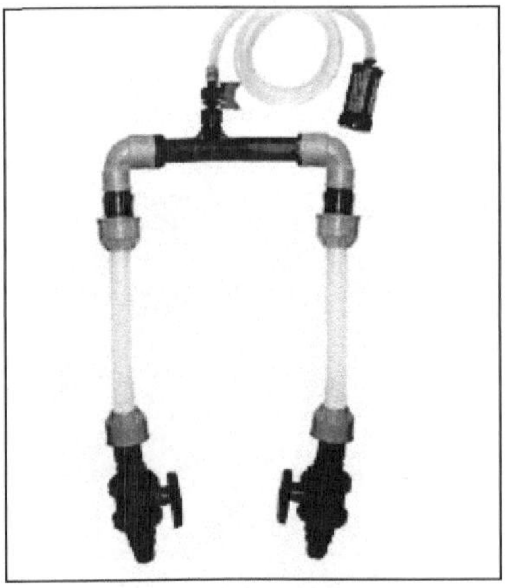

Figure 4.5: Venturi.

Advantages

- ☆ Very simple to operate and no moving parts.
- ☆ Easy to install and to maintain.
- ☆ Injection can be controlled with a metering valve.
- ☆ Suitable for both proportional and quantitative fertigation.

Disadvantages

- ☆ Pressure loss in the main irrigation line.
- ☆ Quantitative fertigation is difficult.
- ☆ Automation is difficult.
- ☆ Very sensitive to pressure variation.

3. Pump Injection

Pumps are used to inject the fertilizer solution from a supply tank into the line (Figure 4.6).

Injection energy is provided by electric motors or hydraulic motors.

Figure 4.6: Pump Injection.

Advantages

- ☆ Very accurate for proportional fertigation
- ☆ No pressure loss in the line
- ☆ Easily adapted for automation

Disadvantages

- ☆ Expensive
- ☆ Complicated design, including a number of moving parts, so wear and breakdown are more likely.

The advantages and disadvantages of using a venturi system for fertigation are given below:

Advantages

- ☆ Simple in operation.
- ☆ Relatively inexpensive in purchasing and maintenance.
- ☆ Good control over the fertilizer concentration with the irrigation water.

Disadvantages

- ☆ High pressure loss (*i.e.* about one third of the total operating pressure).

- ☆ Relatively low discharge rate system.
- ☆ Precise regulation of flow is difficult because the rate of injection is very sensitive to the pressure and rate of flow of water in the main system
- ☆ The irrigation system needs to be operated at full capacity prior to injecting the fertilizer solution.

Methods of Fertigation

When using fertigation, the fertilizer solutions are prepared in advance in stock tanks and the solution is then injected into the irrigation water. The type of fertigation chosen depends on the crop grown, the soil type and the farm management system. There are mainly two types of fertigation.

1. Quantitative Fertigation

It is the application of the plant nutrients in predetermined concentrations to the irrigation system. *e.g.* 20 liter to block A, 40 liter to block B. This method is commonly used in soil. The fertilizer is applied in a pulse after a certain water sheet without fertilizer using a fertilizer tank. Advantages of this method are the low cost and the low maintenance required. Some limitations are also there like, the system is affected by water pressure changes and the concentration of the fertilizer varies during its application.

2. Proportional Fertigation

Nutrients applied in a constant and proportional ratio to the water discharge rate, so that the irrigation water takes a fixed concentration of the applied fertilizer. *e.g.* one liter of fertilizer solution is mixed in, 1000 litres of irrigation water. In this case the fertilizers are applied by direct injection through fertilizer pumps. This method is mostly used in soil less media and sandy soils. Advantage of this method is precise control of the quantity and the injection moment and the disadvantages are high cost, maintenance and complicated operation.

Practices of Fertigation

To capitalize on fertigation benefits, particular care should be taken in fertilizer preparation, dosification, fertilizer injection methods as well in the management and maintenance system.

Fertilizer Preparation

The application of fertilizers is executed by various methods (Tables 4.1 and 4.2).

- ☆ **Stock Solution Preparation**: Mix solid fertilizers as ammonium sulfate, potassium chloride, nitrate and phosphoric acid to prepare a tailor-made stock solution. The stock solution is then injected into the irrigation system @ 2-10 L/m^3

Table 4.1: Acceptable Concentration of different Nutrients

Sl.No.	*Nutrients*	*Acceptable Limits of Concentration, ppm*	*Average Acceptable Concentration, ppm*
1.	Nitrogen	150 - 1000	250
2.	Phosphorus	50-100	80
3.	Potassium	100-400	300
4.	Calcium	100-500	200
5.	Magnesium	50-100	75
6.	Sulphur	200-1000	400
7.	Copper	0.1-0.5	0.25
8.	Boron	0.5-5.0	1.0
9.	Iron	2.0-10	5.0
10.	Manganese	0.5-5.0	2.0
11.	Molybdenum	0.01-0.05	0.02
12.	Zinc	0.5-1.0	0.5
13.	Sodium	20-100	50
14.	Carbonates	20-100	60
15.	Sulfate	200-300	250
16.	Chloride	50-100	70

Table 4.2: Suitable Fertilizers Used for Fertigation

Nutrients	*Fertilizers*	*Per cent Nutrient*
N	Urea	46 per cent N
	Ammonium sulfate	21 per cent N
	Urea ammonium nitrate (L)	32 per cent N
	Ammonium nitrate	34 per cent N
N and P	Mono ammonium phosphate	12 per cent N, 61 per cent P_2O_5
	Urea Phosphate	17 per cent N, 44 per cent P_2O_5
P	Phosphoric acid	52 per cent P_2O_5
P and K	Mono potassium phosphate	52 per cent P_2O_5, 34 per cent K2O
	Potassium chloride	60 per cent K_2O
	Sulphate of Potash	50 per cent K_2O, 17.5 per cent S
K	Potassium nitrate (Multi K)	13 per cent N, 46 per cent K_2O
	Potassium thiosulphate	25 per cent K_2O, 17.5 per cent S
N, P and K	Poly feed	19-19-19 per cent NPK
	Urea Phosphate with SOP	18-18-18 per cent NPK
Mg and Ca	Magnesium nitrate	11 per cent N
	Calcium nitrate	16 per cent N, 19 per cent Ca

- **Compound solid fertilizer mixtures**: Manufactured for use in fertigation, with ratios between the three major elements. Some compositions contain microelements in the form of chelates.
- **Compound liquid fertilizer solutions**: The total nutrient concentration is 5-3-8, 6-6-6, 9-2-8 specified for use in the greenhouses. Some compositions contain microelements in the form of chelates.

Generally, two fertilizer tanks that contain concentrated fertilizer solutions are used to separate those fertilizers that can interact. A possible combination is; tank "A" containing calcium nitrate, potassium nitrate, magnesium nitrate and microelements, whereas tank "B" contains ammonium sulfate, phosphoric acid, nitric acid; in this way P and Ca/Mg are in different tanks to avoid their precipitation. A third tank "C" contains an acid solution to control the pH of the fertilizer solution and to wash the irrigation system to avoid clogging of drippers (Table 4.3).

Precautions to be taken during Fertigation

- Different chemicals like, fertilizers, pesticides, chlorine should not be injected at the same time.
- Selection of most appropriate fertilizer and injection system plays a crucial role.
- Drip irrigation system should be allowed to its working pressure prior to fertilizer injection.
- Be sure the placement of the drip tubing does not interfere with the cultivation system. For example: If planting tomatoes down the center of a raised bed, place the drip tubing about 15 cm off center. This will prevent damaging the tubing when making holes for the transplants and when inserting supporting stakes.
- Make sure the fertilizer solution is compatible with the quality of water into which it is being injected. Some fertilizers can cause a precipitate that will clog the drip system or filter system. For example: Calcium and Phosphorus fertilizers should not be mixed with water containing sulfates in solution.
- A small screen should be put on the end of the suction line to help eliminate solid particles or un-dissolved fertilizer from entering the system and stopping it up.
- Do not inject fertilizers in combination with pesticides or chlorine.
- The injection point must be upstream of the filter system so that the filter will remove any un- dissolved fertilizer or precipitates that occur.
- Before beginning injection of a fertilizer, bring the drip irrigation system up to operating pressure. At this point even the part of the irrigation system farthest from the source should be pressurized.

Table 4.3: Fertigation Requirement and its Response in Major Fruit and Plantation Crops

Sl.No.	*Crop*	*Fertigation Requirement*	*Crop Response*	*References*
1.	Mango (*Mangifera indica* L.) Variety: Dashehari 10 m x 10 m spacing	75 per cent RDF (380 g N-380 g P_2O_5 -380 g K_2O per tree) + mulch + drip irrigation	Increase in growth yield and quality, in addition in V-area of basin in the 16-year-old orchard to soil fertility	Panwar *et al.* (2007)
2.	Mango (*Mangifera indica* L.) Variety: Alphonso 2.5 m x 2.0 m spacing	100 per cent RDF (120 g N- 100 g P_2O_5-100 g K_2O per tree) + 24 litres water per day in 3-year- old orchard	Increase in fruit yield and quality	Prakash *et al.* (2015)
3.	Mango (*Mangifera indica* L.) Variety: Dashehari 10 m x 10 m spacing	75 per cent RDF (750 g N-375 g P_2O_5 -750 g K_2O per tree) + 100 per cent irrigation through drip based on 0.60W/ CPE ratio in 30-yearold orchards	Improvement in nutrition distribution, nutrient use efficiency coupled with better soil moisture distribution and yield	Adak *et al.* (2014)
4.	Guava (*Psidium guajava* L.) Variety: Shweta 2 m x 1 m spacing	75 per cent RDF (30 g N-10 g P_2O_5 -10 g K_2O per tree) + 75 per cent irrigation requirement based on CPE in 3-year-old meadow orchard	Increase in flowering leaf nutrient status, growth, yield and quality	Ramniwas *et al.* (2013)
5.	Guava (*Psidium guajava* L.) Variety: Shweta 2 m x 1 m spacing	100 RDF + surface drip irrigation at1.0 IW/CPE ratio	Increase in growth and fruit yield response	Sharma *et al.* (2013)
6.	Pomegranate (*Punica granatum* L.) Variety: Bhagwa 2 m x 2 m spacing	100 per cent RDF (200 g N-200 g P_2O_5 -500 g K_2O per tree using water-soluble 19: 19: 19 fertilizer) + 100 per cent irrigation at alternate day through drip irrigation in 4-year-old orchard	Better soil moisture distribution, fruit yield and quality	Haneef *et al.* (2014)
7.	Pomegranate (*Punica granatum* L.) Variety: Mridula 2 m x 2 m spacing	50 per cent RDF using water soluble fertilizer+ drip irrigation at 100 per cent CPE in 2-year- old orchard	Improvement in flowering, soil available nutrients and fruit yield	Shanmugasundaram *et al.* (2013)
8.	Almond (*Prunus dulcis* L.) Variety: Waris	75 per cent RDF (330 g N-45 g P_2O_5 -455 g K_2O per plant through water soluble fertilizers) + drip irrigation based on pan evaporation in 8- year-old orchard	Increase in growth response and yield in addition to soil fertility	Dinesh Kumar and Almed (2014)

Sl.No.	Crop	Fertigation Requirement	Crop Response	References
9.	Banana (*Musa paradisiaca* L.)	100 per cent RDF (200 g N- 110 g P_2O_5 - 200 g K_2O plant-1 crop-1) fertigation along with microbial consortium (Azospirillum, phosphate solubilizing bacteria and AM fungi mixed in equal proportions)	Increase in uptake of secondary nutrients and micro nutrients besides fruit yields	Senthil Kumar *et al.* (2014)
10.	Banana (*Musa paradisiaca* L.) Variety: Robusta	100 per cent RDF (13 g N-40 g P_2O_5 -13 g K_2O per plant using urea, DAP and KNO_3 + drip irrigation at 1.0 IW/CPE	Improvement in growth yield and quality	Mahendran *et al.* (2013)
11.	Apple (*Malus domestica* Borkh L.) Variety: Gala on M-26 rootstock 2 m x 4 m spacing	40 g N-17.5 g P_2O_5 using ammonium nitrate and ammonium pyrophosphate (10–15 – 0) in 5-year-old orchard	Improved root distribution and alleviated K-deficiency	Neilsen *et al.* (2000)
12.	Apple (*Malus pumilaalill* L.) Variety: Gala on M-9 rootstock3.5 x 1.25 m spacing	60 per cent RDF (30 g N, 4 g P, 37 g K, 6 g Mn and 4 g Ca per tree) + irrigation based on pan evaporation	Improved growth and yield initially with N-fertigation and later NPK fertigation proved better	Treder (2006)
13.	Strawberry (*Fragaria* x *ananassa* Duch.) Variety: Chandler 25 m x 50 m spacing	75 per cent RDF (112 kg N-75 kg P_2O_5 -90 kg K_2O per ha) + 100 per cent *etc.*	Increase in growth yield and quality	Kachwaya and Chandel (2015)
14.	Papaya (*Carica papaya*) Variety: Taiwan 2.5 m x 2.0 m spacing	100 per cent RDF (250 g N-250 g P2 O5 -500 g K_2O per plant) + 50 mm CPE as100 per cent CPE	Improvement in growth, yield and quality	Deshmukh and Hardha (2014)
15.	Acid lime (*Citrus aurantifolia* Swingle) Variety: Kagzi lime 6 m x 6 m spacing	75 per cent RDF (450 g N-150 g P2 O5 -175 g K_2O per tree) + irrigation at 30 per cent AWC depletion	Improvement in growth, yield, quality and leaf nutrient composition	Shirgure *et al.* (2014)
16.	Nagpur mandarin (*Citrus reticulata* Blanco) Variety: Nagpur mandarin 6 m x 6 m spacing	80 per cent RDF (48 g N-160 g P_2O_5 -240 g K_2O per tree) + irrigation at 20 per cent AWC depletion	Increase in growth, yield, quality, leaf nutrient composition and soil fertility	Shirgure *et al.* (2016)

Sl.No.	*Crop*	*Fertigation Requirement*	*Crop Response*	*References*
17.	Cocoa (*Theobroma cacao* L.)	125 per cent RDF (125 g N+ 50 g P_2O_5 +175 g K_2O per plant per year) with irrigation of 20 L per tree per day	A phenomenal increase in growth parameters such as trunk girth, canopy spread and weight of the pruned branches removed, leaf fresh weight and leaf dry weight as well as levels of NPK	Krishanamoorthy *et al.* (2013)
18.	Arecanut (*Areca catechu* L.)	75 per cent NPK (RDF: 100 g N: 18 g P_2O_5: 117 g K_2O per palm per/year) along with irrigation at IW/CPE ratio of 1 with a 30 mm depth of water through basin irrigation	Increase in leaf water potential, root biomass and organic carbon content in soil based on polynomial regression as well as significant increase in water use efficiency and agronomic nutrient use efficiency	Bhat *et al.* (2007)

Source: Jeyabaskaran *et al.* (2021).

- ☆ After fertigation is completed, irrigate with plain water so that the system is flushed out and fertilizer is washed into the plant beds.
- ☆ Select fertilizer solutions to help adjust water pH if necessary.
- ☆ Never inject any material that is not labelled and recommended for the crop and for injection through the system. Always follow label directives.
- ☆ The use of material that might cause precipitation and clogging of the system should be avoided.

Calculation of Fertiliser Solution Concentrations

The concentration of fertilizer solutions is usually expressed in parts per million (ppm) of nitrogen. To determine how much fertilizer material is required to produce a solution of the desired concentration, use the following formula:

Quantity of fertiliser Required (grams) =

$$\frac{\text{Solution concentration (ppm) x solution volume (litres)}}{\text{10 x (per cent N of fertiliser material)}}$$

For example, to make a 100 ppm solution of 20-10-20 fertilizer in a 500-liter tank, the amount of fertilizer required is

$$\text{Quantity of fertilizer required (grams)} = \frac{(100 \times 500)}{(10 \times 20)}$$

Therefore, the quantity of fertilizer required is 250 grams.

CONCLUSIONS

Fertigation system is an efficient method of applying fertilizers in which the irrigation system is used as the carrier and distributor of the crop nutrients. The combination of water and nutrient leads to efficient use of both by the crop plants. The use of fully soluble fertilizers seems more economical in the first stages when fertigation is adopted, but the use of prepared clear liquid solutions is very convenient in other cases. Fertigation system is very suitable for commercial horticulture where maximization of profitability and yields is a target. Fruit and vegetable crops have been found to be responsive to fertigation due to their wide spacing nature, continuous need for water and nutrients at an optimal rate to give high yields with good quality, and high capital returns on investments. Even though the initial cost of establishing the fertigation system is higher but on long-term basis it is economical compared to conventional methods of fertilization as it brings down the cost of cultivation.

However, to get the desired results, it requires higher management skills at operator level in selection of fertilizers, timing and rate of fertilizer injection, watering schedule, as well as the maintenance of the system. Therefore, to make horticulture sustainable and economically viable and to ensure food and nutritional security of the burgeoning population there is a need to promote the fertigation at large scale by all the concerned stakeholders.

REFERENCES

Adak, T., Kumar, K. and Singh, V. K. (2014). Fertigation regime impacting productivity, moisture, and nutrient distribution in mango under subtropical conditions. *Indian Journal of Soil Conservation* 42(3), 282-292.

Anonymous (2017). PDFCs Research Findings on Fertigation. Deptt. Of Agriculture and Cooperation and Farmers Welfare, MoA and FW, GOI. https: //pmksy.gov.in/pdflinks/PFDCs_Res-earch_Findings_on_Fertigation.pdf.

Bar-Yosef, B. (1992). Fertilization under drip irrigation. In: Fluid Fertilizer, *Science and Technology*. Ed. by D. A. Palgrave. Marcel Dekker, New York., 285-329.

Bhat, R., Sujatha, S. and Balasimha, D. (2007). Impact of drip fertigation on productivity of Arecanut (*Areca catechu* L.). *Agricultural Water Management* 90(1-2), 101-111

Bucks, D. A. and Nakayama, F. S. (1980). In: Proc. Agri-Turi. Irrig. ConI., California, pp 166-180.

Deshmukh, G. and Hardha, M. K. (2014). Effect of irrigation and fertigation scheduling under drip irrigation in papaya. *Journal of Agricultural Search* 1, 216-220.

DeValerio, J., Nistler, D., Hochmuth, R. and Simonne, E. (2012). Fertigation for vegetables: A practical guide for small fields. *EDIS,* (10).

Haneef, M., Kaushik, R.A., Sarrolia, D. K., Mordia, A. and Dhakar, M. (2014). Irrigation scheduling and fertigation in pomegranate cv. Bhagwa under high-density planting system. *Indian Journal of Horticulture* 7, 45-48.

Hagin, J., Lowengart, A. and Sneh, M. (2002). Fertigation: Fertilization through irrigation. International Potash Institute Research Topics No. 23. Ed. by A.E. Johnston. *International Potash Institute*, Basel, Switzerland.

Haynes, R. J. (1985). Principles of fertilizer use for trickle irrigated crops. *Fertilizer research*, *6*(3), 235-255.

Jeyabaskaran, K. J., Shirgure, P. S., Pandey, V., Srivastava, A. K. and Uma, S. (2021). Fertigation in horticulture: a guarantee to economized quality production.

Kachwaya, D. S. and Chandel, J. S. (2015). Effect of fertigation on growth, yield, fruit quality and leaf nutrients content of strawberry (*Fragaria × Ananassa*) cv Chandler. *Indian Journal of Agricultural Sciences* 85, 1319-1323.

Kipp, A. J. (2005). Thirty years fertilization and irrigation in Dutch apple orchards. *Nutrient Cycling in Agroecosystems* 32, 149-156.

Krishnamoorthy, C., Rajamani, K. and Kumar, N. (2013). Effect of fertigation through drip and micro sprinkler on plant biometric characters in cocoa (*Theobroma cacao* L.), *Asian Journal of Horticulture* 8(1), 1-7.

Kumar, D. and Ahmed, N. (2014). Response of nitrogen and potassium fertigation to Waris almond (*Prunus dulcis*) under North Western Himalayan region of India. *The Scientific World Journal*, Article I D 141328, 6 pages.

Kumar, A. and Singh, A. K. (2002). Improving nutrient and water use efficiency through fertigation. *Journal of Water Management* 10, 42-48.

Kumar, S., Kumawat, P. and Choudhary, K. (2020). Irrigation scheduling and fertigation in citrus fruit crops. *Chemical Science Review and Letters* 9(34), 418-424.

Marangoni, B., Toselli M., Zavolloni C. and Flore J. (2001). Nitrogen accumulation and gas exchange in apple trees grafted on M-26 EMLA, M-9 EMLA and Mark rootstocks under low soil temperature and drought stress. *Acta Hort.*, 564, 193-200.

Mahendran, P. P., Yuraj, M., Paremeswari, C., Gurusamy, A. and Krishnasamy, S. (2013). Enchaining growth, yield and quality of banana through subsurface drip fertigation. *International Journal of Chemical, Environmental and Biological Science* 1(2), 391- 394.

Mmolawa, K. and Or, D. (2000). Root zone solute dynamics under drip irrigation: a review. *Plant and Soil* 222, 163-190.

Neilsen, D., Millard, P., Herbert, L. C., Neilsen, G. H., Hogue, E. J., Parchomchuk, P. and Zebarth, B. I. (2001). Remobilization and uptake of N by newly planted apple (*Malus domestica*) trees in response to irrigation method and timing of N application. *Tree Physiology* 21, 513-521.

Neilsen, G. H., Parchoonchuk, P., Neilsen, D. and Zebarth, B. J. (2000). Drip-fertigation of apple trees affects distribution and development of K deficiency. *Canadian Journal of Soil Science* 80(2), 353-361.

Prakash, K., Vijaykumar, R. M., Balamohan, T. N. and Sundhar Singh, S. D. (2015). Effect of drip irrigation regime and fertigation regime on yield and quality of mango cultivar Alphonso under ultra-high-density planting. *Acta Horticulturae*, doi: 10.17660/1066.17.

Panwar, R., Singh, S. K., Singh, C. P. and Singh, P. K. (2007). Mango fruit and quality improvement through fertigation along with mulch. *Indian Journal of Agricultural Sciences* 77, 680-684.

Ranjan, S. and Sow, S. (2021). Fertigation: An efficient means for fertilizer application to enhance nutrient use efficiency.

Kaushik, R. A., Pareek, S., Sarolia, D. K. and Singh, V. (2013). Effect of drip fertigation scheduling on fertilizer use efficiency, leaf nutrient status, yield and quality of 'Shweta' Guava (*Psidium guajava* L.) under meadow orcharding. *National Academy Science Letters*, *36*(5), 483-488.

Sabo M. U., Wailare, M. A., Aliyu, M., Jari, S. and Shuaibu, Y. M. (2013). Effect of NPK fertilizer and spacing on growth and yield of watermelon (*Citrullus lanatus* L.) in Kaltungo Local Government area of Gombe State, Nigeria. *Journal of Agricultural Sciences* 3(8), 325-330.

Shanmugasundaram, T. and Balakrishnamurthy. G. (2013). Effect of fertigation on flowering and yield of tissue culture pomegranate (Punica granatum L.) cv. Mridula grown under ultra-high-density planting. *ischolar*, 8, 601-604

Sharma, S., Patra, S. K., Roy, G. B. and Bera, S. (2013). Influence of drip irrigation and nitrogen fertigation on yield and water productivity of guava. *The Bioscan*, *8*(3), 783-786.

Shirgure, P. S., Srivastava, A. K. and Huchche, A. D. (2014). Water requirements in critical growth stages and fruit productivity of drip irrigated Nagpur mandarin (*Citrus reticulata* Blanco). *Indian Journal of Agricultural Sciences* 84, 317-322.

Shirgure, P. S., Marathe, R. A., Ram, L. and Singh, S. (2000). Irrigation scheduling in acid lime (*Citrus aurantifolia*) as affected by different soil moisture regimes. *Indian Journal of Agricultural Sciences*, 70(3), 173-176.

Senthilkumar, M., Ganesh, S., Srinivas, K. and Panneerselvam, P. (2014). Enhancing uptake of secondary and micronutrients in banana cv. Robusta through intervention of fertigation and consortium of biofertilizers. *Scholars Academic Journal of Biosciences* 2, 472-478.

Treder, W. (2006). Influence of fertigation with nitrogen and a complete fertilizer on growth and yielding of 'Gala' apple trees. *Journal of Fruit and Ornamental Plant Research* 14, 143-154.

Sharma, R., Bhardwaj, S. and Choudhary, M. (2014). Fertigation on Fruit Crops: A Review. *Annals of Biology* 30(1), 48-53.

Shukla, M., Sadhu, A. C., Chinchmalatpure, A. R., Prasad, I., Kumar, S. and Camus, D. Fertigation-Modern Technique of Fertilizer Application. *INDIAN FARMER*, 972, 1062.

Singh, D. and Kumari, N. (2017). Fertigation in Fruit Crops. *International Journal of Economic Plants*, 4(3), 125-130.

Tojnko, S. and Cmelik, Z. (2004). Influence of irrigation and fertilization on performances of young apple trees. *Acta Horticulturae*, 664, 661-664.

Weinbaum, S. A., Johnson, R. S. and De Jong, D. M. (1992). Causes and consequences of over-fertilization in orchards. *Horti Technology* 2, 112-121.

Chapter 5

Foliar Micro-Nutrition of Vegetable Crops

Kushal Sachan[1]**, *Gorantla Prathap Reddy*[2], *Archana Verma*[3] *and Anshul Saxena*[4]

[1]*Department of Soil Science and Agricultural Chemistry, CSAUAT, Kanpur, Uttar Pradesh*

[2]*Department of Agronomy, Institute of Agricultural Sciences, BHU, Varanasi, Uttar Pradesh*

[3]*Department of Agricultural Chemistry, Amar Singh College, Lakhaoti Bulandshahr, Uttar Pradesh*

[4]*Department of Soil Science and Agricultural Chemistry, Sam Higginbottom University of Agriculture, Technology and Sciences, Prayagraj, Uttar Pradesh*

**e-mail: kushalsachan95@gmail.com*

INTRODUCTION

Foliar application of micronutrients is widely used. Most of the micronutrients, after foliar spray, enter the plant body through leaves within a few hours to one day. They mostly enter the leaf via stomata. Since the stomata are mostly present the surface of the leaf, the foliar spray should be applied to both the under and upper surfaces of the leaf as evenly as possible for rapid and complete absorption of the nutrient solution. 2-4 foliar sprays at an interval of 7-10 days are sufficient to correct the deficiency symptoms. This method is most suitable because of the following reasons:

1. Convenience in the application.
2. Requirement of micronutrients in small quantities.
3. Quick correction of deficiency symptoms.
4. Avoidance of fixation of the micronutrients in the soil.

Role of Foliar Nutrition

The initial benefit of a foliar application is an increase in chlorophyll synthesis which can often result in leaves turning a darker green. The increase in photosynthetic activity will stimulate extra root growth; in turn, the root hairs excrete excess sugar which stimulates microbial colonies. These bacterial colonies provide auxins and other root-stimulating compounds. With the increase in cellular activity gas exchange increases the uptake of water. As the roots take up more water they also bring in more nutrients from the soil solution. The foliar application stimulates the entire *"pumping system"* in the plant to increase the uptake of the base applied nutrition. Foliar nutrition will enhance the efficiency of the use of nutrients that plants need to increase ungently for maximum growth and yield. It can also make faster use of the substance and can remedy detected shortcomings in less time than can be achieved with soil applications. Foliar nutrition is the most efficient method for placing fertilizers under problems of soil fixation which normally needs reduced amounts of nutrients compared to soil application. Application of micronutrients in limited quantities and macronutrients, *e.g.*, N, P, or K, without phytotoxicity is the most effective usage of foliar nutritional methods. Foliar spray with micronutrients is one way to improve productivity and reduce environmental risks among other methods of application of plant nutrients. Moreover, it is easy and needs little infrastructure. The potential benefits of foliar nutrition could be concluded as follow (Mikkelsen 2008).

1. Nutrient availability during the maximum demand cycles if an urgent response is required.
2. Provide plants with certain nutrients (zinc and iron) which may not be readily available for root uptake.
3. Providing a source of nutrients during stress when soil applications are not practical.
4. The management of nutrient depletion in high-potential loss environments.
5. Allowing flexibility in supplying nutrients related to improving the quality of the harvest.
6. Providing plants with a nutrient boost while applying other foliar chemicals reducing applications.

Proper Timing of Foliar Nutrition

In relation to the optimal effectiveness of the foliar treatment the timing of the foliar spray, particularly with regard to the growth stage, can be considered crucial and carefully taken. The effectiveness of foliar nutrition is influenced by various endogenous (related to leaf anatomical structure) as well as exogenous (nutrient concentration, soil type and pH) and environmental factors. Including day time, humidity, temperature and wind speed affect the physical as well as biological aspects of foliar application. Plant tissue permeability is critical in nutrient absorption in the plant: warm, moist and calm conditions favor the maximum permeability of tissue, and conditions most frequently seen in late-night hours and often in early morning hours. Rainfall may reduce application effectiveness within 24 to 48 hours of application of foliar since the plant tissue does not absorb every nutrient material immediately (Tables 5.1 and 5.2).

Table 5.1: Meteorological Conditions Favouring Foliar Application of Nutrients (Adapted from MWL (Midwest Lab. 1994)

Time of Day	Late evening; after 6 p.m. or early morning; before 9 a.m.
Temperature	65-85 F; 70 ideal (21C)
Humidity	> 70 per cent relative humidity
Temperature/Humidity Index	140-160
Wind Speed	< 5 mph

Table 5.2: Rate of Foliar Application of Micronutrients

Micronutrient	*Concentration in Foliar Application*
Zinc	0.2-0.6 per cent $ZnSO_4$ + 0.1-0.3 per cent lime
Boron	0.5-0.6 per cent borax
Molybdenum	0.05 per cent ammonium molybdate
Iron	0.4 per cent $FeSO_4$ + 0.2 per cent lime
Manganese	0.4-0.6 per cent $MnSO_4$ + 0.2-0.3 per cent lime
Copper	0.1-0.2 $CuSO_4$ + 0.5 per cent lime

Micronutrients

Nutrients required in relatively smaller quantities, but as essential as micronutrients, are termed micronutrients. These include Fe, Mn, Zn, Cu, Ni, B, Mo, and Cl. Micronutrients are subdivided into cationic micronutrients(Fe, Mn, Zn, Cu, Ni) and anionic micronutrients (B, Mo, and Cl.), and the former are absorbed as the divalent cations(M^{2+}) latter are absorbed in the anionic forms. Boron exception in that it could also be absorbed as neutral H_3BO_3 Molecules. Elements like Na, Al, Si, V, Co, and Se, stimulate plant growth, but are not essential or which are essential only for certain plant species. These are termed *beneficial nutrients*. The term *ultramicronutient* has been proposed for Mo and Co. Zn generally activates enzymes, while Fe, Cu, Mn, and Mo are categorized as redox reagents. The sequence in which these micronutrients are accumulated by plants is as follows: Fe > Mn > Zn > B > Cu > Mo. Depending on the plant species and growth conditions, this order may change (Table 5.3).

Table 5.3: Major Natural Source of Micronutrients

Micronutrient	*Mineral*
Zinc	Sphalerite (ZnS) Smithsonite ($ZnCO_3$)
Copper	Chalcocite (Cu_2S) Malachite [$Cu_2(OH)_2CO_3$] Cuprite [Cu_2O]
Iron	Haematite [Fe_2O_3] Olivine [$(Mg\ Fe)_2SiO_4$] Pyrite [Fe_3O_4]
Manganese	Pyrolusite [MnO_2] Rhodonite [$MnSiO_3$] Manganite [$MnSiO_3$]
Molybdenum	Molybdenite [MoS_2] Wulfenite [$PbMoO_4$] Powellite [$CaMoO_4$]
Boron	Borax [$Na_2B_4O_7.10H_2O$] Tourmaline [$Na\ (Mg,\ Fe)_3Al_6(BO_3)_3Si_6O_8(OH)_4$] Kotoite [$Mg_3(BO_3)_{21}$
Nickel	Serpentine and Olivine
Chlorine	Muriate of Potash [KCL] Sodium Chloride [NaCl]

Needs of Micronutrients

Increased interest in micronutrients as limiting factors in crop growth and yield is due to the following reasons:

- ☆ Long-time cropping has removed measurable amounts of these nutrients.
- ☆ Widespread use of animal manures has decreased and use of high analysis fertilizers low in nutrient impurities has increased.
- ☆ Topsoil has been removed through erosion, land leveling, or terrace construction resulting in a deficiency of certain micronutrients.
- ☆ More information is available on crop responses, availability of nutrients in various soils, soil testing and plant analysis.
- ☆ More attention is being given to crop quality and nutritional value of today's crops.
- ☆ Variability of nutrients across fields has been recognized and the technology to manage variability continues to develop.

- The importance of micronutrients in Indian Agriculture has been realized during the course of The 'Green Revolution in the form of zinc deficiency caused *Khaira* disease of rice. After that the emphasis has been laid on the research and integration of micronutrients in agriculture. Intensive vegetable cultivation and the development of nutrient-responsive improved varieties/hybrids have created a problem of nutrient imbalance in soils due to extra nutrient mining. The imbalance between nutrient addition v/s removals, the latter being on the higher side, is a matter of great concern. Thus, it is imperative to maintain soil health for sustainable olericulture an with adequate supply of micronutrients.

Factors affecting Availability

- Soil pH is the most important factor influencing the availability of micronutirents. As soil pH increases the availability of iron, manganese, copper and zinc decreases, boron availability also decreases as soil pH increases up to above 8.5 where the availability increases due to sodium. Molybdenum is the only element whose availability increases as soil pH increases. The effect of soil Ph on chloride availability is neutral.
- In Coarse textured soils. *i.e.*, sandy soils, tend to be low in boron and copper is often low in leached calcareous soils. Poorly drained soils usually have poor aeration that contributes to iron, zinc and manganese deficiencies.
- Cool weather combined with poorly drained soils accentuates iron and zinc deficiency occurrences because of poor root growth. Under dry soil conditions boron deficiencies are increased and molydenum deficiencies tend to be more severe because less mineralization occurs and less soluble B is available.
- High amounts of phosphorus in soils borderline in copper, zinc, iron and manganese may reduce their availability and/or uptake by plants. Applications of: iron and zinc may reduce copper availability; copper, manganese and zinc may reduce iron availability; and copper, iron and zinc may reduce manganese availability. Applications of potassium may increase boron deficiency on soils low in boron,but high calcium availability increases the capacity of plants for boron. Soils high in iron and aluminum oxides tend to be low in molybdenum, but these soils also are highly weathered and frequently acidic.

Form and Mobility of Micronutrients Absorbed by Plant

The micronutrients are being absorbed by the plants in both ionic and non-ionic forms which depend on the type of element (Table 5.4).

Table 5.4

Micronutrient	*Ionic Form*	*Non-ionic Form*
Boron	$B_4O_7^{2-}$, $H_2BO_3^-$, HBO_3^{2-}	—
Chlorine	Cl^-	—
Copper	Cu^{2+}	$CuSO_4$ with EDTA
Iron	Fe^{2-}, Fe^{3-}	$FeSO_4$ with EDTA
Manganese	Mn^{2+}	$MnSO_4$ with EDTA
Molybdenum	MoO_4^{2-}	—
Zinc	Zn^{2+}	$ZnSO_4$ with EDTA

EDTA: Ethylene diamine tetra-acetic acid.

Mobility of Micronutrients

Bray (1954) taking a comprehensive view of the developments, introduced the concepts of nutrient mobility, where he defined mobility as the "overall process whereby nutrients reach the sorbing root surfaces, thereby making possible their sorption into the plant. The knowledge of the mobility of micronutrients in plants helps in finding the deficiency of particular nutrient. A mobile micronutrient in the plant moves to the growing points in case of deficiency and deficiency symptoms, therefore, appear on the lower leaves while immobile nutrients does not move to the growing points and hence, deficiency symptoms appears on the younger plant parts. Mobility of micronutrients in the soil has considerable influence on availability of nutrients to the plants and method of application.

Mobility in Soil

1. **Mobile:** The mobile micronutrients are highly soluble and not adsorbed on clay complex *e.g.* BO^3, Mn^{2+} and Cl^-
2. **Less mobile:** They are also soluble but adsorbrd on clay complex and so their mobility is reduced *e.g.* Cu^{2+}
3. **Immobile:** these ions are highly reactive and get fixed in the soil *e.g.* Zn^{2+}

Mobility in Plant

Based on mobility of micronutrients in the plant system, they can be grouped as

1. **Moderately mobile:** (Zn),
2. **Less mobile**: Fe, Mn, Cu, Mo, CI)
3. **Immobile**: (B).

An Optimum Level of Micronutrients in Plants

Most of the plant species show a specific ratio of various nutrients in them as an indication of the sufficiency of the concerned elements in the available form in the soil. In addition, there could be a definite quantity of a given nutrient the in the plant which biomass production might suffer with or without any visual deficiency symptoms. The critical nutrient range varies from crop to crop and even from variety to variety of the same crop for any given nutrient, a general guideline can be used for understanding the probable deficiency of nutrients which can be subsequently confirmed by other techniques as below.

Table 5.5: Deficiency, Normal, and Critical Toxicity Levels of Micronutrient Elements in Plants

Micronutrient	*Deficiency*	*Normal*	*Toxicity*
Iron	<20	20-1000	2000 and above
Manganese	<90	90-200	200and above
Copper	<10	10-25	25 and above
Zinc	<10	10-120	120 and above
Boron	<10	10-80	80 and above
Molybdenum	<o.1	0.1-0.90	90 and above

Source: www.FFTC.org

Table 5.6: The Optimum Range of Micronutrients (ppm) in different Vegetables Crops

Vegetables	*Fe*	*Mn*	*Zn*	*Cu*	*B*
Beans	50-300	50-300	20-200	7-30	20-75
Beet root	50-200	50-250	20-200	5-15	—
Brinjal	50-300	40-250	20-250	8-60	25-75
Cabbage	30-200	25-200	20-200	5-15	25-75
Carrot	50-300	60-200	25-250	5-15	30-100
Cauliflower	30-200	25-250	20-250	4-15	30-100
Onion	60-300	50-250	25-100	15-35	22-60
Peas	50-300	30-400	25-100	7-25	–
Radish	50-200	50-250	19-150	5-25	25-125
Tomato	40-200	40-250	20-50	5-20	25-60
Turnip	40-300	40-250	20-250	6-25	40-100

Source: Anjaneyulu, K. and Raja, M.E. (1999). Micronutrient disorders in vegetable crops and their correction. Indian Horticulture (Jan.-March). pp 15-16.

Soil Condition Causes to Micronutrient Deficiency

1. Highly leached acidic sandy soils.
2. Soils with a high-water table
3. Soils with a very high content of organic matter *e.g.* peat and muck soils of Kerela.
4. Calcareous and saline-alkaline soils very high in Ph.
5. Intensively cropped soil with high doses of commercial fertilizers.
6. Application of high doses of lime at one time
7. Maximum application of phosphatic fertilizers decreases the availability of zinc.
8. Application of more N, P and K Fertilizers as it Decreases the availability of micronutrients.
9. Introduction of hybrids and high-yielding varieties of various crops which removes the higher amount of micronutrients from soil increase plant micronutrient demand

Important Functions

1. Zinc

Zinc is indispensable for the normal growth and development of plants. It is effective for the synthesis of plant hormones like auxin and carbohydrate formation **(Pankaj *et al.*, 2018)**. It plays a fundamental role in several critical functions in the cell such as protein metabolism, gene expression, the structural and functional integrity of bio-membranes, and photosynthetic metabolism (Sanju *et al.*, 2003) [61]. Zinc is also a constituent of ribosomes and is essential for their structural integrity ***(Trivedi et al., 2013)***. It promotes starch formation, seed maturation, and production, and enhances seed viability and seedling vigor. Zinc is helpful in the reproduction of certain plants and various enzymatic activities. It also plays a vital role in sulfur and nitrogen metabolism. Critical limit in soil – 0.6 ppm

2. Boron

Boron plays an essential role in the growth and development of new cells in the meristematic region of plants. Boron is necessary for cell wall formation, and the development of fruit and seed. It helps in pollen formation, pollination, and flowering of plants (Malek and Rahim, 2011) [39]. The primary role of boron in plants is to improve the solubility and metabolism of Ca and its mobility and also help in the absorption of nitrogen. It also involves the metabolism and transport of carbohydrates, nucleic acid synthesis, root elongation, photosynthetic activities, and water absorption in plant parts ***(Islam et al., 2018)***. Boron increases the stability of plant cells and is involved in the reproductive phase of plants. Critical limit in soil – 0.5 ppm

3. Iron

Iron is an essential micronutrient required for normal growth and plant function. Iron act as a catalyst in the synthesis of the chlorophyll molecule and helps in the absorption of other elements. It is a structural component of porphyrin molecules like cytochrome, hemes, hematin, ferrichrome, and leg hemoglobin. These substances are involved in oxidation-reduction reactions in respiration and photosynthesis. It also involves DNA synthesis, protein synthesis, and reduction of nitrates and sulphates Further many metabolic pathways are activated by iron and it is a prosthetic group constituent of many enzymes **(*Rout and Sahoo, 2015*)**. Vegetable crops like tomatoes onions, carrots and spinach contain high percentage of Iron. Critical limit in soil – 4.5 ppm or mg/kg

4. Copper

Copper plays pivotal role in regulating multiple biochemical reactions in plants **(*Tripathi et al., 2015*)**. Arnon and Stout (1939) earlier declared copper as an important nutrient for plants in their experiments with tomato. Being the stable cofactor of various enzymes and proteins, copper plays an indispensable role in regulating several metabolic and physiological processes of plants. It helps in the utilization of iron during chlorophyll synthesis **(*Harris, 2016*)**. Copper has been also some indirect effect on nodule formation

5. Manganese

Manganese involve in the activation of many enzymes which helps in photosynthesis and respiration **(*Pankaj et al., 2018*)**. It activates several important metabolic reactions, accelerates germination and maturity while increasing the availability of phosphorus (P) and calcium (Ca). It enhances root growth, development of fruit and develops disease resistance. It also helps in the movement of iron in plants. High concentration of Mn favors the breakdown of indole acetic acid (IAA). Its availability is reduced in high pH calcareous soils but is often very high in the acid soils **(*Jawad et al., 2016*)**. Critical limit in soil – 2.0 ppm

6. Molybdenum

Molybdenum is required for the assimilation of nitrates, as well as, for the fixation of atmospheric nitrogen. It helps in protein synthesis and sulphur metabolism. Low and adequate levels of molybdenum has a positive effect on carotenoid formation. It also helps in absorption and translocation of iron in plants. Critical limit in soil – 0.1 ppm

7. Chlorine

Chlorine is most commonly used as sanitizer, due to its low cost for maintaining the fruit quality like appearance, soluble solids content, acidity, pH, texture and flavor, shelf life and also control microbial growth **(*Rahman et al., 2012*)**. It is essential for photosynthesis (chlorotic tissues), helps in stomatal regulation and raises cell osmotic potential, necessary for shoot apex and root growth. Critical limit in soil – 8.0 ppm

8. Nickel

It is important for activation of urease, an enzyme essential for nitrogen metabolism and also control senescence. It is required for iron uptake and it can substitute for zinc and iron as a cofactor for some enzymes. Critical limit in soil – 0.1ppm

Diagnosis

Micronutrient deficiencies are often blamed when the real cause cannot be identified. Therefore, during diagnosis of micronutrient deficiency symptoms one should have an appropriate knowledge of insect damage, disease symptoms, herbicide injury, drought, water-logging condition, *etc.* for proper identification of micronutrient deficiency symptoms in vegetable crops. Soil tests and plant analysis are excellent diagnostic tools to monitor the micronutrient status of soils and vegetable crops which are as follows:

Soil Analysis

Soil testing has immense value in establishing a sound vegetable cultivation program based on soil fertility. Soil testing is necessary to advocate judicious recommendation of micronutrients. Soil tests and their interpretations are based on the soil samples. Therefore, it is important that the soil samples should be properly collected. After scrapping the surface litter, uniform core of a thin slice of soil from the surface to plough depth (15-22 cm) from 15-20 spots of each sampled field should be taken. About a handful of soil sample should be taken and air-dried on plastic sheet before chemical analysis. The normal range of micronutrients found in mineral soils is as follows

Table 5.7

Micronutrient	*Normal Range (ppm)*
B	5-150
Cu	5-150
Mn	200-10000
Mo	0.2-5
Zn	10-250

Source: Brady, N.C. (1980). The nature and properties of soils (8th ed.). MacMillan Publishing House Inc., USA.

Plant Tissue Analysis

The basic principle of this technique is that the nutrient concentration of plants is related to the amount of nutrient element available in soil. If the nutrient level in the plant tissues falls below the critical level, the soil may be deficient in that element for optimum plant growth. The most recently matured leaf serve well for routine crop monitoring and diagnostic procedures for most micronutrients. However, for the immobile nutrients like Ca and B younger leaves are generally preferred. The plant part to be sampled and growth stage of different vegetable crops for plant, tissue nutrient analysis are as follows:

Table 5.8: Vegetable Crops

Sl.No.	*Crops*	*When to Sample*	*Where to Sample*	*No to Sample*
1	Asparagus	Maturity	Fern, 18=30 inches above the ground line	10-30
2	Beans	Seedling stage or before at bloom	All above-ground portions Recently mature leaf	20-30
3	Broccoli	Before heading	Recently mature leaf	12-20
4	Brussels sprouts	Midseason	Recently mature leaf	12-20
5	Celery	Midseason		12-20
6	Cucumbers	Before fruit set	Recently mature leaf	12-20
7	Head crops (cabbage, cauliflower)	Before heading	Recently mature leaf at centre of the whorl	12-20
8	Leaf crops (lettuce, spinach)	Midseason	Recently mature leaf	12-20
9	Melons	Before fruit set	Recently mature leaf	12-20
10	Peas	B	Leaves from 3rd node from top	40-60
11	Peppers	Midseason	Recently mature leaf	25-50
12	Potatoes	Before or at bloom	3rd to 6th from growing tip	25-30
13	Sweet potatoes	Midseason or before root enlargement	3rd to 6th leaf from tip centre	20-30
14	Roor/bulb crop	Midseason before root or bulb enlargement	Recently mature leaf	20-30
15	Tomatoes (Field)	Mid bloom	3rd to 4th leaf from growing tip	15-20
16	Tomatoes (Indertiminate)	Mid bloom from 1st to 6th cluster stage	Petiole of leaf below or opposite top cluster	12-20

Source: Flynn *et al.*, 2004.

Deficiency Symptoms and its Management

The deficiency of micronutrients can be categorized in the following two ways.

1. Primary Deficiency

It refers to the low total content of micronutrients which depends on the type of soil and other agro-climatic conditions. Generally sandy and calcareous soils contain low amount of micronutrients, whereas clay loam and loam soils contain comparatively more micronutrients. Loose textured soil low in organic matter and can be corrected by the application of chemical compounds containing micronutrient elements. It should be kept in mind that micronutrients should be applied only in deficient conditions otherwise they will cause phytotoxicity.

2. Secondary Deficiency

It implies that total micronutrient content in the soil may be ample but their availability is affected due to soil pH, antagonism, *etc.* The availability of Zn, Mn, Fe, Cu and B is often reduced in soils exhibiting alkaline reaction whereas Mo is unavailable in acidic soils. Under such situation, proper soil management like liming and application of organic matter to the soil should be followed to avoid the deficiency symptoms.

Identification of Deficiency Symptoms

The deficiency symptoms of micronutrients can be distinguished based on **(i)** region of occurrence, **(ii)** presence or absence of dead spots, **and (iii)** chlorosis of entire leaf or interveinal chlorosis.

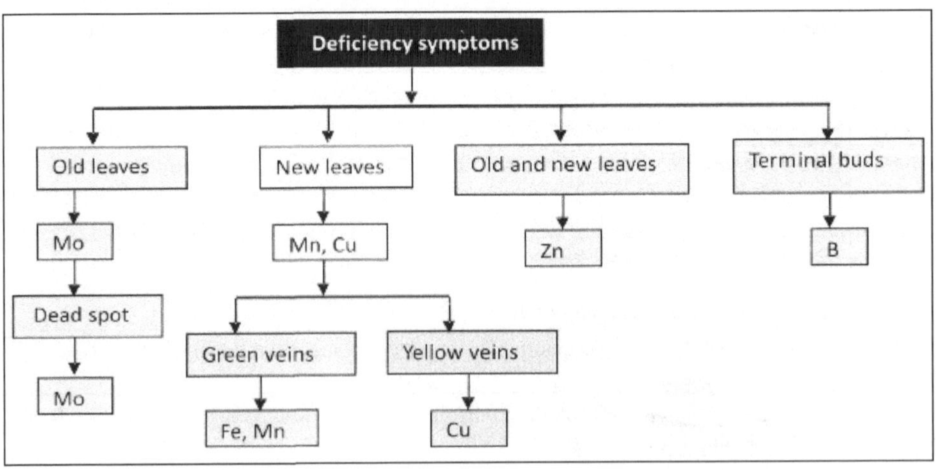

Micronutrient Deficiencies Symptoms and their Management.

Diminishing use of organic manures, adoption of an intensive cropping system, unbalanced use of NPK fertilizers, *etc.*, caused deficiencies of micronutrients. In order to get optimum yield with quality, micronutrient deficiencies have to be detected before visual symptoms are expressed by plants at the hidden hunger stage. Thus, vegetable cultivation needs judicious use of micronutrients to produce a yield of high quality. The most common micronutrient disorders in vegetables are due to the deficiency of B, Mo, Zn, and Fe.

Boron

The deficiency appears first in young tissues, growing points, root tips and developing fruits. Its deficiency may cause sterility, poor fruit set, small fruit size and ultimately lower yield. The shortening of the terminal growth results rosetting occurs due to boron deficiency.

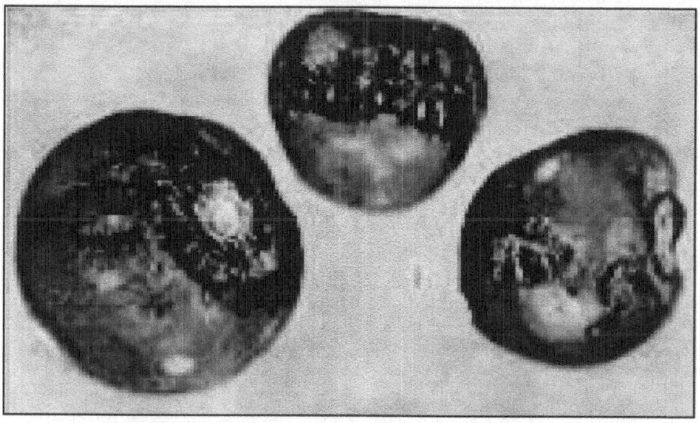

Boron deficiency in Tomato

Typical Symptoms

1. Browning of cauliflower curd.
2. Hollow heart in cauliflower.
3. Black heart of beet.
4. Internal browning of turnip.
5. Brown or black heart of potato.

Management

Boron deficiency is mainly occurring in coarse textured soils and more pronounced during drought periods when root activity is restricted. Boron is highly immobile in the plant so frequent and a constant supply of boron to the young tissue or bud is must for proper growth and development of vegetable crops. Soils are considered deficient if the soil test value of B is less than 0.5 mg boron/kg soil. The most effective method of boron application is basal application in the soil at the time of land preparation or at transplanting. Spray @ 0.5-0.6 per cent at early stage of crop is most effective. The rate of application varies from crop to crop being lowest in beans and highest in cole crops and beet. The common B fertilizers are borax (10.5 per cent B) and boric acid (20 per cent B), and these fertilizers can be applied @ 2-5 kg/ha, although the rate depends on the native availability of boron in the soil.

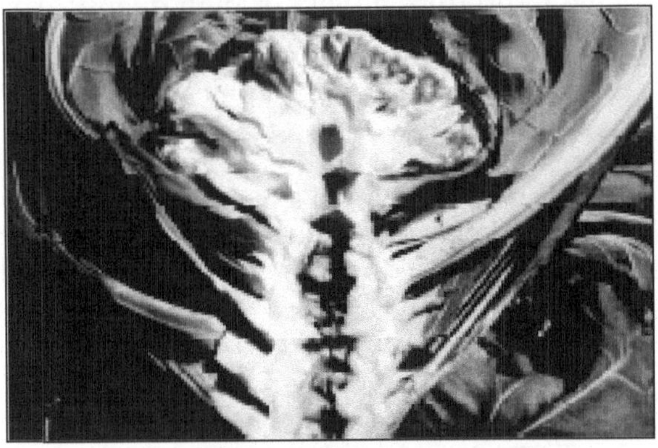

Broccoli: "Hollow-Heart" Disease

Copper

Noticed as stunting of growing leaves, die-back of stems and twigs and light over-all yellowing of leaves giving a withered appearance

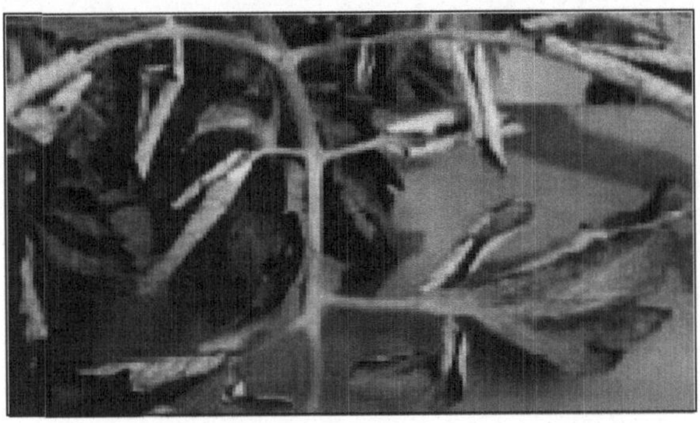

Cu Deficiency in Tomato (Loss of turgor)

Typical Symptoms

1. Bulbs of onion become soft with thin pale young scales.
2. Dieback of stems and twigs, yellowing of leaves, stunted growth and pale green leaves that wither easily.

Management

The level of Cu increase with the increase in clay content of soil. Increased phosphorus and iron availability in soils decreases copper uptake by plants. The residual effect of copper in the soil is very long therefore, one application may be effective for many years. However, soils having high organic matter require annual application of Cu. Some fungicides also contain Cu and their repeated use will increase the level of Cu in the soil. The main Cu containing fertilizers are copper oxide (75 per cent Cu) and copper sulphate (35 per cent Cu). Copper can be applied to the vegetable crops @5-10 kg/ha as soil application and 0.1-0.2 per cent as foliar application.

Iron

Typical symptoms are distinctive green vein colour and yellow interveinal areas on the young leaves. Points and margins of the leaves keep their green colour for long time. In severe cases, the entire leaf, veins and interveinal areas turn yellow which sometimes bleached. Fe deficiency is mainly manifested by yellow leaves due to low levels of chlorophyll. Leaf yellowing first appears on the younger upper leaves in interveinal tissues.

Fe Deficiency in Broccoli.
Chlorosis of leaves beginning as a chlorotic mottling.

Management

Deficiency of iron occurs on soil with pH above 6.8, calcareous soil and soil containing considerable amount of sodium and calcium. Iron chlorosis is mainly corrected through foliar spray of iron containing micronutrients *i.e.* ferrous sulfate @ 0.4 per cent. Soil application of Fe is not recommended as applied Fe quickly reverts to forms unavailable to the plants. Correction of iron chlorosis at early stage of crop growth is essential to prevent yield loss. A surfactant should be added in the solution to improve leaf adherence and absorption of Fe.

Zinc

Its deficiency causes interveinal chlorosis of older leaves then leaves turn grey-white and fall prematurely or die. Stunted growth, distortion in shape and clustering of leaves on short branches known as rosette. The crops like tomato, potato, beans and onion are highly sensitive to Zn deficiency.

Zinc Deficiency in Tomato

Typical Symptoms

1. Interveinal yellowing with marginal burning in beet.
2. Fern leaf disorder in potato.
3. Onion leaves turn yellow.
4. Whole plant turns yellow in garlic.

Yellow Striping, Twisting and Bending of Crops in Onion Due to Deficiency of Zn.

Management

Neutral to alkaline sandy soils and calcareous soils are mostly suffering from Zn deficiency. It has been well documented that at least 50 per cent of the Indian soils suffer from Zn deficiency. Indian soils are generally considered zinc deficient if the test value is less than 0.6 mg DTPA extractable Zn/kg soil. Uptake by plants decreases with increased soil pH and the uptake of zinc also adversely affected by higher levels of available phosphorus and iron into soils. The common fertilizers are zinc sulphate (21 and 33 per cent Zn) and zinc-EDTA (12 per cent Zn). Zinc sulphate (ZnS04) is suitable for soil application @5-25 kg/ha (based on native Zn availability of soil) while Zn-EDTA is suitable for foliar application @ 1 per cent.

Molybdenum

Impairment of growth, deformation of shoot, older leaves show interveinal yellowing with necrotic margins and leaves may crinkle. The symptoms are identical with those of nitrogen deficiency.

Whiptail of Cauliflower-Mo Deficiency.

Typical Symptoms

1. Whiptail leaf (leaf lamina fail to develop) in cauliflower.
2. Mottling and whiptail in turnip.
3. Young leaves turn purple in tomato.
4. Discoloration of leaf and death of growing tip in radish.
5. Decomposition and granulation in veins of chilli.

Management

Ammonium molybdate (54 per cent Mo) and sodium molybdate (38 per cent Mo) are common Mo fertilizers available in the market and can be applied @ 2-5 kg/ha assoil application or 0.03-0.05 per cent as foliar spray

Nickel

The deficiency of nickel causes leaf tip necrosis in nitrogen fixing plants. The deficiency of nickel also delayed nodulation and reduced efficiency of nitrogen fixation in leguminous vegetable crops. Nickel was needed in cowpea at reproductive phase

Manganese

Manganese deficiency causes a light green mottle between the main veins and interveinal chlorotic areas become pale green or dull yellowish colour (Mousavi *et al.*, 2011) [45]. Dry edible beans, cucumbers, lettuce, onions, peas, potatoes, radishes are the most responsive crops.

Spinach: Severe Leaves Chlorosis.

Typical Symptoms

1. Interveinal chlorosis in pea called Marsh spot.
2. Narrow and yellow striping in the leaves of onion

Management

Manganese deficiency is found in soils with pH above 6.7, calcareous and sandy soils. Uptake of Mn decrease with increased soil pH and is adversely affected by high levels of available iron in soils. In acidic soils, apart from low pH and toxicity of Mn may decrease the availability of P and Mo. Vegetable crops are sensitive to excess Mn. Interveinal chlorosis and its relation with over liming show a strong association of manganese deficiency with pH. Deficiency of Mn can be corrected by either soil application (5-10 kg/ha) or foliar spray (0.4-0.6 per cent) of manganese sulphate with an adjuvant at first bloom stage in most of the vegetable crops.

Remission Strategy for Micronutrients Deficiency

Deficiencies in soil may be addressed, and micronutrient stability may be restored, using lits of item to be had available in the marketplace that help in restoring soil stability and improving micronutrient content. Different strategies which can be hired to relieve the micronutrient deficiencies are indexed below:

- ☆ Straight Micronutrient
- ☆ Chelated micronutrient
- ☆ Customized fertilizer
- ☆ Fritted or slow – release fertilizer
- ☆ Nano- micronutrient fertilizer
- ☆ Bio release smart fertilizer
- ☆ Use of organic matrix

- ☆ Biofortification
- ☆ Water soluble fertilizer

Type of Micronutrient Fertilizer

Straight Fertilizer

They come in crystalline and granular forms. They can be disseminated or banded into the soil, foliar sprayed and even applied by irrigation (fertigation). Straight fertilizers can be used in conjunction with primary fertilizers or agrochemicals to reduce application costs. However, there is a chance that nutrients will be separated, resulting in an uneven supply of nutrients to the plant. Typicaly found as sulphate salts of mineras such as zinc sulphate and iron sulphate which are water soluble. Oxysulpahtes are the oxides of copper manganese and zinc, which are partially acidulated with sulphuric acid. The solubility of oxysulphates in water increases with increasing acidulation.

Types of Micronutrients	*Trade name of Micronutrients*	*Composition (per cent)*	*Manufacture*
Straight	Copsomin 25	24 per cent Cu as copper sulphate	Greencare Biosciences, Hyderabad
Straight	Zealbor	20 per cent (Boron)	Agri Search (I) Pvt. Ltd
Straight	Anmol	52 per cent (Ammonium molybdate)	Agri Search (I) Pvt. Ltd
Straight	Uttam Zinc	33 per cent (Zinc sulphate monohydrate)	Chambal fertilizer

Customized Fertilizers

Customized fertilizers manufacture basically involves mixing and crushing of urea, DAP, MOP, ZnS, bentonite sulphur and boron granules for obtaining the desired proportion of N, P, K, S and micronutrients. The objective behind the customized fertilizer is to provide site specific nutrient management for achieving maximum fertilizer use efficiency for the applied nutrient in a cost-effective manner. Customized fertilizers are combination of micro nutrients like sulphur, zinc, boron added to the key items such as urea and diammonium phosphate (DAP) and potash, in a proportion that suits specific crops and soil patterns. A fertilizer formulated according to specifications that are furnished by/for a consumer prior to mixing, usually based on the results of soil tests. The efficiency of customized fertilizers depends on soil properties, crop, water and specific nutrients. They are generally multi-nutrient carriers that contain macro, secondary, and/or micronutrients from both inorganic and organic sources and are developed for certain crops and regions.

e.g., TRADE NAME- Grasim by Indo-Gulf fertilizer

Constituents (per cent) - N (14 per cent), P (27 per cent), K (10 per cent), Zn (4 per cent) and B (0.5 per cent)

Chelated Micronutrients

Chelated micronutrients are fertilizers where the micronutrient ion (for example Fe or iron) is surrounded by a larger molecule called a ligand or chelator. Ligands can be natural or synthetic chemicals. These compounds combined with a micronutrient form a chelated micronutrient. Chelated micronutrients are protected from oxidation, precipitation, and immobilization in a certain condition. There are also many naturally occurring chelating agents such as amino acids, organic acids, humic and fluvic acids, lignin sulfonates, ligninipolycarboxylates, sugar acids, phenols, polyphosphates, flavonoids, and siderophores. The most popular synthetic chelating agent is EDTA, which is used to chelate both soil and foliar nutrients DTPA is primarily utilized in chelate for alkaline soils. The most efficient iron fertilizers on high-pH soils are iron chelates produced with HEDTA and EDDHA, but they are also the most expensive **(*Shukla and Behera, 2017*).** Chelated micronutrients are more effective per unit of micronutrients, but they are also more expensive, limiting their use to high-value crops such as vegetable crops.

Micronutrient	*Trade Name of Micronutrient*	*Constituent*	*Constituent*
Chelated	Lohadrip	6 per cent (FeEDDHA)	Agri Search (I) Pvt. Ltd
Chelated	Amoga	12 per cent (ZnEDTA)	Agri parie -Commerce Pvt., Ahmedabad
Chelated	Volgamin-12	11 per cent (FeEDTA)	Greencare Biosciences, Hyderabad

Singh *et al.*, 2022.

Slow Release Fertilizer

Slow-release feature ensures a constant supply of the micronutrient source with no waste. Frits is more suited to maintenance applications than to alleviate the deficiency. These products were intended to be used on course-textured soils in high-rainfall locations to prevent leaching. Boron glass frits, boron phosphate, colemanite, and ulexite are some examples of fritted fertilizers.

Advantage

- ☆ Fewer applications
- ☆ Low burn potential
- ☆ Release rates vary depending on fertilizer characteristic
- ☆ Comparatively slow release rate

Disadvantage

- ☆ Unit cost is high Availability limited
- ☆ Release rate goverened by factor other than plant needed
- ☆ Regulated by temperature

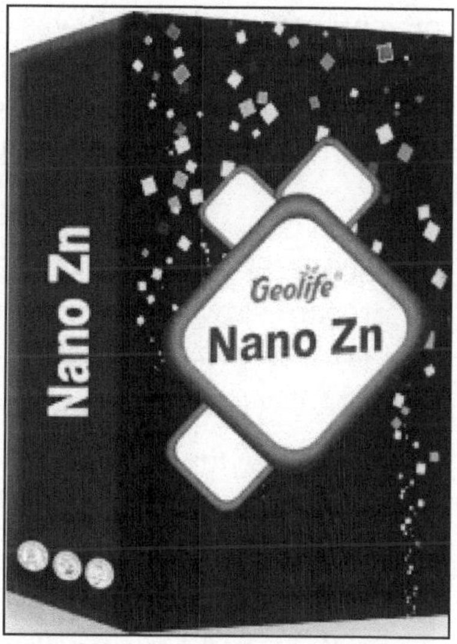

Nano Fertilizer

Nano-fertilizers are nutrient carriers of nano-dimensions ranging from 30 to 40 nm (10-9 m or one-billionth of a meter) and capable of holding bountiful of nutrient ions due to their high surface area and release it slowly and steadily that commensurate with crop demand. ***Subramanian et al.* (*2008*)** reported that nano-fertilizers and nanocomposites can be used to control the release of nutrients from the fertilizer granules so as to improve the NUE while preventing the nutrient ions from either getting fixed or lost in the environment. Nano-fertilizers have high use efficiency andcan be delivered in a timely manner to a rhizospheric target. ***Laware and Rakar* (*2014*)** reported that the usage of ZnO nanoparticles can impact flowering time,for example, onion flowering time is reduced by 12-14 days.

Micronutrient	*Trade Name of Micronutrient*	*Constituent*	*Constituent*
Nano fertilizer	Geolife Nano Fe	12 per cent	Geolife Agritech IndiaPvt. Ltd.
Nano fertilizer	Geolife Nano boron	20 per cent	Geolife Agritech IndiaPvt. Ltd.
Nano fertilizer	Geolife Nano Mn	26.6 per cent	Geolife Agritech IndiaPvt. Ltd.
Nano fertilizer	Geolife Nano Zn	12 per cent	Geolife Agritech IndiaPvt. Ltd.

Fertilizer Based on Biological Material

Biological materials with a natural tendency to bind micronutrient ions could be a potential alternative to routinely used fertilizers. Functional groups (hydroxyl, phosphate, or amino groups) have an affinity for metal cations containing microelements, producing metal complexes, which has been discovered in the development of slow-release fertilizers (***Arslanoglu*, 2019**). As a result, biosorption of micronutrient cations on bio-based materials provides an intriguing technique for producing control release fertilizer. Micronutrients would be delivered *via* a biodegradable carrier in this fertilizer have better bioavailability and a slower release of nutrients into the soil. This sort of bio-based fertilizer is completely biodegradable, reducing the environmental impact of fertilizers. When comparing the efficiency of biosorption fertilizers to conventional fertilizers, it was discovered that biosorption fertilizers are easily absorbed by plants and have a steady release of nutrients. In comparison to plants fed with traditional reference fertilizers (inorganic salts and chelate), the crop yielded by zinc-fertilized plants was higher (Tuhy *et al.*, 2014).

REFERENCES

Arslanoglu, H. (2019) Adsorption of micronutrient metal ion onto struvite to prepare slow release multielement fertilizer: copper (II) doped-struvite. *Chemosphere* 217, 393-401.

Brady, N. C. (1980). The nature and properties of soils (8th ed.). MacMillan.

FFTC, (2002). *Nitrates in Vegetables*, 5F. Food and Fertilizer Technology Centre, 14 Wenchow St., Taipei- 10616 Taiwan, (http: / /www.agnet.org.).

Flynn, R., Ball, S. T. and Baker, R. D. (2004). *Sampling for plant tissue analysis*, Guide A-123. New Mexico State University, New Mexico.

Harris, K. D. and Lavanya, L. (2016). Influence of foliar application of boron, copper and thei combinations on the quality of tomato (*Lycopersicon esculentum* Mill.). *Research journal of Agriculture and Forestry Sciences.* 4(7): 1-5.

Islam, M. Z., Mele, M. A., Choi, K. Y. and Kang, H. M. (2018). The effect of silicon and boron foliar application on the quality and shelf life of cherry tomatoes. Zemdirbyste-Agriculture. 105(2): 159-164.

Laware, S. L. and Raskar, S. (2014). Influence of zinc oxide nanoparticles on growth, flowering and seed productivity in onion. *International Journal of Current Microbiology Science* 3(7), 874-881.

Mani, P. (2011). Fertilizers in Vegetable Production.

Midwest Lab, M. W. L. (1994). Foliar nutrition. Midwest Laboratories, Inc., Omaha, NE.

Mikkelsen, R. L. (2008). Where does foliar fertilization fit in? Plant Nutrition TODAY, International Plant Nutrition Institute (IPNI), Spring.3.

Oosterhuis, D. M. (1995). Potassium nutrition of cotton in the USA, with particular reference to foliar fertilization.In: G.A. Constable and N.W. Forrester (Eds.). Challenging the Future: Proc. World Cotton Conference Brisbane Australia. CSIRO, Melbourne. 133-146.

Oosterhuis, D. M. and Weir, B. L. (2010). Foliar fertilization of cotton. In: J. M. D. Stewart (Eds.), Physiology of Cotton. Springer Science + Business Media B.V. 272–288.

Pandav, A. K., Nalla, M. K., Aslam, T., Rana, M. K. and Bommesh, J. C. (2016). Effect of foliar application of micronutrients on growth and yield parameters in Eggplant cv HLB 12. *Environment and Ecology*.35(3):

Pankaj, P., Kujar, P. K. and Saravanan, S. (2018). Effect of different micronutrient on plant quality of broccoli (*Brassica oleracea* var. *italica*) CV green magic. *Journal of Pharmacognosy and Photochemistry*. 1: 2825-2828 Publishing House Inc., USA.

Rahman, M. H., Halder, B. C. and Khan, M. A. (2011). Effects of zinc and manganese on growth and yield of potato. *Journal of Bangladesh Society for Agricultural Sciences and Technology*. 8(1-2): 17-22.

Rahman, M. M., Miaruddin, M. D., Chowdhury, M. D. G. F., Khan, M. D. H. H. and Martin, M. A. (2012). Effect of different system and chlorination on the quality and shelf life of green chilli. *Bangladesh Journal of Agricultural Research*. 37(4): 729-736.

Rout, G. R. and Sahoo, S. (2015). Role of iron in plant growth and metabolism. *Reviews in Agricultural Science*. 3: 1-24.

Shukla, A. K. and Behera, S. K. (2017) Micronutrients research in India: Retrospect and prospects. FAI Annual Seminar.

Singh, K. P. and Rajput, C. B. S. (2008). Effect of Molybdenum on Cauliflower (*Brassica oleracea* L. var *Botrytis*) in Sand Nutrient Culture. *Experimental Agriculture*. 12(2): 195-199.

Singh, S., Pandey, A., Bahuguna, Ayush, M., Kumar, K., Patra, A. Sathyanarayana, E., Jatav, H., Jatav, S. and Rajput, V. (2022). Current State of Knowledge in Diagnosis and Mitigation of Micronutrients Deficiency in Crop Production in Indian Prospective.

Subramanian, K. S., Paulraj, C. and Natarajan, S. (2008). Nanotechnological approaches in nutrient management: Nanotechnology applications in agriculture, TNAU technical bulletin. TNAU, Coimbatore, pp 37-42.

Tripathi, D. K., Singh, S., Singh, S., Mishra, S., Chauhan, D. K. and Dubey, N. K. (2015). Micronutrients and their diverse role in agricultural crops advances and future prospective. *Acta Physiologiae Plantarum*. 37(7): 1-14.

Trivedi, A. P. and Dhumal, K. N. (2013). Effect of soil and foliar applications of zinc and iron on the yield and quality of onion (*Allium cepa* L.). *Bangladesh Journal of Agricultural Research*. 38(1): 41-48.

Tuhy, L., Samoraj, M., Michalak, I. and Chojnacka, K. (2014) The application of biosorption for production of micronutrient fertilizers based on waste biomass. *Applied biochemistry and biotechnology 174*, 1376-1392.

Zahed, Z., Kumar, S. S., Mahale, A. G., Krishna, J. R. and Mufti, S. (2021). Foliar Micro-nutrition of Vegetable Crops - A Critical Review. *Current Journal of Applied Science and Technology, 40*(7), 1-12. https: //doi.org/10.9734/ cjast/2021/v40i731322.

Chapter 6

Fruit Crops in the World of Genome Editing

Homeshvari[1] and Devid Kumar Sahu[2]*

[1]Ph.D. Scholar, Department of Horticulture (Fruit Science),

[2]Ph.D. Scholar, Department of Soil Science and Agricultural Chemistry,

JNKVV, Jabalpur, Madhya Pradesh

**e-mail: homeshvari1997@gmail.com*

INTRODUCTION

Fruit growing is one of the oldest and most important practices in the world. Fruits are essential for a healthy diet, being a substantial source of nutrients and antioxidants and therefore quality improvement in these crops has gained permanent interest from producers and researchers. Valuable cultivars and varieties of many fruit crops have been developed by introducing desirable traits through conventional breeding and genetic transformation. Despite their improved qualities, genetically modified (GM) plants have been accepted with restrictions on the market. Even if GM fruits are free of pesticide residues and have more flavors and are lower in fat, consumers are reluctant and biotech companies should find a compelling case to sell GM foods. In many countries, fruits are not considered staple foods. Therefore, the development of new varieties of GM fruit crops with a range of novel characteristics has gained consumer acceptance mainly as luxury products. The recent development of high-throughput sequencing technologies provided information on genomes and valuable qualities in fruit crops. In addition, the genomes of many plant species have been sequenced (Bolger *et al.*, 2014), contributing to the decipherment of the molecular mechanisms of physiological processes, including flowering, youth, maturation and lifespan. In addition to social obstacles, the genetic transformation of fruit crops has some technical drawbacks, such as multiple restriction sites in the genome guaranteed by endonucleases, low insertion efficiency of modified constructs, low efficiency of correct insertion into the target chromosomal site, much time, laborious selection/detection strategies, and possible adverse mutagenic effects (Capecchi, 2005) RNA interference (RNAi) was developed as a valuable gene knockout technology to overcome some drawbacks of existing methods. Unfortunately, it also exhibited disadvantages such as incomplete and transient knockdown of genes and unpredictable off-target effects (McManus and Sharp, 2002). The last decade has been marked by the emergence of a new approach that allows direct manipulation of any gene in various cell types and organisms. This technology, known as "genome editing", is based on the use of engineered nucleases composed of sequence-specific DNA-binding domains fused to a non-specific DNA-cleavage cassette (Urnov *et al.,* 2010; Carrol, 2011). These engineered nucleases allow efficient and precise genetic modifications via double-strand breaks (DSBs) in target DNA. As effective technologies in genetic engineering, genome editing techniques are used for the insertion, substitution, deletion or interruption of DNA sequences.

Concept of Genome Editing

The terms 'gene editing' and 'genome editing' are used interchangeably and generally mean the same thing. Literally, 'gene editing' refers to the editing of 'genes', which are the units within hereditary material that are translated into functional components such as proteins. The word "genome" refers to the hereditary material of an organism. A gene can be compared to a sentence in a text, and the genome to the full text. Genes determine heritable characteristics, but it is the genome that determines the organism. Spontaneous changes in sentences are the basis of evolution.

Editing is the activity of deliberately altering sentences in a text with the goal of adding meaning, changing meaning, removing errors, or increasing readability. It involves changing, adding or removing letters and words, and sometimes even removing or adding entire sentences. This is also what genome editing is all about. It is about deliberately substituting, removing or adding some DNA letters (Figure 6.1) within the existing genetic blueprint of an organism, with the aim of altering its genetic properties. Genome editing is a specific approach in which one or more predetermined sites in the genome are edited to achieve a predetermined characteristic. In recent decades, our knowledge of plant biology and genes has grown exponentially, and genome editing allows researchers to use that knowledge more efficiently.

Tools for Plant Genome Editing

The genome editing toolbox consists of various methods that can be used to introduce specific, targeted changes into an organism's genome (Songstad *et al.,* 2017). Currently, the CRISPR (Clustered Regularly Interspaced Short

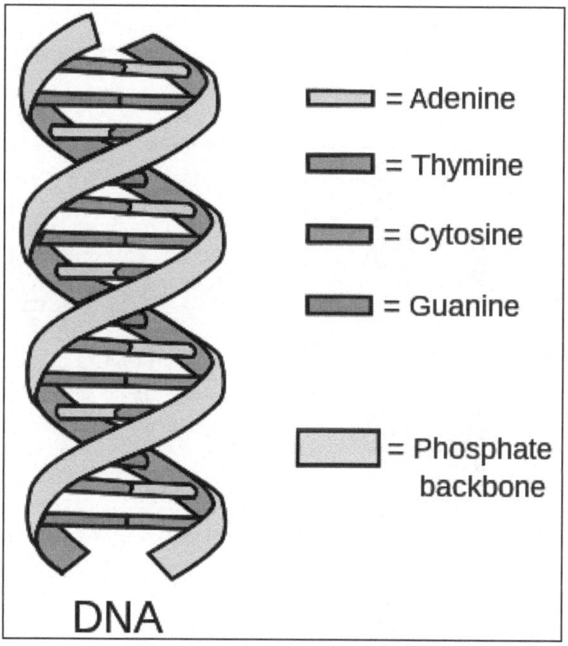

Figure 6.1: The DNA Helix.

Palindromic Repeats) system is by far the most widely used genome-editing tool (Menz *et al.*, 2021; Parisi and Rodríguez-Cerezo, 2021). It is based on a natural mechanism and was introduced to the scientific community as a genome-editing tool about a decade ago. CRISPR has swept aside most other previously developed genome-editing tools, such as meganucleases, Zinc Finger Nuclease (ZFN) technology, and transcription activator-like effector nucleases (TALENs). Similar to CRISPR, these editing tools make use of three biological mechanisms (Songstad *et al.*, 2017):

1. The ability to find a specific sequence of DNA letters in the genome;
2. The ability to cleave DNA at that location, and
3. The activity of the innate DNA repair machinery.

All of these tools make use of a nuclease, which is an enzyme that cleaves DNA. Alongside these nuclease-based technologies, there is also oligo-directed mutagenesis (ODM), which can be used to change a DNA letter at a desired location (Sauer *et al.*, 2016). CRISPR has become the dominant genome editing method because it is much more versatile than the other systems and can very easily be programmed to direct the nuclease to the desired location in the genome.

New Tools for Plant Genome Editing

Based on the high diversity of Cas proteins, the CRISPR/Cas systems have

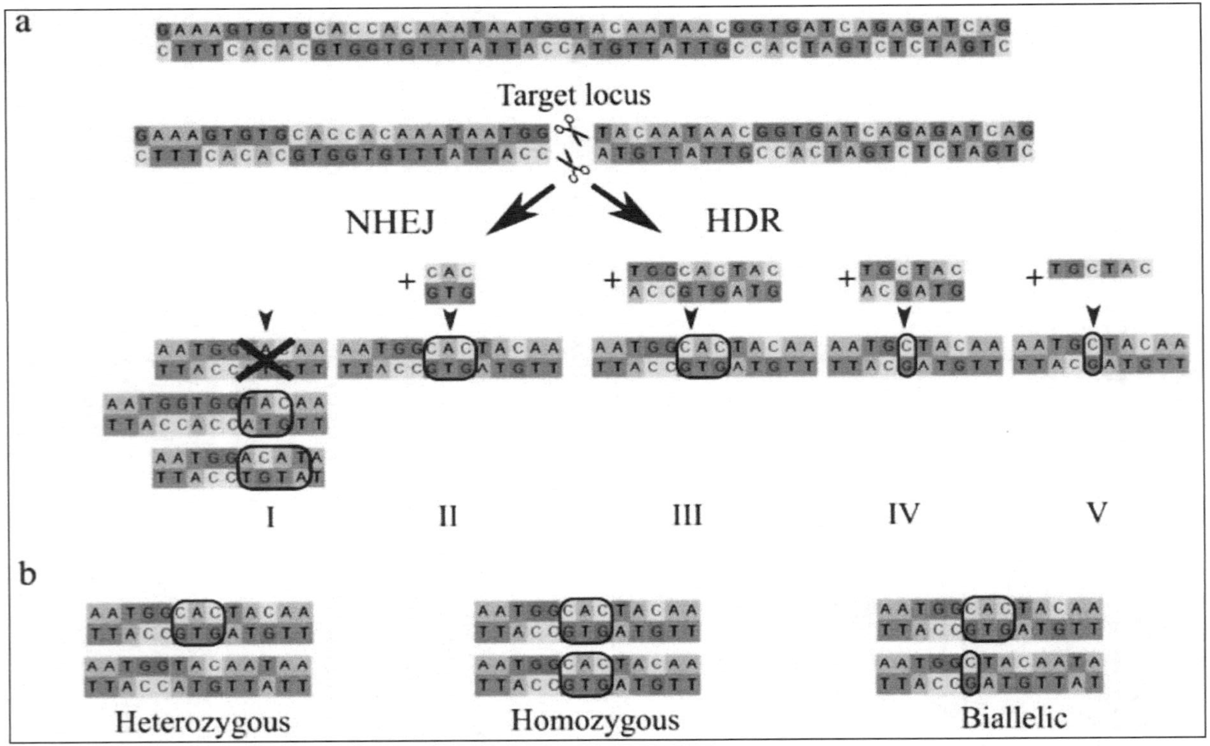

Figure 6.2: Genome Editing at Target Locus.

(a) Site-specific nucleases introduce double-strand breaks where gene modification is acquired by two repair pathways. Non-homologous end joining (NHEJ) generates gene knockout (I) by deletion, insertion or inversion, in the absence of donor DNA, and gene insertion (II) when integrates donor DNA by compatible ends. Homology dependent repair (HDR) results in gene insertion (III) when integrates donor DNA based on homology regions and gene correction when assimilates a small change provided as either double-stranded (IV) or single-stranded DNA (V); (b). Gene modifications in diploid plants. Adapted from Zhu *et al.*, 2017.

been classified into two classes and six types with several subtypes. Types I, III and IV belong to class 1, while types II, V and VI belong to class 2 (Koonin and Makarova, 2019). Discovered in Prevotella and Francisella, class II CRISPR has a type V effector (Cpf1 or Cas12a) (Malzahn *et al.*, 2019) that can be designed to cleave specific DNA sequences (Ma *et al.*, 2018). It targets T-rich motifs and does not require the tracrRNA to form a mature crRNA. Cpf1 represents a valuable alternative to Cas9 due to its ability to induce DSBs and process RNA and DNA (Safari *et al.*, 2019). As a valuable nuclease, Cpf1 generates displaced ends (Ding *et al.*, 2018), enabling genome manipulation (Li *et al.*, 2018b). Cfp1 allows precise gene knockout (Gaudelli *et al.*, 2018), insertion or deletion of DNA sequences, base substitutions, and the development of "prime editing" that can insert new sequences into a DNA site, expanding the applications of genome editing (Anzalone *et al.*, 2019). Another CRISPR/Cas system was recently identified in bacteriophages, suggesting that the CasÖ enzyme could also be used for genome editing in plants. It shows higher target recognition properties compared to Cas9 and Cas12a and has half their molecular weight. Furthermore, the CasÖ enzyme uses the same active site for processing mature crRNA and cleavage of foreign nucleic acids (Pausch *et al.*, 2020).

Mechanisms of Genome Editing

The main difference between current technologies and conventional breeding practices based solely on recombination and, to a small extent, on genetic recombination is that genomic editing achieves strict specificity towards the intended DNA target. At first, the artificially engineered nuclease enzymes called molecular scissors (Punwar *et al.*, 2014) such as zinc-finger nucleases (ZFNs) (Carroll, 2011) and the transcription activator-like effector nucleases (TALENs) (Mahfouz *et al.*, 2011; Li *et al.*, 2012) that are capable of generating desired genomic modifications (Shan *et al.*, 2013) have been deployed. The most recent system developed for genome-editing is the clustered regulatory interspaced short palindromic repeat (CRISPR)/Associated Protein-9 Nuclease (CRISPR/Cas9), based on RNA-guided DNA endonucleases that allow precise modification, insertion, or replacement of genes at specific sites. CRISPR/Cas technology is considered the most efficient, cheap, and user-friendly among the genome editing tools (Kaul *et al.*, 2020). All these new technologies for crop improvement allow the modification of any kind of genomic trait (Jaganathan *et al.*, 2018).

1. Zinc-Finger Nucleases (ZFNs)

ZFNs are engineered nucleases consisting of the DNA-binding zinc-finger (ZF) motifs and the FokI endonuclease. The recognition target sites consist of two ZF binding sites flanking up to 5-7 bp spacer sequence recognized by the Fokl endonuclease cleavage domain. Each ZF recognizes short sequences (3 bp), but it is possible to increase the recognition sequence up to 20 bp by combining 6-8 ZFs with specific recognition sites. 3 to 4 ZF domains together compile a ZFN, with each ZF domain containing approximately 30 amino acid residues organized in ββα motifs (Petolino, 2015). The editing method based on ZFN is based on the protein dimer composed of two DNA-binding proteins (each weighing 3-6 ZF) with the catalytic domain of the FokI endonuclease, which cleaves the double-stranded DNA. The two ZF proteins recognize two specific DNA sequences and bring the two FokI domains closer together. The dimerization of FokI is obligatory for nuclease activity and is followed by increased specificity of DNA recognition. Furthermore, FokI nucleases have been modified to function only as heterodimers to increase recognition specificity (Urnov *et al.*, 2010). Due to their efficiency, minimal off-target effects and high specificity, ZFNs are valuable genome-editing tools suitable for editing various crops of interest (Kamburova *et al.*, 2017).

2. Transcription Activator-like Effector Nucleases (TALENs)

These artificial nucleases contain a binding domain (TALE) that consists of a series of approximately 32- 34 amino acid residue repeats and a Fok I DNA cleavage domain. Each repeat is conserved, except the amino acid in positions 12 and 13, variable di-residues (RVDs), which establish the DNA binding site of TALE. These binding domains can be designed to bind any DNA sequence. The origin of the binding domain is in TAL effectors from *Xanthomonas* spp. TALENs can create DSBs at the target site that can be repaired by NHEJ, introducing small insertions or deletions (Pérez-Quitero *et al.*, 2013). TALENs also require dimerization of the FokI domain, which is similar to ZFNs, but, conversely to a ZF domain, which identifies a DNA triplet, a TALE protein only recognizes a single bp (Dheer *et al.*, 2020).

3. Clustered Regularly Interspaced Short Palindromic Repeats/ Associated Protein (CRISPR/Cas)

The newest technology of genome editing consists of a specialized RNA sequence and a Cas9 enzyme working as molecular scissors to cleave the DNA (Figure 6.3). The CRISPR/Cas system confers immunity against viral DNA and RNA in bacteria and archaea, and the mechanism is described in detail by several authors (Charpentier *et al.*, 2015; Rath *et al.*, 2015; Jiang and Doudna, 2017). The CRISPR/Cas system used for genome editing is comprised of chimeric RNA molecules crRNA (CRISPRassociated RNA) and tracrRNA (transcribed trans-activating crRNA) that are transcribed in a single guide RNA (sgRNA), and the Cas9 protein (Jinek *et al.*, 2012). Different sequences in the genome can be targeted by designed sequences of gRNA (Jiang and Doudna, 2017). The break is repaired by two mechanisms: non-homologous end-joining (NHEJ) and homology-directed repair (HDR). NHEJ, also known as the "non-homologous" mechanism, uses different enzymes that join break ends without the need for a homologous template. In most cases, the NHEJ pathway causes indel mutations (insertions/deletions), which often cause the loss of gene function. In contrast, the HDR mechanism requires a homologous sequence for reparation by recombination at the breakpoint (Zhu *et al.*, 2017). The main advantage of the system is the specificity that relies on the complementarity between the gRNA and the target sequence. However, off-target activity may occur in some loci with partial complementarity to the gRNA (Sledzinski *et al.*, 2020). NHEJ repair mechanism induces reparations by direct ligation of the broken ends, leading to insertions, deletions, or substitutions at the DSB site. HDR acts in the presence of a donor DNA sequence and corrects the existing modifications or inserts new sequences of interest (Puchta, 2017). The integrated transgene is functional in the plant genome and can be expressed (Jaganathan *et al.*, 2018). Different types of CRISPR/Cas systems have distinct molecular mechanisms for DNA targeting (Makarova *et al.*, 2011; Chylinski *et al.*, 2014).

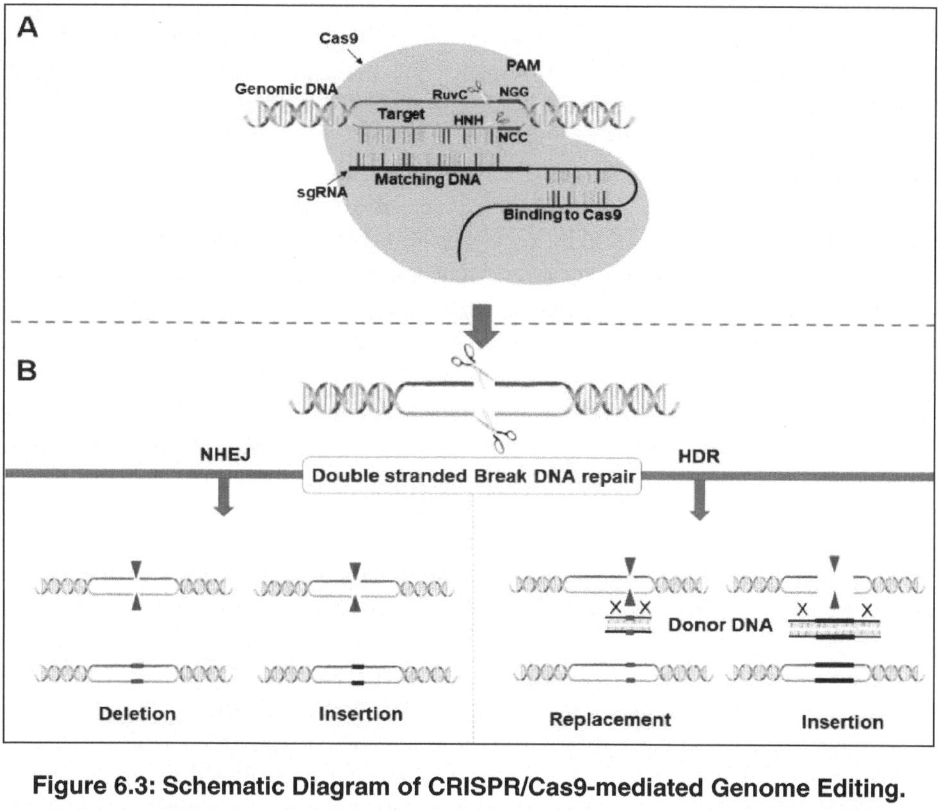

Figure 6.3: Schematic Diagram of CRISPR/Cas9-mediated Genome Editing.

(A) Representation of the Cas9/sgRNA/target DNA complex. The sgRNA is represented by the black line. The target sequence is indicated by the blue strand of genomic DNA. The protospacer adjacent motif (PAM) sequence is highlighted in red. The enzyme Cas9 is represented by the gray background. The catalytic domains RuvC and HNH responsible for non-complementary and complementary strand break, respectively, are indicated by scissors. (B) Repair of the DNA double-strand breaks. Two mechanisms may be involved in the repair: (i) the non-homologous end joining (NHEJ) or (ii) the homology-directed repair (HDR).

Recently, a new Cas protein named Cpf1 (also known as Cas12a), identified in Prevotella and Francisella, has also been used as a nuclease in plant genome editing (Tang *et al.*, 2017, Zetsche *et al.*, 2015). Unlike CRISPR/Cas9, the CRISPR/Cpf1 system requires a single crRNA that does not need to couple with a tracrRNA, allowing the use of a smaller gRNA. Cpf1 recognizes an alternative PAM sequence (50 -TTTN-30) upstream of the target site, increasing the range of targets for forming DSBs. Additionally, unlike Cas9, Cpf1 generates staggered cuts with 50 overhangs, analogous to traditional restriction enzyme cloning, that are 18–23 bp distal from the PAM sequence (Zetsche *et al.*, 2015). Expression of Cas9 with sgRNA leads to introduction of DSBs, which can be repaired via two major pathways: non-homologous end joining (NHEJ) and homology-directed repair (HDR; Figure 6.1B) (Puchta, 2007); DSB repair via NHEJ is the most frequent and is characterized by the fusion of the break ends where insertions or deletions (indels) of one or more random nucleotides can occur, leading to imperfect repair and mutations (Chiruvella *et al.*, 2013). When DSBs are repaired by NHEJ, it typically results in gene knockout or loss of protein function. Thus, NHEJ can be used as an alternative to RNAi and other gene silencing techniques (Unniyampurath *et al.*, 2016). Alternatively, if an artificial repair template is provided, HDR is induced and larger exogenous DNA fragments can be inserted, or single nucleotides can be exchanged in a specified genomic location, resulting in a more precise gene editing. However, the target has a low mutation frequency since it is difficult to insert an artificial DNA repair template into the plant cells (Branzei and Foiani, 2008; Puchta *et al.*, 1996). In plants, efficient CRISPR/Cas9-based genome editing comprises of the following steps as depicted in Figure 6.3. Initially, it is necessary to choose a region in the gene of interest that is unique in the genome and located upstream of a 50 -NGG-30 PAM sequence (Jiang *et al.*, 2013). The sgRNA must have homology to the sequence of interest (approximately 20 bp sequence) and should also contain a "scaffold" sequence necessary for Cas9 binding (Jiang *et al.*, 2013; Doudna and Charpentier, 2014). When expressed, the Cas9 protein and sgRNA form a complex that can bind to any genomic sequence with a PAM but will only cleave the target if it is sufficiently homologous to the sgRNA (Kleinstiver *et al.*, 2016). sgRNAs are solely responsible for recruiting Cas9 to specific genomic loci, and their optimal design is critical for successful gene editing. It is also important to minimize off-target possibilities, *i.e.*, when sgRNA matches additional sites that are similar to the target sequences, unwanted mutations could occur (Kleinstiver *et al.*, 2016; Doench *et al.*, 2016). If the goal is to induce HDR repair mechanism, it is necessary to provide a DNA repair template containing the desired mutation in addition to the sgRNA(s) and the Cas9 nuclease. To facilitate this process, there must be an additional homologous sequence immediately upstream and downstream of the target (named left and right homology arms, respectively). The repair template can be delivered as single or double-stranded oligonucleotide as well as a double-stranded DNA plasmid (Li *et al.*, 2016; Sun *et al.*, 2016). Then, the Cas9 and sgRNA expression cassettes are incorporated into the plant's genome,

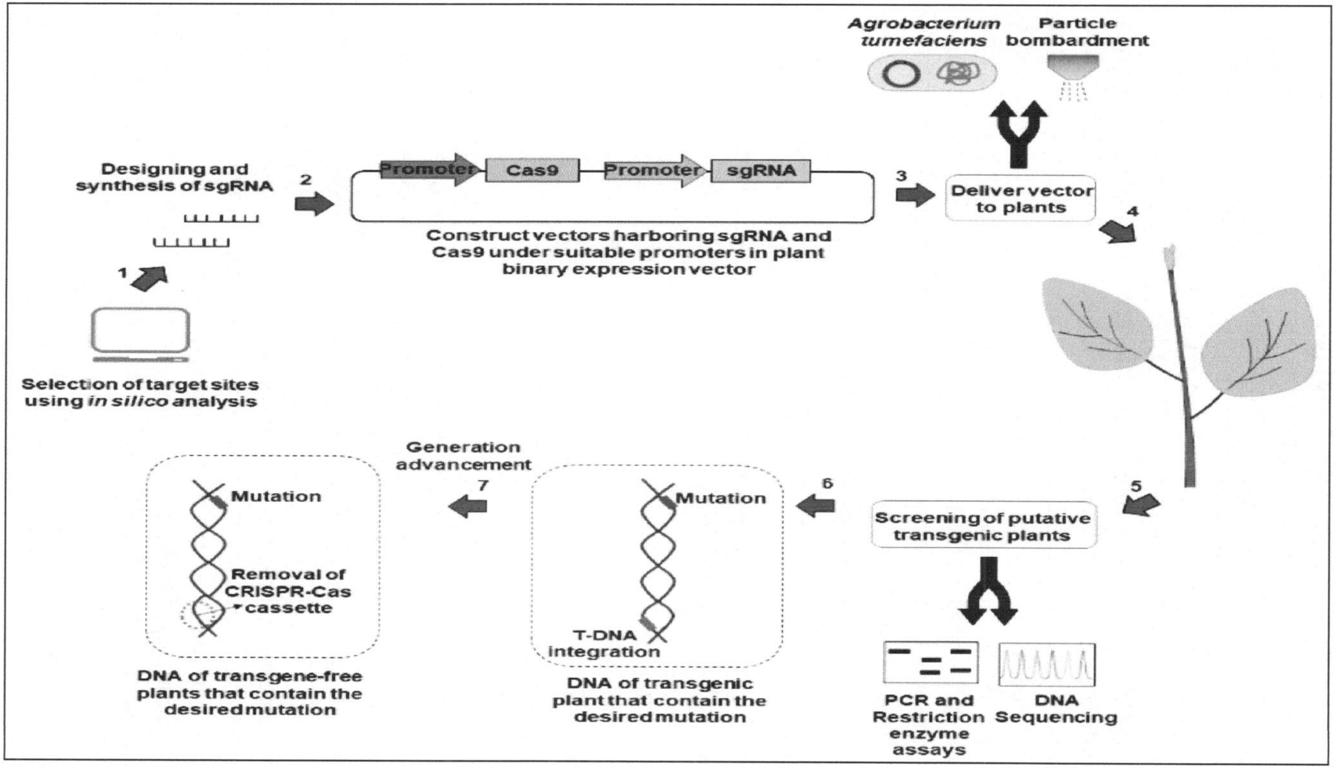

Figure 6.4: Overview of the Basic Flow of the CRISPR/Cas9 System in Plant Genome Editing.

1. Select a genomic target where the mutation is to be introduced. 2. Design the sgRNA complementary to the expected target sequence. 3. The sgRNA and Cas9 under suitable promoters are cloned in plant binary expression vectors. 4. The components of the CRISPR/Cas9 system construct are delivered into the plants, via *Agrobacterium*-mediated transformation or particle bombardment. 5. The mutations in regenerated transgenic plants are identified using restriction enzyme assays and sequencing. 6 and 7. The removal of the CRISPR/Cas9 cassettes is possible in subsequent generations of plants.

normally via *Agrobacterium*-mediated transformation (Fan *et al.*, 2015; Zhou *et al.*, 2014). Expression of Cas9 and sgRNA introduces DSBs at the target site (Yin *et al.*, 2017). After being repaired via NHEJ or HDR, desired mutations in regenerated transgenic plants are identified by restriction enzyme assays or sequencing (Soda *et al.*, 2018). Usually, the transgenic population contains a wide variety of mutations or edits within the target gene. Often, the following four genotypes are generated in T0 lines: homozygote (same mutation in two alleles), heterozygote (mutation of a single allele), biallelic (different mutations in two alleles) and chimera (more than two different mutations) (Yang *et al.*, 2017). Some plants remain unmodified because of the absence of Cas9/sgRNA expression or insufficient target cleavage. Even when a DNA repair template is provided to induce repair via HDR, some of the DSBs are still repaired by NHEJ, leading to undesirable mutations (Xingliang and Yaoguang, 2015). Since the Cas9/sgRNA transgene and the target region are often at different locations in the genome, removal of the Cas9/sgRNA transgene through segregation is possible in the subsequent generations.

Thus far, the majority of published studies with the CRISPR/Cas9 system involve NHEJ-mediated single-gene knockout or down-regulation (Khatodia *et al.*, 2016; Rani *et al.*, 2016). However, when Cas9 is combined with multiple sgRNAs specific for different target genes (Ma *et al.*, 2015), NHEJ may also introduce targeted chromosome deletion (Zhou *et al.*, 2014) and knock out the whole gene family, which provides the possibility to introduce multiple traits in an elite variety (Peterson *et al.*, 2016). Additionally, HDR can be employed to generate a specific sequence change for gene correction, gene replacement, and gene knock-in (Endo *et al.*, 2016; Svitashev *et al.*, 2015).

Genome Editing in Fruit Crops

In plants, genome editing using the CRISPR/Cas9 system was first reported in 2013 (Li *et al.*, 2013, Shan *et al.*, 2013). In addition to the pioneer studies in the model plants *Arabidopsis* (Chen *et al.*, 2017; Mao *et al.*, 2013) and *Nicotiana* (Li *et al.*, 2013; Nekrasov *et al.*, 2013; Gao *et al.*, 2015), the CRISPR/Cas9 system has been successfully used for genome editing in several fruit crops, including apple, banana, cacao, citrus, grape, kiwifruit, and pear. Several approaches have been evaluated to optimize the CRISPR/Cas9 technique for its use within a fruit cultivar. In most of these studies, the phytoene desaturase (PDS) gene, which encodes an enzyme in the carotenoid biosynthesis pathway was targeted. The disruption of this gene impairs chlorophyll and carotenoid production resulting in an albino phenotype (Qin *et al.*, 2007) and is an easy target for manipulation to confirm the efficacy of the genome modification system. For instance, transgenic expression of Cas9 guided by 19-bp sgRNA designed to target the conserved region of two PDS genes in the banana genome resulted in complete albino and variegated phenotype among regenerated plantlets and mutation efficiency of 59 per cent (Kaur *et al.*, 2018). Higher editing efficiency (100 per cent) targeting the same gene via polycistronic gRNAs was also described in banana (Naim *et al.*, 2018). Mutations induced in the PDS gene in diploid and octoploid strawberry resulted in a clear albino phenotype at a high frequency (Wilson *et al.*, 2019). Similar results were seen following the editing of the PDS gene of *Carrizo citrange*. When the Cas9 gene was driven by the Arabidopsis YAO gene promoter instead of the commonly used cauliflower mosaic virus 35S promoter, the mutation efficiency increased and in several instances was reported to be 100 per cent (Zhang *et al.*, 2017). Mutations in the PDS gene introduced by CRISPR/Cas9 have also been shown to confer an albino phenotype in apple (Charrier *et al.*, 2019; Nishitani *et al.*, 2016), grapes (Nakajima *et al.*, 2017) kiwifruit (Wang *et al.*, 2018), pear (Charrier *et al.*, 2019), and kumquat (Zhu *et al.*, 2019).

Recently, CRISPR/Cas9 was also utilized to induce mutations in the MaGA20ox2 gene. This gene regulates dwarfism in banana (Chen *et al.*, 2016). Seven mutant lines were detected with semi-dwarf phenotype following genome editing. In this study, endogenous GA levels in the different tissues were quantified and the results were consistent with the phenotype observed. There was a significant change in GAs levels in the mutants compared to the non-transformed control in both leaves and roots (Shao *et al.*, 2019). This strategy can be subsequently utilized to develop the much-needed semi-dwarf or dwarf banana cultivars. Other crops that have been modified using the CRISPR/Cas9 system are outlined in Table 6.1.

Table 6.1: List of Targeted Gene(s) and Trait Modified via the CRISPR/Cas9 System in different Fruit Crops

Species	*Target Gene*	*Target Trait*	*Delivery Method*	*Reference*
Apple	DIPM-1,DIPM-2,DIPM-4	Fire blightdisease resistance	PEG-mediated protoplast transfection	(Malnoy *et al.*, 2016)
Apple	PDS	Albino phenotypes	*Agrobacterium*-mediated leaf discs transformation	(Nishitani *et al.*, 2016)
Apple and Pear	PDSTFL1	Albino phenotypes Early flowering	Vacuum-infiltration in a suspension of *Agrobacterium* tumefaciens	(Charrier *et al.*, 2019)
Banana	PDS	Albino phenotype	*Agrobacterium*-mediated suspension cells transformation	(Aim *et al.*, 2018)
Banana	PDS	Albino phenotypes	*Agrobacterium*- mediated embryogenic cell suspension cultures transformation	(Kaur *et al.*, 2018)
Banana	MaGA20ox2	Semi-dwarfing size	*Agrobacterium*-mediated suspension cells transformation	(Shao *et al.*, 2019)

Species	*Target Gene*	*Target Trait*	*Delivery Method*	*Reference*
Banana	Ebsv	Control of virus pathogenesis	*Agrobacterium*-mediated suspension cells transformation	(Tripathi *et al.*, 2019)
Cacao	TcNPR3	*Phytophthora tropicalis* resistance	*Agrobacterium*-mediated transient leaf transformation	(Fister *et al.*, 2010)
Citrus (Carrizo Citrange)	PDS	Albino phenotypes	*Agrobacterium*-mediated epicotyl transformation	(Jia *et al.*, 2017)
Citrus (Grapefruit)	CsLOB1	Canker disease resistance	*Agrobacterium*-mediated epicotyl transformation	(Jia *et al.*, 2017)
Citrus (Grapefruit)	PDS	Albino phenotype	*Agrobacterium*-mediated epicotyl transformation	(Jia *et al.*, 2017)
Citrus (Kumquat)	PDS	Albino phenotype	*Agrobacterium*-mediated epicotyl transformation	(Zhu *et al.*, 2019)
Citrus (Sweet Orange)	CsLOB1	Canker disease resistance	*Agrobacterium*-mediated epicotyl transformation	(Peng *et al.*, 2017)
Citrus (Sweet Orange)	CsWRKY22	Canker disease resistance	*Agrobacterium*-mediated epicotyl transformation	(Zhang *et al.*, 2019)
Citrus (Sweet Orange)	DMR6	Huanglongbing resistance	*Agrobacterium*-mediated epicotyl transformation	(Zhang *et al.*, 2018)
Grape	PDS	Albino phenotypes	*Agrobacterium*- mediated callus transformation	(Nakajima *et al.*, 2017)
Grape	PDS	Albino phenotypes	*Agrobacterium*-mediated suspension cells transformation	(Ren *et al.*, 2019)
Grape and Apple	IdnDH	Biosynthesis of tartaric acid	*Agrobacterium*-mediated suspension cells transformation	(Osakabe *et al.*, 2018)
Grape	L-idonate dehydrogenase gene (IdnDH)	Tartaric acid content	*Agrobacterium*-mediated suspension cells transformation	(Ren *et al.*, 2016)
Grape	VvWRKY52	*Botrytis cinerea* resistance	*Agrobacterium*-mediated somatic embryos transformation	(Wang *et al.*, 2018)
Kiwifruit	PDS	Albino phenotype	*Agrobacterium*-mediated transformation	(Wang *et al.*, 2018)
Strawberry	APETALA3 (AP3)	Flowering control	*Agrobacterium*-mediated leaf disk	(Martín-Pizarro *et al.*, 2018)

Species	*Target Gene*	*Target Trait*	*Delivery Method*	*Reference*
Strawberry	Auxin Response Factor 8 (FvARF8) and Auxin biosynthesis gene (FveTAA1)	Auxin biosynthesis	*Agrobacterium*-mediated transformation	(Zhou *et al.*, 2018)
Strawberry	PDS	Albino phenotypes	*Agrobacterium*-mediated leaf and petiole transformation	(Wilson *et al.*, 2019)

Overview of Fruit Crop Genome Editing

Genome Editing in Banana

Bananas are one of the most economically important fruits in the tropics. It is also the fourth largest food crop in the world (Perrier *et al.*, 2011). Traditional breeding is difficult due to its triploid genome and sterility (Dash *et al.*, 2016). Genetic engineering is perhaps the only way to improve this important staple food and fruit. A single guide RNA was designed to target the conserved domain of two RASPDS genes (RAS-PDS1 and RAS-PDS2) in banana embryonic cell suspension cultures cv. Rasthali (Kaur *et al.*, 2018). Fully albino and variegated phenotypes were observed among the regenerated seedlings with a mutation rate of 59 per cent. More recently, a multiplexed approach via the tRNA-gRNA polycistronic system was used to target PDS exon 1 in cultivar Cavendish; A 100 per cent mutation rate and triallelic deletions or insertions were found among the 19 regenerated plants whose albino grade correlated with genotype (Naim *et al.*, 2018). The study demonstrates that CRISPR/Cas9-mediated genome editing occurs efficiently in banana. Targeting endogenous genes will be the next necessary step towards banana breeding.

Genome Editing in Citrus

Citrus is one of the most economically important fruit crops in the world however, genetic modification through gene breeding or manipulation is extremely difficult due to polyembryonic seeds, pollen incompatibility, parthenocarpic fruits, a prolonged juvenile phase, and resistance to transformation by traditional means (Talon and Gmitter, 2008). In 2014, Jia and Wang successfully used *Xanthomonas citri* subsp. *citri* (Xcc)-facilitated agro infiltration to deliver Cas9 and a synthetic sgRNA targeting CsPDS in sweet orange (*Citrus* X *sinensis*) and Duncan grapefruit leaves, leading to successful but low mutation rates in infiltrated leaves (Jia and Wang, 2014). Later, the same group applied *Agrobacterium*-mediated transformation to Duncan grapefruit epicotyl explants and created six transgenic lines containing a CRISPR/Cas9 construct targeting the CsLOB1 gene that controls susceptibility to citrus canker, a disease caused by citrus canker. *Xanthomonas axonopodis* bacterium. The resulting plants showed varying degrees of resistance to citrus canker

(Jia *et al.*, 2017b). Another effort successfully edited the CsLOB1 promoter, rendering the resulting homozygous plants resistant to citrus canker (Peng *et al.*, 2017). Therefore, genome editing of the promoter region proves to be an effective method to disrupt the expression of the target gene. These encouraging developments illustrate that CRISPR/Cas9-mediated editing will help combat devastating diseases affecting citrus, including citrus greening.

High Efficiency Editing in Kiwifruit is Achieved with the PTG/Cas9 System

Kiwi (*Actinidia* Lindl.) is an economically important fruit species that contains a high content of vitamin C and dietary fiber (Huang *et al.*, 2013). Genome editing in kiwi was achieved using the standard CRISPR/Cas9 system and the Cas9 system (Xie and Yang 2013; Xie *et al.*, 2015). The kiwi PDS gene was edited in callus-derived kiwi seedlings with the PTG/Cas9 system with 10-fold higher efficiency than the traditional CRISPR/Cas9 system (Wang *et al.*, 2018b). In addition, deletions of large chromosomal fragments were generated in kiwi with the paired sgRNAs of the PTG/Cas9 system, suggesting that PTG/Cas9 is a powerful tool in kiwi genome editing and quality improvement.

Genome Editing in Grape and Apple through Direct Delivery of RNPs into Protoplasts

Grape (*Vitis vinifera* L.) is another economically important fruit crop that can be eaten fresh or used for making wine, jam, juice, and even medicine. Whole genome sequence of grape (*V. vinifera* L.) was reported and released in 2007 (Jaillon *et al.*, 2007), and grape transformation using embryonic suspensions was established as early as 1993 (Hebert *et al.*, 1993). In 2016, a group successfully edited the Lidonate dehydrogenase gene (IdnDH) in the biosynthesis of tartaric acid (Ren *et al.*, 2016). A CRIPSR/Cas9 construct containing two sgRNAs was transformed into the 'Chardonnay' embryonic suspension cells via *agrobacterium*, yielding 100 per cent mutational efficiency. A second report targeting *Vitis vinifera* phytoene desaturase (VvPDS) showed albino leaves at efficiencies ranging from 0–86 per cent (Nakajima *et al.*, 2017). Interestingly, older leaves possess higher ratio of mutant cells, probably due to the accumuation of DSBs or decreased DSBs repair in older leaves. A very recent study edited the VvWRKY52 transcription factor, which plays a role in biotic stress response (Wang *et al.*, 2018a). Proembryonal masses derived from embryo calli were infected with *agrobacterium* containing the CRISPR/Cas9 construct. Fifteen of 22 first generation transgenic lines carried biallelic mutations, and four of these showed increased resistance to Botrytis cinerea. Transgene-free genome editing is highly desirable to facilitate commercialization of genetically modified grapes. To knockout MLO-7, a susceptible locus to powdery mildew, direct delivery of purified CRISPR/Cas9 ribo nucleoproteins (RNPs) into grapevine protoplasts was demonstrated (Malnoy *et al.*, 2016). The RNPs were introduced into protoplasts via the PEG-mediated transformation and subsequently led to the editing of MLO-7 without involving any foreign DNA. However, the mutagenesis efficiency was only 0.1 per cent to 6.9 per cent. Recently, an improved protocol was established for the direct delivery of CRISPR/Cas9 RNPs into grapevine protoplasts (Osakabe *et al.*, 2018). The complete protocol of the direct delivery takes only 2–3 weeks, and whole plants can be subsequently generated from edited protoplasts. Therefore, grape is becoming a pioneering fruit crop in the arena of transgene-free genome editing. Apple is another major fruit crop in the world. Efficient genome editing was shown in first generation young transgenic plants of semi-dwarfing rootstock cultivar JM2. The PDS gene was precisely modified and a variety of albino phenotypes from clear white to pale green, to variegated young shoots were observed (Nishitani *et al.*, 2016). The white sectors can be used to indicate when and where successful editing occurs. Direct delivery of RNPs into apple leaf protoplasts was shown possible, and edited mutations in the DIPM-1, -2, and -4 genes were detected by deep sequencing of pooled protoplast DNA (Malnoy *et al.*, 2016). However, microcalli generation from apple protoplasts will need to be established before transgene-free apple plants are attainable (Osakabe *et al.*, 2018).

Transgene-Free Edited Plants in Strawberry through Segregation of the T-DNA Vector

Diploid wild strawberry (*Fragaria vesca*) has emerged as a new model system for the garden strawberry (*Fragaria x ananassa*), which is octoploid (Shulaev *et al.*, 2011). Recently, Zhou *et al.* (2018) reported successful application of CRISPR/Cas9 in *F. vesca*. In this study, dual sgRNA cassettes were driven by an Arabidopsis U6 promoter (AtU6-26) and a *Fragaria vesca* U6 promoter (FveU6-2), respectively. T_0 generation transgenic *F. vesca* lines exhibited a mutation efficiency of 49 per cent –75 per cent in the two targeted genes FveARF8 and FveTAA1, both acting in auxin pathway. Interestingly, analysis of T_1 progeny plants revealed new mutations generated in T_1 with even a higher efficiency (83 per cent) than the T_0 generation (49 per cent). The increased editing rate in T_1 resulted in half of the T_1 plants harboring large fragments deletion between the two sgRNA target sites. Some of these T_1 homozygous mutant plants lost the transgene and became transgene-free. Further, the arf8 mutant seedlings showed faster growth than the wild type seedling (Zhou *et al.*, 2018). Another recent study targeted the auxin biosynthesis gene FveYUC10, showing 100 per cent efficiency in the mutation rate (Feng *et al.*, 2019). However, homozygous Fveyuc10 mutants did not exhibit any morphological phenotype. Garden strawberry *F. ananassa* is octoploid and a hybrid. CRISPR/Cas9 was used to edit the FaTM6 MADS box gene, encoding a putative ortholog of Arabidopsis APETALA3 (Martin-Pizarro *et al.*, 2018). The mutant FaTM6 knockout lines failed to develop any fruit due to a lack of fertile anthers (Figure 6.5), indicating that the FaTM6 MADS box gene plays a key role in anther development in strawberry. This result is very exciting as editing octoploid genome was not as difficult as previously anticipated. As such, the result suggests that the CRISPR/

Figure 6.5: Phenotype of CRISPR knockout Lines of FaTM6 in Garden Strawberry (*Fragaria x ananassa*) Revealing a Key Role of FaTM6 in Anther Development.

Top panels: adult plants of wild type control and three tm6 mutant lines tm6-1, tm6-7, and tm6-9. Bottom panels: close up images of the flower and fruit. While the wild type control plants develop wild-type berries at different stages, the mutant tm6 plants failed to develop any fruits due to a lack of fertile anthers. The figure is modified from Martýn-Pizarro *et al.* (2018) under Creative Commons CC BY license.

Cas9 technology will be invaluable for the genetic improvement of commercial strawberry.

Challenges for Genome Editing in Fruit Crops

With the completion of genome sequencing of more and more fruit crop species, understanding of their genomic structure, gene pathways and gene function is paving the way for genome editing of economically important traits. Particular interest for fruit crops are the characteristics of the response to biotic or abiotic stress, shelf life, flavor and nutritional value, plant architecture and flowering time. However, polyploidization is found in many fruit species (Chen 2007; Emshwiller *et al.*, 2009; Wood *et al.*, 2009) such as apple, banana, strawberry, watermelon, plum, and kiwi. The complexity of the genome, including heterozygosity and polyploidy, contributes to the difficulties in applying genome editing to these species, since many more copies of the genes need to be mutated to have the desired phenotype. Second, many fruit crops lack efficient transformation methods, severely limiting the delivery of CRISPR/Cas9 constructs. Third, the low editing efficiency and somatic mosaics of CRISPR/Cas9-induced mutations may make it difficult to identify heritable mutations. Fourth, the long juvenile stage of many fruit trees makes experimentation and optimization of genome editing difficult. In addition, it takes a long time to obtain homozygous germline-transmitted or segregating next-generation mutations. Fifth, public resistance to GMOs suggests that robust non-transgene genome editing methods need to be developed for fruit crops.

Scope

Genome editing is a revolutionary technology for basic and applied plant research. The CRIPSR/Cas9 system is currently being developed and applied to trait engineering of fruit crops. However, due to transformation difficulties, complex genomes, slow growth cycles and lack of genomic information, it remains a long and arduous task to apply CRISPR/Cas9 to most fruit crops. The rapid development of new and improved CRISPR/Cas9-based tools and delivery systems may help overcome some of the barriers in horticultural crops. Methods that reduce the need for multigenerational cycles or that increase editing efficiency are also highly desirable. Rapid and low-cost detection of mutants is an area of urgent need. Targeting cis elements to alter gene regulation is a clever way to generate allelic sequences as demonstrated in the citrus LOB1 gene and tomato CLV3 gene. Strategies that aid in the production of GM-free edited fruit crops will be vital to consumer acceptance and commercialization. Furthermore, the importance of public dialogue about genome editing and investment in consumer-friendly features such as healthier products and longer shelf life can never be overemphasized. Given what has already been demonstrated in tomato, the de novo domestication and designer fruit is not far from reality.

REFERENCES

Anzalone, A.V., Randolph, P.B., Davis, J.R., Sousa, A.A., Koblan, L.W. (2019). Search-and replace genome editing without double strand breaks or donor DNA. Nature 576: 149-157.

Barrangou, R., Fremaux, C., Deveau, H., Richards, M., Boyaval, P., Moineau, S., Romero, D.A., Horvath, P. (2007). CRISPR provides acquired resistance against viruses in prokaryotes. Science, 315, 1709-1712.

Bolger, M.E., Weisshaar, B., Scholz, U., Stein, N., Usadel, B. (2014). Plant genome sequencing applications for crop improvement. *Current Opinion in Biotechnology*. 26: 31-37.

Bortesi, L., Fischer, R. (2015). The CRISPR/Cas9 system for plant genome editing and beyond. Biotechnol. Adv. 33, 41-52.

Branzei, D., Foiani, M. (2008). Regulation of DNA repair throughout the cell cycle. *Nat. Rev. Mol. Cell Biol.* 9: 297-308.

Capecchi, M.R. (2005). Gene targeting in mice: functional analysis of the mammalian genome for the twenty-first century. *Nature Review Genetics.* 6: 507-512.

Carroll, D. (2011). Genome engineering with zinc-finger nucleases. Genetics. 188: 773-782.

Charpentier, E., Richter, H., Van, D. J., White, M.F. (2015). Biogenesis pathways of RNA guides in archaeal and bacterial CRISPR-Cas adaptive immunity. FEMS Microbiology Reviews. 23: 428-441.

Charrier, A., Vergne, E., Dousset, N.J.-P., Richer, A., Petiteau, A., Chevreau, E. (2019). Efficient targeted mutagenesis in apple and first time edition of pear using the CRISPR-Cas9 system. Front. *Plant Sci.* 10, 40.

Chen, J., Xie, J., Duan, Y., Hu, H., Hu, Y., Li, W. (2016). Genome-wide identification and expression profiling reveal tissue-specific expression and differentially-regulated genes involved in gibberellin metabolism between Williams banana and its dwarf mutant. *BMC Plant Biol.* 16: 123.

Chen, Y., Wang, Z., Ni, H., Xu, Y., Chen, Q., Jiang, L. (2017). CRISPR/Cas9-mediated base editing system efficiently generates gain-of-function mutations in Arabidopsis. *Sci. China Life Sci.* 60: 520-523.

Chen, Z.J. (2007). Genetic and epigenetic mechanisms for gene expression and phenotypic variation in plant polyploids. *Annu Rev Plant Biol.* 58: 377-406

Chiruvella, K.K., Liang, Z., Wilson, T.E. (2013). Repair of double-strand breaks by end joining. Cold Spring Harb. *Perspect. Biol.* 5: a012757.

Chylinski, K., Makarova, K.S., Charpentier, E., Koonin, E.V. (2014). Classification and evolution of type II CRISPR-Cas systems. *Nucleic Acids Research.* 42: 6091-6105.

Dash, P.K., Rai, R. (2016). Translating the "banana genome" to delineate stress resistance, dwarfing, parthenocarpy and mechanisms of fruit ripening. *Front Plant Sci.* 7: 1543.

Dheer, P., Rautela, I., Sharma, V., Dhiman, M., Sharma, A. (2020). Evolution in crop improvement approaches and future prospects of molecular markers to CRISPR/Cas9 system. Gene 753: 144795.

Ding, D., Chen, K., Chen, Y., Li, H., Xie, K. (2018). Engineering introns to express RNA guides for Cas9- and Cpf1-mediated multiplex genome editing. *Molecular Plant*. 11: 542-552.

Doench, J.G., Fusi, N., Sullender, M., Hegde, M., Vaimberg, E.W., Donovan, K.F., Smith, I., Tothova, Z., Wilen, C., Orchard, R. (2016). Optimized sgRNA design to maximize activity and minimize off-target effects of CRISPR-Cas9. Nat. Biotechnol. 34, 184-191.

Doudna, J.A., Charpentier, E. (2014). The new frontier of genome engineering with CRISPR Cas9. Science, 346, 1258096.

Emshwiller, E., Theim, T., Grau, A., Nina, V., Terrazas, F. (2009). Origins of domestication and polyploidy in oca (*Oxalis tuberosa*; Oxalidaceae). *Am J Bot*. 96: 1839-1848.

Endo, M., Mikami, M., Toki, S. (2016). Biallelic gene targeting in rice. *Plant Physiol*. 170: 667-677.

Fan, D., Liu, T., Li, C., Jiao, B., Li, S., Hou, Y., Luo, K. (2015). Efficient CRISPR/Cas9 mediated targeted mutagenesis in Populus in the first generation. *Sci. Rep*. 5: 12217.

Feng, J., Dai, C., Luo, H., Han, Y., Liu, Z., Kang, C. (2019). Reporter gene expression reveals precise auxin synthesis sites during fruit and root development in the wild strawberry. *J Exp Bot*. 70: 563-574.

Gao, J., Wang, G., Ma, S., Xie, X., Wu, X., Zhang, X., Wu, Y., Zhao, P., Xia, Q. (2015). CRISPR/Cas9-mediated targeted mutagenesis in *Nicotiana tabacum*. *Plant Mol. Biol*. 87: 99-110.

Gaudelli, N.M., Komor, A.C., Rees, H.A., Packer, M.S., Badran, A.H. (2018). Publisher correction: Programmable base editing of A_T to G_C in genomic DNA without DNA cleavage. Nature 559: E8.

Ishino, Y., Shinagawa, H., Makino, K., Amemura, M., Nakata, A. (1987). Nucleotide sequence of the iap gene, responsible for alkaline phosphatase isozyme conversion in Escherichia coli, and identification of the gene product. *J. Bacteriol*. 169: 5429-5433.

Jaganathan, D., Ramasamy, K., Sellamuthu, G., Jayabalan, S., Venkataraman, G. (2018). CRISPR for crop improvement: An update review. *Frontiers in Plant Science*. 9: 985.

Jaganathan, D., Ramasamy, K., Sellamuthu, G., Jayabalan, S., Venkataraman, G. (2018). CRISPR for crop improvement: An update review. *Frontiers in Plant Science*. 9: 985.

Jaillon, O., Aury, J.M., Noel, B., Policriti, A., Clepet, C., Casagrande, A., Choisne, N., Aubourg, S., Vitulo, N., Jubin, C., Vezzi, A., Legeai, F., Hugueney, P., Dasilva, C., Horner, D., Mica, E., Jublot, D., Poulain, J., Bruyere, C., Billault, A., Segurens, B., Gouyvenoux, M., Ugarte, E., Cattonaro, F., Anthouard, V., Vico, V., Del, C., Alaux, M., Gaspero, G., Dumas, V., Felice, N., Paillard, S., Juman, I., Moroldo, M., Scalabrin, S., Canaguier, A., Clainche, I., Malacrida, G., Durand, E., Pesole, G., Laucou, V., Chatelet, P., Merdinoglu, D., Delledonne, M., Pezzotti, M., Lecharny, A., Scarpelli, C., Artiguenave, F., Pe, M.E., Valle, G., Morgante, M., Caboche, M., Adam-Blondon, A.F., Weissenbach, J., Quetier, F., Wincker, P., Public, F.I. (2007). The grapevine genome sequence suggests ancestral hexaploidization in major angiosperm phyla. Nature 449: U463-U465.

Jansen, R., van Embden, J.D.A., Gaastra, W., Schouls, L.M. (2002). Identification of genes that are associated with DNA repeats in prokaryotes. *Mol. Microbiol.* 43: 1565-1575.

Jia, H., Wang, N. (2014a). Xcc-facilitated agroinfiltration of citrus leaves: A tool for rapid functional analysis of transgenes in citrus leaves. *Plant Cell Rep.* 33: 1993-2001.

Jia, H., Wang, N. (2014b). Targeted genome editing of sweet orange using Cas9/ sgRNA. *PLoS* one 9: e93806.

Jia, H., Zhang, Y., Orbovic, V., Xu, J., White, F.F., Jones, J.B., Wang, N. (2017b). Genome editing of the disease susceptibility gene CsLOB1 in citrus confers resistance to citrus canker. *Plant Biotechnol J*. 15: 817-823.

Jiang, F., Doudna, J.A. (2017). CRISPR–Cas9 structures and mechanisms. *Annual Review of Biophysics*. 46: 505-529.

Jiang, W., Bikard, D., Cox, D., Zhang, F., Marraffini, L.A. (2013). RNA-guided editing of bacterial genomes using CRISPR-Cas systems. *Nat. Biotechnol.* 31: 233-239.

Jinek, M., Chylinski, K., Fonfara, I., Hauer, M., Doudna, J.A. (2012). Programmable dual RNA–guided DNA endonuclease in adaptive bacterial immunity. Science. 337: 816-821.

Jinek, M., Chylinski, K., Fonfara, I., Hauer, M., Doudna, J.A., Charpentier, E. (2012). A programmable dual-RNA-guided DNA endonuclease in adaptive bacterial immunity. Science, 337: 816-821.

Kamburova, V.S., Nikitina, E.V., Shermatov, S.E., Buriev, Z.T., Kumpatla, S.P. (2017). Genome editing in plants: An overview of tools and applications. *International Journal of Agronomy*. ID 7315351.

Karvelis, T., Gasiunas, G., Miksys, A., Barrangou, R., Horvath, P., Siksnys, V. (2013). crRNA and tracrRNA guide Cas9-mediated DNA interference in *Streptococcus thermophilus*. *RNA Biol*. 10: 841-851.

Kaul, T., Sony, S.K., Verma, R., Motelb, K.F.A., Prakash, A.T. (2020). Revisiting CRISPR/Cas-mediated crop improvement: Special focus on nutrition. *Journal of Biosciences*. 45: 137. PMID: 33361628.

Kaur, N., Alok, A., Kaur, N., Pandey, P., Awasthi, P., Tiwari, S. (2018). CRISPR/Cas9 mediated efficient editing in *phytoene desaturase* (PDS) demonstrates precise manipulation in banana cv. Rasthali genome. *Funct Integr Genomics*. 18: 89-99.

Khatodia, S., Bhatotia, K., Passricha, N., Khurana, S., Tuteja, N. (2016). The CRISPR/Cas genome-editing tool: Application in improvement of crops. *Front. Plant Sci*. 7: 506.

Kleinstiver, B.P., Pattanayak, V., Prew, M.S., Tsai, S.Q., Nguyen, N.T., Zheng, Z., Joung, J.K. (2016). High-fidelity CRISPR–Cas9 nucleases with no detectable genome-wide off-target effects. 529, 490-495.

Li, J., Meng, X., Zong, Y., Chen, K., Zhang, H., Liu, J., Li, J., Gao, C. (2016). Gene replacements and insertions in rice by intron targeting using CRISPR–Cas9. Nat. 2, 16139.

Li, J.-F., Norville, J.E., Aach, J., McCormack, M., Zhang, D., Bush, J., Church, G.M., Sheen, J. (2013). Multiplex and homologous recombination–mediated genome editing in *Arabidopsis* and *Nicotiana benthamiana* using guide RNA and Cas9. *Nat. Biotechnol*. 31: 688-691.

Li, L., Piatek, M.J., Atef, A., Piatek, A., Wibowo, A. (2012). Rapid and highly efficient construction of TALE-based transcriptional regulators and nucleases for genome modification. *Plant Molecular Biology*. 78: 407-416.

Li, S., Zhang, X., Wang, W., Guo, X., Wu, Z. (2018b). Expanding the scope of CRISPR/Cpf1-mediated genome editing in rice. *Molecular Plant*. 11: 995-998.

Ma, X., Zhang, Q., Zhu, Q., Liu, W., Chen, Y., Qiu, R., Wang, B., Yang, Z., Li, H., Lin, Y. (2015). A robust CRISPR/Cas9 system for convenient, high-efficiency multiplex genome editing in monocot and dicot plants. Mol. Plant. 8: 1274-1284.

Mahfouz, M.M., Li, L., Shamimuzzaman, M., Wibowo, A., Fang, X. (2011). De novo engineered transcription activator-like effector (TALE) hybrid nuclease with novel DNA binding specificity creates double-strand breaks. Proceedings of the National Academy of Sciences of the USA 108: 2623-2628.

Makarova, K.S., Haft, D.H., Barrangou, R., Brouns, S.J.J., Charpentier, E. (2011). Evolution and classification of the CRISPR-Cas systems. *Nature Review Microbiology*. 9: 467-477.

Malnoy, M., Viola, R., Jung, M.H., Koo, O.J., Kim, S., Kim, J.S., Velasco, R., Kanchiswamy, C.N. (2016). DNA-free genetically edited grapevine and apple protoplast using CRISPR/Cas9 ribonucleoproteins. *Front. Plant. Sci.* 7: 1904.

Malzahn, A.A., Tang, X., Lee, K., Ren, Q., Sretenovic. (2019). Application of CRISPR Cas12a temperature sensitivity for improved genome editing in rice, maise, and Arabidopsis. *BMC Biology*. 17: 9.

Mao, Y., Zhang, H., Xu, N., Zhang, B., Gou, F., Zhu, J.-K. (2013). Application of the CRISPR-Cas system for efficient genome engineering in plants. *Mol. Plant*, 6: 2008-2011.

Martin-Pizarro, C., Trivino, J.C., Pose, D. (2018). Functional analysis of TM6 MADS-box gene in the octoploid strawberry by CRISPR/Cas9 directed mutagenesis. *J Exp Bot*.70: 885-895.

McManus, M.T., Sharp, P.A. (2002). Gene silencing in mammals by small interfering RNAs. *Nature Reviews Genetics*. 3: 737-747.

Mojica, F.J., Diez-Villasenor, C., García-Martínez, J., Almendros, C. (2009). Short motif sequences determine the targets of the prokaryotic CRISPR defence system. *Microbiology*. 155: 733-740.

Mojica, F.J., Rodriguez-Valera, F. (2016). The discovery of CRISPR in archaea and bacteria. FEBS J. 283, 3162-3169.

Naim, F., Dugdale, B., Kleidon, J., Brinin, A., Shand, K., Waterhouse, P., Dale, J. (2018). Gene editing the phytoene desaturase alleles of Cavendish banana using CRISPR/Cas9. *Transgenic Res*. 27: 451–460.

Nakajima, I., Ban, Y., Azuma, A., Onoue, N., Moriguchi, T., Yamamoto, T., Toki, S., Endo, M. (2017). CRISPR/Cas9-mediated targeted mutagenesis in grape. PLoS ONE 12: e0177966.

Nekrasov, V., Staskawicz, B., Weigel, D., Jones, J.D., Kamoun, S. (2013). Targeted mutagenesis in the model plant Nicotiana benthamiana using Cas9 RNA-guided endonuclease. *Nat. Biotechnol*. 31: 691–693.

Nishitani, C., Hirai, N., Komori, S., Wada, M., Okada, K., Osakabe, K., Yamamoto, T., Osakabe, Y. (2016). Efficient genome editing in apple using a CRISPR/Cas9 system. *Sci Rep*. 6: 31481.

Osakabe, Y., Liang, Z., Ren, C., Nishitani, C., Osakabe, K., Wada, M., Komori, S., Malnoy, M., Velasco, R., Poli, M., Jung, M.H., Koo, O.J., Viola, R., Kanchiswamy, N.C. (2018) CRISPR-Cas9- mediated genome editing in apple and grapevine. *Nat Protoc*. 13: 2844-2863.

Pausch, P., Al-Shayeb, B., Bisom-Rapp, E., Tsuchida, C.A., Li, Z. (2020). CRISPR/CasU from huge phages is a hypercompact genome editor. Science, 369: 333-337.

Peng, A.H., Chen, S.C., Lei, T.G., Xu, L.Z., He, Y.R., Wu, L., Yao, L.X., Zou, X.P. (2017). Engineering canker-resistant plants through CRISPR/Cas9-targeted editing of the susceptibility gene CsLOB1 promoter in citrus. *Plant Biotechnol J.* 15: 1509-1519.

Perrier, X., De, Langhe. E., Donohue, M., Lentfer, C., Vrydaghs, L., Bakry, F., Carreel, F., Hippolyte, I., Horry, J.P., Jenny, C., Lebot, V., Risterucci, A.M., Tomekpe, K., Doutrelepont, H., Ball, T., Manwaring, J., Maret, P., Denham, T. (2011). Multidisciplinary perspectives on banana (*Musa* spp.) domestication. Proc Natl Acad Sci, USA 108: 11311-11318.

Peterson, B.A., Haak, D.C., Nishimura, M.T., Teixeira, P.J., James, S.R., Dangl, J.L., Nimchuk, Z.L. (2016). Genomewide assessment of efficiency and specificity in CRISPR/Cas9 mediated multiple site targeting in Arabidopsis. PLoS ONE, 11, e0162169.

Petolino, J.F. (2015). Genome editing in plants via designed zinc finger nucleases. In Vitro Cellular and Developmental Biology - Plant. 51: 1.

Puchta, H. (2004). The repair of double-strand breaks in plants: Mechanisms and consequences for genome evolution. *J. Exp. Bot.* 56: 1-14.

Puchta, H. (2017). Applying CRISPR/Cas for genome engineering in plants: The best is yet to come. *Current Opinion in Plant Biology*. 36: 1-8.

Puchta, H., Dujon, B., Hohn, B. (1996). Two different but related mechanisms are used in plants for the repair of genomic double-strand breaks by homologous recombination. *Proc. Natl. Acad. Sci.* 93: 5055-5060.

Punwar, B.S., Ram, C., Singh, A., Vala, A. (2014). Genome editing: conceptual introduction. Technical Report. doi: 10.13140/RG.2.1.1169.5203.

Qin, G., Gu, H., Ma, L., Peng, Y., Deng, X.W., Chen, Z., Qu, L.J. (2007). Disruption of phytoene desaturase gene results in albino and dwarf phenotypes in Arabidopsis by impairing chlorophyll, carotenoid, and gibberellin biosynthesis. *Cell Res.* 17: 471-482.

Rani, R., Yadav, P., Barbadikar, K.M., Baliyan, N., Malhotra, E.V., Singh, B.K., Kumar, A., Singh, D. (2016). CRISPR/Cas9: A promising way to exploit genetic variation in plants. *Biotechnol. Lett.* 38: 1991-2006.

Rath, D., Amlinger, L., Rath, A., Lundgren, M. (2015). The CRISPRCas immune system: Biology, mechanisms and applications. Biochimie. 117: 119-128.

Ren, C., Liu, X., Zhang, Z., Wang, Y., Duan, W., Li, S., Liang, Z. (2016). CRISPR/Cas9 mediated efficient targeted mutagenesis in Chardonnay (*Vitis vinifera* L.). *Sci Rep.* 6: 32289.

Safari, F., Zare, K., Negahdaripour, M., Barekati-Mowahed, M., Ghasemi, Y. (2019). CRISPR Cpf1 proteins: Structure, function and implications for genome editing. *Cell and Bioscience*. 9: 36.

Shan, Q, Wang Y., Li, J., Zhang, Y., Chen, K. (2013). Targeted genome modification of crop plants using a CRISPR-Cas system. *Nature Biotechnology*. 31: 8.

Shan, Q., Wang, Y., Li, J., Zhang, Y., Chen, K., Liang, Z., Zhang, K., Liu, J., Xi, J.J., Qiu, J.L. (2013). Targeted genome modification of crop plants using a CRISPR-Cas system. *Nat. Biotechnol*. 31: 686–688.

Sledzinski, P., Nowaczyk, M., Olejniczak, M. (2020). Computational tools and resources supporting CRISPR-Cas experiments. Cells 9: 1288. doi: 10.3390/cells9051288.

Soda, N., Verma, L., Giri, J. (2018). CRISPR-Cas9 based plant genome editing: Significance, opportunities and recent advances. Plant Physiol. *Biochem*. 131: 2-11.

Steinert, J., Schiml, S., Fauser, F., Puchta, H. (2015). Highly efficient heritable plant genome engineering using Cas9 orthologues from *Streptococcus thermophilus* and *Staphylococcus aureus. Plant J*. 84: 1295-1305.

Sun, Y., Zhang, X., Wu, C., He, Y., Ma, Y., Hou, H., Guo, X., Du, W., Zhao, Y., Xia, L. (2016). Engineering herbicide-resistant rice plants through CRISPR/Cas9-mediated homologous recombination of acetolactate synthase. Mol. Plant, 9, 628-631.

Svitashev, S., Young, J., Schwartz, C., Gao, H., Falco, S.C., Cigan, A.M. (2015). Targeted mutagenesis, precise gene editing and site-specific gene insertion in maize using Cas9 and guide RNA. *Plant Physiol*. 169: 931-945.

Talon, M., Gmitter, F.G., Jr. (2008). Citrus genomics. *Int J Plant Genomics*. 2008: 528361.

Tang, X., Lowder, L.G., Zhang, T., Malzahn, A.A., Zheng, X., Voytas, D.F., Zhong, Z., Chen, Y., Ren, Q., Li, Q. (2017). A CRISPR–Cpf1 system for efficient genome editing and transcriptional repression in plants. *Nat. Plants*, 3: 17018.

Unniyampurath, U., Pilankatta, R., Krishnan, M. (2016). RNA interference in the age of CRISPR: Will CRISPR interfere with RNAi. *Int. J. Mol. Sci*. 17: 291.

Urnov, F.D., Rebar, E.J., Holmes, M.C., Zhang, H.S., Gregory, P.D. (2010). Genome editing with engineered zinc finger nucleases. *Nature Review Genetics*. 11: 636-646.

Wang, X., Tu, M., Wang, D., Liu, J., Li, Y., Li, Z., Wang, Y., Wang, X. (2018a) CRISPR/Cas9-mediated efficient targeted mutagenesis in grape in the first generation. *Plant Biotechnol J*. 16: 844-855.

Wang, Z., Wang, S., Li, D., Zhang, Q., Li, L., Zhong, C., Liu, Y., Huang, H. (2018). Optimized paired-sgRNA/Cas9 cloning and expression cassette triggers high-efficiency multiplex genome editing in kiwifruit. *Plant Biotechnol. J*. 16: 1424-1433.

Wilson, F.M., Harrison, K., Armitage, A.D., Simkin, A.J., Harrison, R.J. (2019). CRISPR/Cas9-mediated mutagenesis of phytoene desaturase in diploid and octoploid strawberry. *Plant Methods*. 15: 45.

Wood, T.E., Takebayashi, N., Barker, M.S., Mayrose, I., Greenspoon, PB., Rieseberg, L.H. (2009). The frequency of polyploid speciation in vascular plants. *Proc Natl Acad Sci*.106: 13875-13879.

Xie, K., Minkenberg, B., Yang, Y. (2015). Boosting CRISPR/Cas9 multiplex editing capability with the endogenous tRNA-processing system. *Proc Natl Acad Sci*, USA 112: 3570-3575.

Xie, K.B., Yang, Y.N. (2013). RNA-guided genome editing in plants using a CRISPRCas system. *Mol Plant*. 6: 1975-1983.

Xing, H.L., Dong, L., Wang, Z.P., Zhang, H.Y., Han, C.Y., Liu, B., Wang, X.C., Chen, Q.J. (2014). A CRISPR/Cas9 toolkit for multiplex genome editing in plants. BMC Plant Biol. 14: 327.

Xingliang, M., Yaoguang, L. (2015). CRISPR/Cas9-based genome editing systems and the analysis of targeted genome mutations in plants. Hereditas (Beijing), 38, 118-125.

Yang, H., Wu, J.J., Tang, T., Liu, K.D., Dai, C. (2017). CRISPR/Cas9-mediated genome editing efficiently creates specific mutations at multiple loci using one sgRNA in *Brassica napus*. *Sci. Rep*. 7: 7489.

Yin, K., Gao, C., Qiu, J.L. (2017). Progress and prospects in plant genome editing. *Nat. Plants*, 3: 17107

Zetsche, B., Gootenberg, J.S., Abudayyeh, O.O., Slaymaker, I.M., Makarova, K.S., Essletzbichler, P., Volz, S.E., Joung, J., Van Der Oost, J., Regev, A. (2015). Cpf1 is a single RNA-guided endonuclease of a class 2 CRISPR-Cas system. 163, 759-771.

Zhang, F., LeBlanc, C., Irish, V.F., Jacob, Y. (2017). Rapid and efficient CRISPR/Cas9 gene editing in Citrus using the YAO promoter. *Plant Cell Rep*. 36: 1883-1887.

Zhou, H., Liu, B., Weeks, D.P., Spalding, M.H., Yang, B. (2014). Large chromosomal deletions and heritable small genetic changes induced by CRISPR/Cas9 in rice. *Nucleic Acids Res*. 42: 10903-10914.

Zhou. J., Wang, G., Liu, Z. (2018). Efficient genome-editing of wild strawberry genes, vector development, and validation. *Plant Biotechnol J*. 16: 1868-1877.

Chapter 7

Grafting in Vegetable Crops: An Excellent Technique of Agriculture

Khursheed Alam[1], Muzeev Ahmad[2], Astha Vishwaraj[3], Amit Kumar[4] and Shalini Singh[4]*

[1]Ph.D. Research Scholar, [3]M.Sc., Student, Department of Vegetable Science,

[4]Ph.D. Research Scholar, Department of Fruit Science,

Sardar Vallabhbhai Patel University of Agriculture and Technology, Meerut – 250 110, Uttar Pradesh

[2]Assistant Professor, Department of Agriculture, Integral Institute of Agricultural Sciences and Technology, Lucknow – 226 026, Uttar Pradesh

**e-mail: alamk1996@gmail.com*

INTRODUCTION

Grafting is the process of using tissue regeneration to unite two plant parts (a rootstock and a scion) so that the resulting combination of plant parts achieves physical reunion and grows as a single plant. Grafting is the process of using a robust plant to replace the root system of a cultivar that is economically valuable but vulnerable to one or more stress conditions. This technology is of great importance in controlling soil pathogens, and as a result, it increases the efficiency of plants in absorbing water and nutrients, improving the quantity and quality of the crop, and also extending the production season (**Oda, 2002; Hang *et al.*, 2005**). Most of the continuous agricultural losses in the production of vegetable crops in greenhouses are due to soil diseases and nematodes. As an anti-measure against such damages such as fusarium wilt and bacterial wilt, seedlings are grafted into different types and varieties of origins. This technology has evolved into equipping large fields with grafted seedlings and for various crops **(Kubota *et al.*, 2008).**

There is a great need to increase the area of cultivation with vegetable crops and also to increase the productivity per unit area in order to supplement the local market with the vegetable yield for human consumption, thus the need arises to get rid of the diseases (bacterial, fungal, and nematodes) that are endemic in the soil and that cause a lack of production and the death of plants. One of the most important global common methods for eliminating these endemic diseases is the use of methods of sterilizing the soil with chemical compounds, including methyl bromide, but then it was found that these compounds have significant negative effects on human health and the environment, thus prompting researchers to study new alternatives to soil sterilization, including sterilization based on Sun heat (solar sterilization), heat sterilization, evaporation, and use of resistant varieties **(Edelstein and Ben-Hur, 2006 and Ellis *et al.*, 2008)**. Grafting technology is a quick way to take advantage of disease-resistant genotypes resulting from plant breeding programs while minimizing the use of chemicals and using high-yielding varieties sensitive to soil diseases to contribute to integrated disease management programs and help achieve alternative cultivation **(Leonardi and Romano, 2004, Mudge *et al.*, 2009). Uysal *et al.* (2010)** indicate that grafting improves plant growth and production depending on the inherent genotype used. **Sawalha (2012)** also found that the production of a dunum of uplifting in Palestine in the Sanur Plain was 2.7 tons for uplifting grafted on the origin of pumpkins and 2.4tons for uplifting grafted on the origin of gourds. Grafting also had a significant effect in reducing the number of damaged fruits and dead plants due to soil fungi compared to the control. **Ibrahim *et al.* (2016)** indicated that the grafting of cucumber hybrids (Emir F_1) on five origins of RS 841 F_1, large honey squash, musk squash, pumpkin, and fiber in addition to unvaccinated plants (comparison) for Fusarium wilt disease. The results showed poor plant growth. The treatment of vaccination on the pumpkin was significant compared to the rest of the treatment. No significant differences were found between the comparison treatment and the vaccination treatments on pumpkin origins in growth traits.

The rootstock is the new root system, and the scion is the upper section or cultivar of interest. Watermelon (*Citrullus lanatus* L.) plants were grafted onto Lagenaria siceraria L. to prevent Fusarium wilt in the late 1920s in Japan, according to the first account of vegetable grafting. Grafting has become popular all around the world since then. Grafted seedlings are used in around 92 per cent and 95 per cent of the area cultivated with watermelon in Japan and Korea, respectively. **Lee *et al.* (2010)** and interest in the technology of grafting in vegetable crops increased in 1927. when watermelon was grafted onto some varieties of the Gourd vegetable group in mainly East Asian countries **(Ashita, 1927)**. And in some European countries with limited areas due to the lack of the possibility of following agricultural rotations, which in turn causes the concentration of many soil diseases that affect the root system of vegetable crops, unlike what is found in countries with large agricultural areas such as the United States of America, which follow a cycle system. Therefore, the method of producing vegetables with grafting technology is not widespread. It has been found that more than 92 per cent of watermelon and melon production in Japan and Korea are grafted before seedlings, and the total grafted seedlings in Korea are about 540million seedlings, and Japan with about 750 million seedlings **(Oda, 1999).**

The rootstock used is a hybrid root obtained from several parents, which are often wild stock. Resistance inherent in origin and its ability to increase and absorb nutrients from the soil and provide them with flavor is one of the most important technologies that help grafted plants resist soil diseases **(Guan *et al.*, 2012).**

Grafting Technique has many Benefits for Vegetable Crop Production

Although the benefits of using grafted transplants are now fully recognized worldwide, the need to enlighten the scientific basis of rootstock-scion interactions under variable environmental pressures remains vital for extracting grafting-mediated crop improvement.

1. **To manage soil borne diseases:** Various vegetable crops are serious threaten by soil borne diseases. Since 1930s, grafting has been using to create resistance against the soil borne diseases in vegetable crops. It was found the best alternate after the phase out of Methyl Bromide Fumigation which was usually used to control soil-borne diseases. Thus, this technique is exclusively using to manage mainly fusarium, bacterial, and verticillium wilt in tomato and melons. Furthermore, this is one of the effective techniques to control root knot nematodes in Solanaceaeous and cucurbitaceous crops by grafting susceptible commercial cultivars onto resistant rootstocks. For this, various wild species has been exploiting for disease resistance.
2. **Higher adaptability to environmental stresses:** These are highly susceptible to environmental changes. The change in temperature and erratic rainfall cause drought, flooding, and salinity which led to serious loss of vegetable crop production. It can be said that production of vegetable crops is challenging under changing climate. At commercial level, development of tolerant cultivars using breeding is limited due to genetically and physiologically complexity of the abiotic stresses. Many vegetable crops do not have resistant crossing compatible wild resistant sources. In this situation, grafting a plant surgical technique has effectively been used to mitigate broad spectrum of various environmental pressures. Particularly, in tomato and melons, grafting of high yielding commercial sensitive cultivars onto

rootstock having resistance and vigorous root system has showed great success for improving many environmental stresses.

3. **Enhancement of yield with better fruit quality:** The grafted vegetable plants with resistance/tolerance of abiotic stressors has increased yield and quality in the plants, but this improvement depends on the genotype of scion and rootstock, their compatibility, and reciprocal effect on each other. Moreover, root system of rootstock plays a great role to effect epigeous part of grafted plants under rhizosphere stresses. Tomato, variability for abiotic stresses especially salt tolerance is limited. But, wild species such as *S. pimpinellifolium, S. peruvianum, S. cheesmaniae, S. habrochaites, S. chmielewskii* and *S. pennellii* in the genus Solanum possessed sources of tolerance and the same have been exploited as rootstock to enhance productivity under wide environmental stress conditions, whereas, introgression with commercial tomato is difficult due to the problem of crossibility.

Vegetable Grafting Techniques

Currently, vegetable grafting is done on fruit vegetables from the Solanaceae and Cucurbitaceae families Tables 7.1 and 7.2 and Figure 7.1, but the method chosen is based on the type of crop grown **(Lee, 1994)**, the grower's experience, personal preference, grafting purpose **(Lee *et al.*, 2010)**, and the cost paid **(Maurya, 2019)**. Splice and cleft grafting are more common with Solanaceous vegetables, but tongue approach grafting ensures that Cucurbitaceous crops survive better when grafted **(Yassin and Hussen, 2015)**.

Table 7.1: Use of different Rootstocks for Grafting in Vgetables

Scion	*Rootstocks*
Tomato	*Solanum pimpinellifolium*
Cucumber	*Cucurbita moschata, Cucurbita ficifolia Cucurbita maxima*
Melon (for open field)	*Cucurbita Spp, C. moschata × C. maxima*
Melon (for Greenhouse)	*Benincasa hispida, Cucurbita Spp. C. moschata × C. maxima*
Watermelon	*Cucurbita maxima, Cucurbita moschata, C. moschata × C. maxima, Lagenaria siceraria*
Bitter gourd	*Cucurbita moschata, Lagenaria siceraria*
Eggplant	*Solanum nigrum, S. torvum, S. integrifolium*

Hole Insertion Grafting

When the root stock and scion have hollow hypocotyls, this approach is especially popular with Cucurbits. The scion is put into a hole drilled in the rootstock. When watermelon seedlings are grafted onto bottle gourd or squash rootstocks, it is especially popular **(Maurya *et al.*, 2019)**. Although this procedure needs more skill, commercial growers prefer it since it eliminates the need for additional labour for clipping, transplanting, cutting, and removing the grafting clip **(Lee *et al.*, 2010).**

Table 7.2: Rootstock and Method of Grafting in different Vegetable Crops

Sl.No.	*Vegetables*	*Rootstocks*	*Method of grafting*
1.	Watermelon	Squash and bottle gourd	Cleft and tongue approach grafting
2.	Cucumber	Squash and fig leaf gourd	Splice and tongue approach grafting
3.	Melon	Squash for oriental melon and melon for other melon	Splice and tongue approach grafting
4.	Tomato	Wild species of tomato	Splice and tongue approach grafting
5.	Brinjal	Brinjal and wild species	Splice and tongue approach grafting
6.	Pepper	Pepper and wild relative	Splice grafting

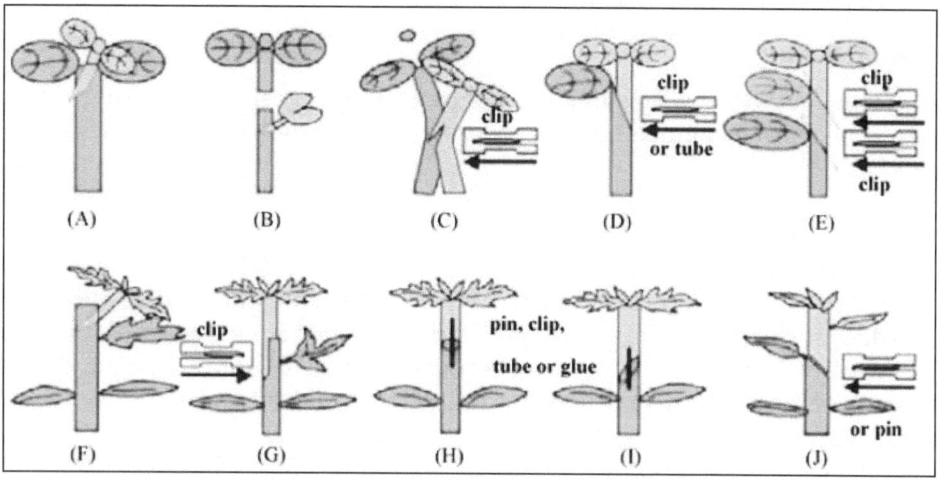

Figure 7.1: Major Grafting Methods in Cucurbits and Solanaceous Vegetables. (A and B) hole insertion grafting; (C) tongue approach grafting; (D, E and J) splice grafting; (F, G) cleft grafting; (H and I) pin grafting.

Tongue Approach Grafting/Side Grafting

This method can be used on both Solanaceous and Cucurbitaceous crops, however it is most commonly used on the latter. In this approach, a downward slanting cut is made on the rootstock stem and an upward slanting cut is made on the scion stem at a 30 - 45-degree angle. At the place of cut, the root stock and scion are joined and held together with grafting clips. Because this method necessitates equal-diameter rootstock and scion seedlings, the scion and rootstock seedling planting times should be modified accordingly.

Splice Grafting

One of the most frequent grafting methods for solanaceous vegetables is top grafting, often known as tube grafting. It is slightly modified in cucurbits and referred to as one cotyledon splice grafting **(Lee *et al.*, 2010)**. Both the rootstock and the scion are cut at a 450 angle with matching cuts and clipped together with a grafting clip, according to **Johnson *et al.* (2011)**. The main advantage of this approach is that it produces sturdy and healthy grafted seedlings because the vascular bundles are properly attached **(Lee and Oda, 2003)**.

Cleft Grafting

It's also known as apical grafting or wedge grafting, and it's usually utilised for solanaceous vegetables. After decapitating the rootstock, a vertical incision of about 0.5 cm is created in the centre, and a scion with a wedge-shaped end is inserted at the centre of the incision before tying the graft union with a plastic clip or parafilm **(Johnson *et al.*, 2011)**.

Pin Grafting

This method is very similar to splice grafting, except instead of grafting clips, specifically engineered pins are used to secure the grafted seedlings **(Lee *et al.*, 2010)**.

Tools, Clips and Grafting Aids

Several agricultural companies have created grafting instruments to accomplish automated grafting and hold the graft union intact **(Lee and Oda, 2003)**. Unfortunately, commercial producers have not utilised the most of them extensively. Simple grafting aids including grafting clips, tubes, tapes, and pins have been utilised for grafting in a limited but widespread fashion (Figure 7.2). The most common grafting clips used for tongue approach grafting in cucurbits and other crops are the standard grafting clips, which are formed of a circular plastic spring (Figure 7.2A). The clips are expensive, easy to use and handle for varied stem diameters, and may be reused numerous times despite having somewhat different sizes and shapes depending on the maker. Many commercial growers also frequently use a variety of alternative clips, particularly elastic tube-shaped clips with or without attachment for a supporting pole for the grafted seedlings (Figures 7.2B and C), for both manual and automated grafting. Many nations, like Israel and the Netherlands (Figure 7.2E), employ much smaller elastic slit-tubes for grafting tomatoes and peppers. Ceramic pins (Figure 7.2D) are a very useful and effective tool for fixing the grafted interface and are ideal for grafting by machines or robots. It can be operated with bare hands, a basic pencil-shaped tool, a machine, or a robot. Another method of keeping the grafted equivalents in place uses adhesive tape, glue, or occasionally aluminium foil. Growers around the world utilise specialised knives and gimlets for grafting that have been created. The Netherlands and

Korea have developed a unique knife with a self-feeding connection for skim milk or alcohol to inactivate some powerful viruses. A consistent wedge and receptacle can be concurrently formed in the stem of Phaseolus vulgaris using a hand-held grafting tool made of replaceable stainless steel, single-edge razor blades. With the advancement of grafting technology and the launch of new and effective grafting robots, it is clear that significant progress will be achieved on these devices. Experienced growers are keen to generate consistent, healthy seedlings for effective grafting by utilising superior quality seeds, preferably primed seeds, as uniformly small seedlings are unquestionably preferred for grafting, especially for machine grafting.

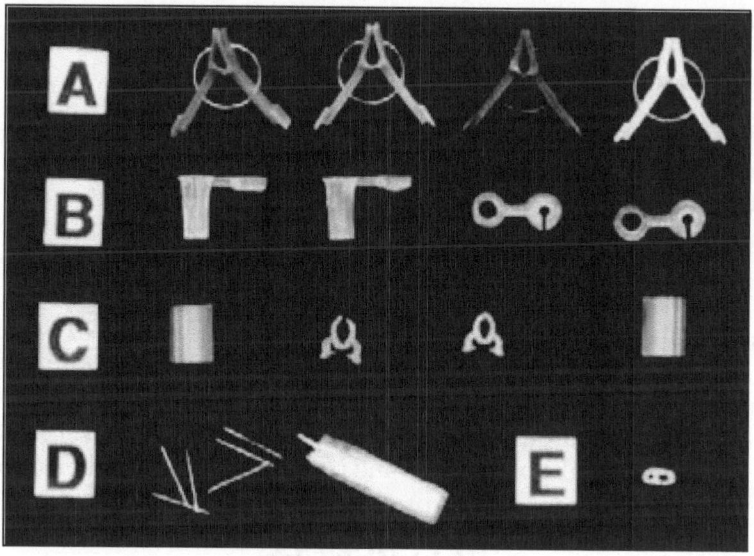

Figure 7.2: Grafting Clips, Tubes, Pins, and other Aids.

Acclimatization

In order for grafted plants to survive, proper acclimation is essential. Healing the wounded surface and hardening for field or greenhouse survival are both involved in acclimatisation **(Lee and Oda, 2003)**. To produce consistent grafted seedlings, appropriate moisture content must be maintained both before and after grafting. Simply placing the rootstock and scion in a black plastic bag (to prevent heat buildup) until the union is formed may be sufficient to achieve acclimatisation. Acclimatization is typically accomplished by growers using plastic film coverings. A single layer of semi-transparent high-density polyethylene film (0.01 mm or thinner) is used to seal the cell trays in many commercial nurseries, where the grafted plants are typically placed in cell trays with 32–72 cells. This reduces moisture loss and allows the plants to be kept sealed for 5-7 days without additional irrigation. To prevent an excessive buildup of heat during the day, partial shade may be required. Commercial plug seedling producers in Japan and Korea have created and utilised a variety of acclimatisation chamber types **(Lee *et al.*, 2008).**

Advantage of Vegetable Grafting

1. Yield increase and disease resistance
2. Shoot growth promotion
3. Low and High temperature tolerance
4. Enhanced nutrient and water uptake
5. High salt tolerance and Wet soil tolerance
6. Quality changes and Extended harvest period

CONCLUSION

Grafted vegetable seedlings have been employed in vegetable production over the world to maximize yield and overcome various biotic and abiotic stress. It fits well into the organic and integrated crop production system. Grafting and other propagation techniques has a potential in promotion of cultivation in non-traditional and fragile agro- ecosystem. Since grafting gives increased diseased tolerance and vigor to crops, it will be useful in the low-input **sustainable horticulture** of the future. The usage of grafted vegetables is rapidly increasing due to an increase in protected vegetable farming, intensive cropping cycles, paucity of production resources, and changing climate. For India's low vegetable productivity, this technique could be a useful technique for increasing the overall Solanaceous and Cucurbitaceous vegetable production system by overcoming numerous production constraints.

REFERENCES

Ashita, E. (1927). Grafting of watermelons in Japanese. *Korea* (*Chosun*) *Agri. Nwsl* 1: 9-11.

Bakir (2008). Grafting vegetables to improve production in greenhouses, *Journal of the Agricultural Engineer,* Year 35, Issue 85, April. the Hashemite Kingdom of Jordan.: 66-69.

Edelstein, M. and Ben-Hur, M. (2006). Use of grafted vegetable to minimize toxic chemical usage and damage to plant growth and yield quality under irrigation with marginal water. *ISHS Acta Horticulturae, 699: 159-168.*

Ellis, S., Boehm, M. and Mitchell, T. (2008). Fungal-like diseases of plant, Fact sheets, agriculture and natural resources, Ohio State University: 401-407.

Guan, W., Zhao, X., Hassell, R. and Thies, J. (2012). Defence mechanisms involved in disease resistance of grafted vegetable. *HortScience,* 47 (2): 164-170.

Hang, S. D., Zhao, Y. P. Wand, G. Y. and Song, G. Y. (2005). Vegetable Grafting, Beijng, China. China Agriculture Press.

Ibrahim, Alaa, Hammoudi, O., Asmar, G. and Suleiman, N. S. (2016). Vegetative growth of cucumber hybrids (Emir F1) grafted on different origins and tolerated to Fusarium wilt disease under protected cultivation conditions. *The Jordanian Journal of Agricultural Sciences,* 12 (3): 949-961.

Kubota, C., Michael, A. M., Burelle, N. K., Bausher, M. G. and Rosskopf, E. N. (2008). Vegetable Grafting: History, Use and Current Technology Statue in North America. *HortScience,* 43(6): 1664-1669.

Lee, J. M., Kubota, C., Tsao, S. J., Bie, Z., Echevarria, P. H., Morra, L. and Oda, M. (2010). Current status of vegetable grafting: Diffusion, grafting techniques, automation. *Scientia Horticulturae,* 127(2), 93-105.

Lee, J. M., Kubota, C., Tsao, S. J., Vinh, N. Q., Huang, Y. and Oda, M. (2008). Recent progress in vegetable grafting. In: International Workshop on Development and Adaptation of Green Technology for Sustainable Agriculture and Enhancement of Rural Entrepreurship, IRRI, Los Banos, Laguna, Philippines, September 28–October 2,2009, 21 pp.

Lee, J. M., Oda, M. and Janic, J. (2003). Horticultural Reviews. Grafting of herbaceous vegetable and ornamental crops, 28, 61-124.

Leonardi, C. and D. Romano (2004). Recent Issues on vegetable grafting. *Acta Hort. (ISHS),* 631: 163-174.

Mudge, K., Janick, J. Scofield, S. and Goldschmidt, E. (2009). A history of grafting. Horticultural Reviews, John Wiley and Sons Inc. 35: 437-493.

Oda, M. (2002). Grafting of vegetable crops. Sci. Rep. Agric. Biol. Sci. Osaka Pref. Univ., 35(9) : 1-5.

Sawalha, H. (2012). The use of watermelon (Baladi cultivar) grafting on bottle gourd and pumpkin rootstocks to control the soil-borne fungi casing vascular wilt disease of watermelon in the plain of Sanoor in Palestine. *Al-Aqsa University Journal (Natural Sciences Series),* 16(1): 39-54.

Uysal, N., Tuzel, Y. Oztekin, G. B. and Tuzel, H. I. (2010). Effect of different rootstocks onplant growth, yield, fruit quality and water consumption of cucumber in greenhouse conditions. *IHC. Lispoa,* S03. 052: 191.

Chapter 8

Growing of Vegetables in Kitchen Garden for Household Food and Nutritional Security

Astha Vishwaraj[1], Khursheed Alam[2], Muzeev Ahmad[3], Vibhu Pandey[4] and Amit Kumar[4]*

[1]M.Sc., Student, [2]Ph.D. Research Scholar, Department of Vegetable Science,

[4]Ph.D. Research Scholar, Department of Fruit Science,

Sardar Vallabhbhai Patel University of Agriculture and Technology, Meerut – 250 110, Uttar Pradesh

[3]Assistant Professor, Department of Agriculture, Integral Institute of Agricultural Sciences and Technology, Lucknow – 226 026, Uttar Pradesh

**e-mail: asthavishwaraj9925@gmail.com*

INTRODUCTION

India has a score of 29.1 in the 2022 global hunger index, ranking it the 107th-worst-performing countries out of 121 **(GHI, 2022).** In fact, managing the high rate of undernutrition that the nation continues to face is still a huge challenge. According to the National Family Health Survey 4, stunting rates are 38.4 percent and underweight rates are 35.8 percent **(GHI, 2019).** Over time, there has only been a slight improvement. Long-term consequences of inadequate nutrition include cognitive and development deficiencies as well as lowered infection resistance. In India, it is the primary factor in approximately half of all deaths of children under the age of five. One of the most important things a developing country like India should focus on is food security and nutritional diversity. To address the problem of food production and food security, numerous techniques are needed.

By 2050, the population of the world is predicted to have increased to nearly 9 billion people **(Ghanem, 2010).** It is imperative that we find ways to feed the world's expanding population. According to projections, by 2050, each person's daily energy requirements could reach 3050 kcal, necessitating a 70 per cent increase in world food production **(Grofova and Srnec 2012).** This has led to a rise in interest in kitchen gardens as a technique to promote nutrition and household food security.

A kitchen garden is a system that integrates the family home, a space for pleasure, and a garden to produce a range of foods, including fruits, vegetables, and medicinal plants for use at home or for sale **(Galhena *et al.*, 2013).** The biggest issues with kitchen gardening in other nations have been a lack of understanding of their importance, climatic challenges, a lack of water, and a lack of management expertise **(SDSN, 2013).** The agricultural landscape and local food systems of emerging nations around the world both include kitchen gardens, though. They have endured through time and could lessen hunger **(Galhena *et al.*, 2013).**

In essence, kitchen gardens play a significant role in enhancing household income, nutritional status, and food security **(SDSN, 2013).** Improved dietary diversity ratings, increased consumption of fruits and vegetables high in vitamin A, pulses, other fruits and vegetables, and better access to complementary foods have all been linked to kitchen gardens **(Girard *et al.*, 2012).** Changes in household food production and consumption, parental and child intake of target foods and micronutrients, and an increase in the overall dietary diversity of rural low-income households are all ways that kitchen gardens have an impact on nutrition **(Ruel and Alderman, 2013).**

Vegetables are the main source of vitamins, minerals, and fibre, and it is generally known that they have beneficial health effects on humans. Tomato, brinjal, bottle gourd, snake gourd, Knolkhol, potato, cauliflower, spinach, methi, and coriander are the main vegetable crops. As they include a variety of nutrients necessary for many bodily functions, vegetables are a crucial component of a healthy diet. Additionally, these vegetables give us taste, palatability, better digestion, and an increase in hunger. Since vegetables are often short-duration crops, kitchen gardens are ideal places to produce them. The family's nutritional garden, which is typically close to the house, is where short-term vegetables, fruits, herbs, medicinal plants, and other plants are grown. When a compost pit is dug in the side of the kitchen garden, it helps us recycle kitchen waste. Gardening promotes the health of the neighborhood and the individual by providing benefits to both. The advantages of food production go beyond an individual's physical, mental, and emotional well-being to have a lasting impact on others and on the social and physical environment of the community **(Armstrong, 2000).**

The space available in the backyard of the house, on the roof, or in containers can be used to cultivate a kitchen garden. Kitchen gardening techniques provide several societal advantages, including improved health and nutrition, higher income, more work opportunities, and household food security. According to the ICMR, a person needs about 300 g of vegetables each day, yet there aren't many of them around. Many rural families used to produce vegetables in their backyards for their own consumption, but due to the unorganized fashion in which they were grown, they still did not consume enough vitamins and minerals.

With the use of locally accessible planting materials, green manures, "live" fencing, and traditional insect control techniques, gardening must be done with almost no financial resources. So, to some extent, home gardening is a production system that the underprivileged can easily access **(Marsh and Talukder, 1994).**

Advantage

- ☆ Provide nutrient-dense fresh produce, especially fruits and vegetables.
- ☆ provide organic vegetables and fruits free of harmful chemicals.
- ☆ help reduce the cost of purchasing vegetables and therapies.
- ☆ increases children's awareness of the value of labour.
- ☆ The taste of vegetables harvested from a home garden is superior to those purchase at the market.

Principles of Kitchen Gardening

The following are key guidelines that must be complied with in order to enhance production and ensure a consistent year-round supply of nutrient-dense fruits and vegetables on a small plot of land:

- ☆ Choosing fruit and vegetable crops based on the agroecology that is currently in place.
- ☆ Storing vegetable seeds for the new season.
- ☆ Growing a variety of healthy fruit and vegetable crops that are rich in vitamins and minerals.
- ☆ Year-round availability of a continuous supply of fruits and vegetables.
- ☆ Ensures the households' nutritional and food security.
- ☆ Returns on investment from the selling of surplus goods.
- ☆ Helps in the growth of organic vegetables and fruits.
- ☆ Helps in the preservation of traditional vegetable seed varieties.
- ☆ Reduces costs for other food items, which increases household income.
- ☆ A balanced diet improves health and lowers household medical costs.
- ☆ If accessible, a good way for women to spend their free time.

☆ Gives them the ability to pay for minor personal expenses on their own and enhances their planning, problem-solving, and management skills.

Benefits of Kitchen Garden

Vitamins, proteins, and minerals like calcium and iron are abundant in fruits and vegetables. In addition, they are a fantastic source of nourishment that is high in calories. The various fruits and vegetables that are rich sources of vitamins, iron, and calcium are listed in Tables 8.1 and 8.2.

Table 8.1: The Vitamins and Nutrients that are Present in Various Fruits and Vegetables

Sl.No.	*Nutrients/Vitamins*	*Fruit and Vegetable Crops*
1.	Calories	Tapioca, sweet potato, yam, potato, colocasia, onion, plantain, breadfruit, jackfruit, pumpkin, banana, peas *etc.*
2.	Proteins	Peas, cowpea, drumstick *etc.*
3.	Vitamin A	Carrot, spinach, drumstick, pumpkin, mango, papaya, passion fruit, tomato *etc.*
4.	Vitamin B complex	Chillies, peas, broad beans, tomato, banana, grape, bhindi, capsicum, bael *etc.*
5.	Vitamin C	Aonla, leafy vegetables, tomato, orange, lemon, guava, mango *etc.*
6.	Calcium	Curry leaves, drumstick, spinach, custard apple and all the other leafy vegetables.
7.	Iron	Karonda, date *etc.*

Source: Adam, D. M *et al.* (2020).

Table 8.2: Nutritional Information of Kitchen Gardening Vegetables

Sl. No.	*Vegetable*	*Energy (kcal)*	*Moisture (per cent) (wb)*	*Protein (per cent)*	*Fat (per cent)*	*Total CHO's (per cent)*	*Dietary Fabre (per cent)*	*Minerals (per cent)*
1	Tomato	18.67	93.62	0.9	0.47	2.71	1.77	0.52
2	Brinjal	22.88	90	1.48	0.32	3.52	3.98	0.7
3	Ridge Guard	11.78	94.99	0.91	0.14	1.72	1.81	0.44
4	Cluster bean	37.17	84.65	3.55	0.37	4.91	4.83	1.68
5	Lady's finger	24.78	89.06	2.08	0.22	3.62	4.02	0.94
6	Bottle guard	10.01	95.17	0.53	0.13	1.68	2.12	0.36
7	Field bean	31.64	85.57	3.71	0.6	2.85	6.19	1.08
8	Green chilies	14.86	93.89	1.11	0.34	1.84	2.06	0.76
9	Amaranth leaves	28.13	86.85	3.29	0.65	2.28	4.41	2.52

Sl. No.	Vegetable	Energy (kcal)	Moisture (per cent) (wb)	Protein (per cent)	Fat (per cent)	Total CHO's (per cent)	Dietary Fabre (per cent)	Minerals (per cent)
10	Gogu leaves	33.49	87.42	1.86	1.09	4.06	4.59	0.98
11	Spinach	22.52	90.31	2.14	0.64	2.05	2.38	2.47
12	Fenugreek leaves	30.87	86.73	3.68	0.83	2.17	4.9	1.69
14	Drumstick leaves	62.88	75.65	6.41	1.64	5.62	8.21	2.46

Source: Longvah *et al.* (2017).

Sites Selection and Size

- ✰ Due to the scarcity of land on the homestead, there are few options for choosing a location for a kitchen garden.
- ✰ A kitchen garden is typically created in the rear of a home, close to a water source in a spacious area that receives lots of sunlight.
- ✰ Size and design of a vegetable garden are determined by the amount of area available, the size of the household, and the free time available for maintenance.
- ✰ Vegetables may be produced year-round for a family of five member on about 200 M^2 of land, or about five cents. Instead of a square yard or a broad stretch of ground, a rectangle garden is preferable.

Layout of Kitchen Garden

- ✰ The agro-climatic conditions available in the area determine the design of the garden and the choice of crops appropriate for each season. Layout and crop allocation (Figure 8.1) adjustments must be made in response to climatic and seasonal changes.
- ✰ In a kitchen garden, long-lasting and consistent yielding variety of various crops should be preferred over high-yielding ones that call for ongoing maintenance.
- ✰ A bed should only be 3 to 4 feet wide to make it simple to reach the centre without walking on the dirt. Beds can be set down at soil level or raised.
- ✰ For efficient use of farm and kitchen waste, one or two compost pits could be placed on one corner of the kitchen garden.
- ✰ To make it simple to weed and harvest, make access roads between beds at least 2 feet wide.
- ✰ Three annual crops can be grown on the same plot by using scientific crop rotation, which is required to help prevent illnesses and avoid plant-hungry pests that overwinter in the soil.

Goose Berry	Drumstick	Papaya		Manure pit		Coccinea
Acid lime	Pomegranate			Nursery beds		
Lab lab	Curry leaf					Bitter Gourd
	Checkurrminas			Herbal plants (Gymnea, Thuthuvalai, Oma valli, Thulsi)		
	Mint	Amaranthus				
Snake gourd	Water Amaranth	Okra		Greater yam	Elephant foot yam	Ribbed Gourd
	Coriander	Chillies		Lesser yam	Cluster bean	
	Fenugreek	Brinjal		Colacasia	Cow pea	
Small Onion	Bellary Onion	Tomato	ENTRANCE GATE	Tapioca	Radish	Beet root

Figure 8.1: Layout Plan of Kichen Garden.

- For efficient use of space, companion cropping, accession cropping, intercropping, and mixed cropping can be used.
- Care should be given to plant varieties or crops at the appropriate time of planting or season while allocating or organizing crops in each sub-plot.

Land Preparation

- First, a depth of 30 to 40 cm is dug with a thorough spade.
- The removal of perennial weeds, bushes, and stones.
- The soil is amended with 100 kg of well decomposed farmyard manure or vermicompost.
- Depending on the need, ridges and furrows are constructed at spacing of 45 or 60 cm.
- Ridges and furrows can be replaced by flat beds.

Vegetable Cultivation in Containers or Raised Beds on Terraces

It is getting more challenging to own and maintain traditional types of house gardens, even on a small quantity of land, as a result of rising urbanization and the resulting strain on land. Depending on the available space, vegetables can be grown in terraces either in pots or in beds as shown below:

Boxes or Pots

Vegetables grow well in standard earthen pots with diameters of 30 to 40 cm. Large 40 x 40 x 30 cm cement pots can also be used. The ideal dimensions for the wooden boxes are 75 cm in length, 45 cm in width, and 50 cm in height. These are convenient for growing any type of vegetable.

Beds

On the roof, small, rectangular cement beds might be built along the wall. The beds should be 75 cm wide and at least 45-60 cm high. The entire roof should not be covered in soil because of the possibility of damage from large soil loads and regular irrigation.

Sowing and Planting

Maximum production and a year-round supply of fruits and vegetables for the table are the major goals of a kitchen garden. This goal is simply accomplished by keeping to specific processes. Spacing requirement of fruits and vegetables recommended for the kitchen garden is given in Tables 8.3 and 8.4.

- On one side of the ridges, direct-sown crops including okra, cluster beans, and cowpea can be planted at a spacing of 30 cm. After combining one part of seeds with twenty parts of fine sand, Amaranthus (meant for whole plant pull out and clipping) can be distributed in the plots. Along the bunds of plots, little onions, mint, and coriander can be planted or sowed.

- By drawing lines, it is possible to sow the seeds of transplanted crops like tomato, brinjal, and chilli one month in advance in nursery beds or pots.

Table 8.3: Spacing, Sowing/Transplanting, Sowing Time and Seed Rate of Vegetable Crops in Kitchen Garden

Vegetables	*Row to Row Spacing (cm)*	*Plant to Plant Spacing (cm)*	*Depth of Sowing/ Transplanting (cm)*	*Sowing Time*	*Seed Rate (10 M²)*
Okra	60	30	2.0-3.0	Feb-March, June-July	15 g.
French Bean	60	15	2.5-3.5	May-June	40 g.
Cowpea	45	15	2.5-3.5	March to Mid-July	30 g.
Pea	20	5	2.0-3.0	Oct-Nov	150 g.
Cabbage	45	45	Transplanting	Sept-Oct	0.4 g.
Carrot	45	30	1-2	Aug-Sept-Oct	4 g.
Cauliflower	60	45	Transplanting	Aug-Sept-Oct	0.4 g.
Cucumber	150	50	1-2	Feb-March, June-July	2 g.
Brinjal	90	60	Transplanting	June to July November April	0.4 g.
Onion	45	5-0	Transplanting	Oct-Nov, May-June	8 g.
Chili	60	30	Transplanting	Nov-Jan, May-June	0.8 g.
Pumpkin	150	50	3-4	Jan-Mar Sept-Dec May-Jun	4 g.
Radish	30	2.5 - 3.5	1-2	April-Aug, Sept-Oct, Nov-Jan	8 g.
Spinach	30	5-10	1-2	Sept-Nov Feb	20 g.
Tomato	60	30	Transplanting	June-Aug, Nov-Dec	0.4 g.

Table 8.4: Different Variety and Spacing of Fruit Species for Kitchen Garden

Sl.No.	*Fruit Species*	*Variety*	*Spacing*
1.	Jackfruit (grafted)	Gumless	10m x 10m
2.	Mango	MallikaAmrapali	8m x 8m5m x 5m
3.	Mango	Dashehari	10m x 10m
4.	Kagzi lime	Seedlings	4m x 4m
5.	Pomegranate	Bhagwa	5m x 5m
6.	Guava	Lalit	5m x 5m
7.	Guava	L-49/Sardar	5m x 5m
8.	Litchi	Rose scented	8m x 8m
9.	Bael	NB-5, NB-7	5m x 5m
10.	Drumstick	PKM-1	4m x 4m
11.	Peach	Shan-e-Punjab	5m x 5m
12.	Plum	Green Gauge	5m x 5m

Source: Adam, D.M. *et al.* (2020).

- ☆ To protect the seeds from ants, sow the seeds, cover with top soil, and then sprinkle with 250 grammes of neem cake.
- ☆ The seedlings are removed from the nursery and transplanted along one side of the ridges at spacings of 30-45 cm for tomato, brinjal, and chilli and 10 cm on both sides of the ridges for big onion, depending on the variety of plant being grown. After planting, water the plants right away, and then again, the third day. In the early stages, the seedlings can be watered once every two days, and thereafter once every four days.
- ☆ The perennial plants should be placed on one side of the garden, typically at the back, so they won't shadow out other crops or compete with other vegetable crops for nutrients.
- ☆ Coriander, spinach, fenugreek, alternanthera, mint, and other short-season green vegetables can be grown next to the sidewalk all around the garden and on the central walkway.

Selection of Crops

There is a long list of crops that are quite simple to cultivate at home, and that too without assistance from a professional. Gardening will be much more enjoyable for you after you figure out how to cultivate Fruits, vegetables, Spices, Medicinal plant, Flower and trees on your own. A few vegetables that are simple to cultivate in your kitchen garden are listed Table 8.5.

Table 8.5. Different fruit and vegetable Suitable for Kitchen Garden

Fruits	*Vegetable*	*Spices*	*Medicinal Plants*	*Flowers*	*Trees*
Mango	Tomato	Turmeric	Aloe	Rose	Bottle brush
Banana	Brinjal	Coriander	Solanum	Sombaruthi	Pagoda tree
Sapota	Chilly	Fenugreek	Mint	Jasmine	Mandarai
Guava	Onion		Basil	Nerium	
Papaya	Okra		Thulsi		
Acid lime	Bitter gourd		Omavalli		
Amla	Snake gourd		Karisilanganni		
Pomegranate	Ribbed gourd		Keelanelli		
Anona	Bottle gourd		Pirandai		
Date Pam	Amaranthus		Thoothuvelai		
	Lab lab		Ponnanganni		
	Beetroot		Poduthalai		
	Radish		Manathakali		
	Curry leaf		Vettiver		
	Moringa				
	Spinach				
	Cluster bean				
	Cowpea				
	Moringa				
	Tapioca				

Selection of Vegetable Variety for Kitchen Garden

It is suggested to cultivate just those vegetables, such as lettuce, green vegetables, radish, coriander, fenugreek, *etc.*, that cannot be purchased from the market in fresh condition. When growing vegetables in pots and boxes, it is best to choose those that bear more fruits per plant, such as tomatoes, brinjal, chillies, beans, and cucurbits.

Vegetables come in a wide range of varieties and hybrids that can be grown in a backyard garden, as seen Table 8.6.

Constraints/Gaps Identified

Five categories were used to group the constraints farm families face (Adam, D. M. *et al.*, 2020). A. Input constraints B. Technical constraints C. Social and cultural limitations D. Postharvest restrictions, and E. Overall restrictions.

Table 8.6: Different Vegetable Variety for Kitchen Garden

Groups	*Crop*	*Varieties/Hybrids*	*Growing Season*
Solanaceous crops	Brinjal	Pusa Shyamla (long purple), Small round-Pusa Bindu, Pusa Ankur, Arka Kusmukar, Pusa Safed Baingan-1, Arka Shirish, Kashi Himani	Kharif (June/July-Oct/Nov)
	Tomato	Pusa Sheetal, Pusa Sadabahar, Pusa Hybrid-2, Pusa Hybrid-4, Pusa Hybrid-8, Pusa Rohini, Arka Samrat	Autumn-Winter
	Chilli	Pusa Sadabahar, Kashi Gaurav	Kharif
Leafy vegetables	Amaranth	Pusa Lal Chaulai, Pusa Kiran, Kashi Suhaavani	Spring summer and Kharif
	Chenopodium	Pusa Bathua-1, Kashi Bathua-2	Winter
	Beet leaf	All Green, Pusa Harit, Pusa Bharati	Winter
	Fenugreek	Pusa Early Bunching, Pusa Kasuri	Winter
	Vegetable mustard	Pusa Sag-1	Winter
Cole crops	Cabbage	Golden Acre	Pusa Mukta, Pusa Cabbage Hybrid-1
	Cauliflower	Pusa Meghna, Pusa Kartik Sankar, Pusa Deepali, Pusa Sharad, Pusa Hybrid-2, Pusa Paushja, Pusa Shukti, Pusa Snowball K-1, Pusa Snowball Kt-25	Early-June/July-Oct/Nov Mid early-July/August-Nov/Dec Mid late- August/September-December/January Late-October/Nov-Feb/March
	Knolkhol	White Vienna, Pusa Virat, Palam Tender knob	Winter
Bulb and root crops	Onion	Pusa Red, Pusa Madhvi, Pusa Riddhi	Winter
	Radish	Pusa Chetki, Rapid Red White Tipped, Pusa Mridula (breakfast radishes) Pusa Jamuni (pink fleshed), Pusa Gulabi Pink fleshed), Pusa Vidhu, Kashi Lohit	Mar-Sept Oct-Nov Sept-Nov

Groups	Crop	Varieties/Hybrids	Growing Season
	Carrot	Tropical- Pusa Vristi (red, heat tolerant);	Kharif
		Pusa Meghali (orange), Pusa Rudhira (red), Pusa Asita (black)	Winter
		Temperate: Pusa Yamdagni (orange, temperate), Pusa Nayanjyoti (orange, temperate)	Winter and Spring summer
	Turnip	Pusa Sweti,	Kharif (August)
		Purple Top White Globe, Pusa Chandrima, Pusa Swarnima	Winter
Legumes	Garden Pea	Arkel, Pusa Pragati, Pusa GP 17	Winter
	Cowpea	Pusa Sukomal, Ankur Gomti	Spring summer and Kharif
	French bean	Bush type (Contender, Pusa Parvati), Pole type (Kentucky Wonder, PusaHimlata), Arka Anoop	Autumn and spring summer
	Dolichos	Pusa Sem-2, Pusa Sem-3, Kashi Haritima	Kharif and autumn winter
	Cluster bean	Pusa Navbahar	Spring summer and Kharif
Malvaceous	Bhindi	Pusa A-4, Kashi Lalima	Spring summer and Kharif
Cucurbits	Bottle gourd	Pusa Sandesh (round fruit), Pusa Naveen (long fruit), Pusa Hybrid-3, Pusa Santushti (pear shaped), Pusa Samridhi (Long), Kashi Kirti	Spring summer and Kharif
	Bitter gourd	Pusa Vishesh, Pusa Do Mousami, Pusa Hybrid 1, 2, Pusa Aushadhi	Spring summer
	Pumpkin	Pusa Vikas, Pusa Viswas, Pusa Hybrid-1	Spring summer
	Sponge gourd	Pusa Sneha, Kashi Divya	Spring summer and Kharif
	Ridge gourd	Pusa Nutan, Kashi Shivani	Spring summer and Kharif
	Cucumber	Pusa Uday, Pusa Barkha	Spring summer and Kharif
Exotic vegetables	Bunching onion	Pusa Soumya	All the year round
	Leek	Palam Paustik	Winter
	Celery	Ford Hook Emperor	Winter
	Lettuce	Great Lakes, Chinese Yellow	Winter
	Parsley	Moss Curled	Winter

Groups	Crop	Varieties/Hybrids	Growing Season
	Broccoli	**Green Sprouting:** Pusa Kts-1, Palam Samridhi **Yellow heading:** Palam Kanchan **Purple heading:** Palam Vichitra	Winter
	Kale	Pusa Selection	Winter
	Brussels Sprout	Hilds Ideal	Winter
	Cherry tomato	Pusa selection	Autumn winter
Perennial tree	Drumstick	CO-1, Co-2	-

Source: Selvakumar, R. and Ganghadhar, K. (2018).

A. Input Constraints

1. Lack of quality planting materials (saplings and seeds) of fruits and vegetables
2. Due to a lack of irrigation facilities caused by water constraint
3. Lack of appropriate land near residences for a kitchen garden
4. Availability of plant protection agents when needed

B. Technical Constraints

1. Awareness about better varieties, seed rates, and sowing dates
2. Lack of understanding regarding the choosing of nutrient-dense fruit and vegetable crops
3. Lack of understanding regarding the identification and control of important pests and diseases
4. Insufficient knowledge of important irrigation stages
5. A lack of awareness of seed propagation, nursery, and seedling techniques

C. Socio-cultural Constraints

1. Fear of farm produce being stolen
2. Farmers habit of non-adoption until other farmers in the social system start doing it
3. Farmers' preoccupation with a set of traditional practices
4. Migration of rural youth to urban areas in search of job

D. Postharvest Constraints

1. Having trouble selling a little quantity of extra produce
2. Insufficient storage space for excess produce
3. There is no market in the village.
4. Lack of understanding of methods for processing and preserving excess produce

E. General Constraints

1. Wild animal menace
2. Monoculture of just a few vegetables in the home garden and a lack of variety
3. Open grazing by cattle and goat
4. The importance of the kitchen garden is lower than that of other farm activities
5. Excess rain during kharif season

Tools for Gardening

The following implements are considered to be required for helping to a small or moderate-sized garden: a spade or four-pronged digging fork, transplanting trowel, sickle, garden knife, khurpi, hand sprayer, rake, hand hoe, trench hoe, hand cultivator, hand duster, water cane, measuring stick, string, stakes, and irrigation equipment. The need for a wheel cultivator, a small sprayer, and a wheelbarrow may go beyond this basic list. One may easily perform the majority of gardening activities with these items at hand (Figure 8.2).

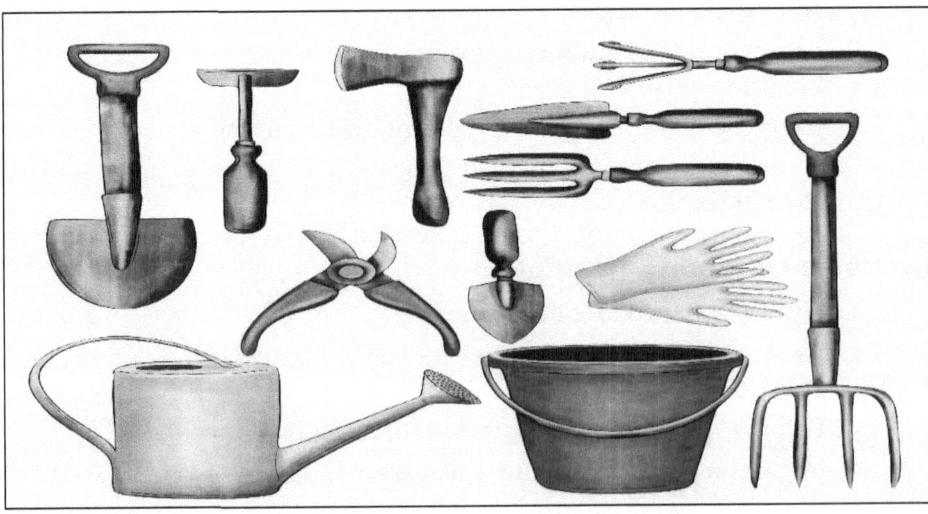

Figure 8.2: Use of Different Tools for Kitchen Garden.

In reality, the following fruits and vegetables should be grown in a home garden because their edibility and nutritional worth depend greatly on their freshness.

CONCLUSION

It is possible to come to the conclusion that growing kitchen gardens is extremely important for addressing the issue of malnutrition and micronutrient deficiencies in rural communities. The cheapest and most straightforward strategy to maintain excellent health is to increase fruit and vegetable eating. Backyard kitchen gardening provides direct access to food that can be gathered, prepared, and given to family members, frequently on a daily basis, thereby contributing to household food security. In the human diet, fruit and vegetable crops are essential as supplies of fresh food, calories, vitamins, dietary fibre, and nutraceuticals. Such crops that are ideal for urban production, processing, and value addition have a rich genetic heritage in India. A large portion of the population's health problems are caused by malnutrition and undernutrition. Fruits and vegetables are crucial in ensuring that our meal is tasty, simple to digest, well-balanced, and nutritious. Even those who are extremely impoverished or landless can grow their own food on little amounts of land, in containers, or elsewhere. As a result, kitchen gardening is an effective way to increase household food security and should be promoted and implemented across the nation. An option for farm women to work and earn money while also making good use of space and water is provided by a kitchen garden.

REFERENCES

Armstrong, A. and Donna (2000). A survey of community gardens in upstate New York: Implications for health promotion and community development. Health and Place 6.

Galhena, D. H., Freed, R. and Maredia, K. M. (2013). Home gardens: A promising approach to enhance household food security and wellbeing. Agriculture and Food Security, 2(8), 1–13.

Ghanem, H. (2010). The State of Food Insecurity in the World 2010: Addressing Food Insecurity in Protracted Crises. 2010: Food and Agriculture Organization of the United Nations.http: //www.fao.org/docrep/013/i1683e/i1683e.pdf.

Girard, A. W., Self, J. L., McAuliffe, C. and Olude, O. (2012). The effects of household food production strategies on the health and nutrition outcomes of women and young children: A systematic review. Paediatric and Perinatal Epidemiology, 26(S1), 205–222.

Global Hunger Index (2019). https: //www.globalhungerindex.org/results.html

Global Hunger Index (2022). https: //www.globalhungerindex.org/india.html

Grofova, S. and K. Srnec. (2012). Food crisis, food production and poverty. Agricultural Economics.http: //www.agriculturejournals.cz/publicFiles/61881.pdf.

Longvah, T., Anathan, R., Bhaskarachary, K. and Venkaiah, K. (2017). Indian Food Composition Tables 2017. Published by National Institute of Nutrition, ICMR, India.

Marsh, R. and Talukder, A. (1994). Production and consumption effects of the introduction of home gardening on target, interaction and control groups: a case study from Bangladesh. In Proceedings of the International Symposium on Systems-Oriented Research, 83-85.

Kadam, D. M., Bihari, B., Kumar, S., Bishnoi R., Roy, T., Singh, M., Maurya, U. K., Muruganandan, M., Shrimali, S. S., Kumar, R., Sharma, S. K. and Malik, A. (2020). Kitchen garden for nutritional and livelihood security in North Western Himalayan Region. Technical bulletin published by ICAR-IISWC, Dehradun under Farmer FIRST Programme. pp. 28.

Ruel, M. T. and Alderman, H. (2013). Nutrition-sensitive interventions and programmes: How can they help to accelerate progress in improving maternal and child nutrition? The Lancet, 382 (9891), 536–551.

Systems of the Sustainable Development Solutions Network (2013). Solutions for sustainable agriculture and food systems: technical report for the post-2015 development agenda/prepared by the Thematic Group on Sustainable Agriculture and Food. - New York: SDSN.

Selvakumar, R. and Ganghadhar, K. (2018). Vegetable Varieties for Urban Home Garden. *Indian Farming,* 5(10): 1260-1268.

Chapter 9

Hydroponics: A Futuristic Technological Advancement in Vegetable Farming

Rehan[1]*, Ankita Sharma[1], Akhilesh Tiwari[2] and Reena Nair[3]

[1]Research Scholar, [2]Professor, [3]Assistant Professor,

Department of Horticulture, College of Agriculture, Jawaharlal Nehru Krishi Vishwa Vidhyalaya, Jabalpur, Madhya Pradesh

**e-mail: rehanazmi477@mail.com*

INTRODUCTION

India presently necessitates food security, which implies that every individual seemed to have both physically and financially access to adequate and nourishing food to satisfy dietary demands. Lack of adequate water for agriculture diminishes food production, which in eventually stimulate a large number of people in our nation to be hungry and suffer from malnutrition. The proportion of arable land is shrinking daily as a result of the population boom, urbanisation, and industrialisation. Due to unusual environmental behaviour, the traditional techniques of agricultural production are also experiencing a variety of challenges. Therefore, in order to sustainably feed the world's expanding population, innovative and cutting-edge techniques for generating enough food must be developed.

A change in the growth medium can be an alternate strategy for sustainable crop production and the preservation of water and land resources that are already quickly running out. Therefore, there is a critical need for the use of technology in agriculture that may help save water and have a favourable effect on food availability and production. That's why this 'hydroponics' systems is becoming increasingly widespread over the world and according to the most recent report, it is expected to reach a world growth of 18.8 per cent from 2017 to 2023, corresponding to a global hydroponic market USD 490.50 Million by 2023 (Jan *et al.*, 2020).

'Hydroponics' is one such soilless growing technology, and its water consumption efficiency is substantially higher than that of a traditional system. Owing to its extensive efficient resources management and high-quality food production, hydroponic agriculture is currently growing increasingly popular on a global scale. This method offers a number of benefits, notably year-round production, less disease and insect infestations, and the removal of labor-intensive agricultural activities like weeding, spraying, and watering. Crops also develop more quickly than typical crops grown in soil.

Hydroponics

The technique of hydroponics involves growing plants in nutrient solutions with or without the use of an inert medium, such as gravel, vermiculite, rockwool, peat moss, sawdust, coir dust, coconut fibre, *etc.* to offer mechanical support. The name "hydroponics" literally translates to "water work" since it comes from the Greek words "hydro" for water and "ponos" for labour.

Hydroponics is the technique of growing plants in soil-less condition with their roots immersed in nutrient solution (Maharana and Koul, 2011). Terrestrial plants can be cultivated either with their roots just in the mineral fertiliser solution or in an inert media like perlite, gravel, or mineral wool. This method aids in managing the production system for effective use of natural resources and reducing hunger in addition to assisting in addressing the difficulties posed by climate change.

Other similar terms related to hydroponic are 'aqua (water) culture', 'hydroculture', 'nutriculture', 'soilless culture', 'soilless agriculture', 'tank farming', or 'chemical culture'. A 'hydroponicist' is defined as one who practices hydroponics and 'hydroponicum' defined as a building or garden in which hydroponics is practiced (Jones, 2014). Most hydroponics systems operate automatically to control the amount of water, nutrients and photoperiod based on the requirements of different plants. (Resh, 2013). Various commercial and special crops can be grown using hydroponics including leafy vegetables, tomato, cucumber, pepper, strawberry and many more.

A mineral nutrient solution, an inert substrate like gravel, sand, wool, or perlite, or even rice husks, can all be used as the media. Aeroponics, however, is extremely similar to hydroponics and employs a thin mist of liquid nutrient solution instead of a solid or liquid substrate. Without the drawbacks of soil, vertical farming's advantages might be used to increase productivity on the same amount of area. This is so that plants may be grown in shelves or layers, one layer being grown on top of the other. These layers may be piled on top of one another, allowing for expansion in the vertical direction.

Figure 9.1: Hydroponics Technique for Growing Vegetables.
(*Source*: https: //www.rimolgreenhouses.com/)

History of Hydroponics

The word hydroponics was coined by Professor William Gericke in the early 1930s; describe the growing of plants with their roots suspended in water containing mineral nutrients. Researchers at Purdue University developed the nutriculture system in 1940. During 1960's and 70's, commercial hydroponics farms were developed in Arizona, Abu Dhabi, Belgium, California, Denmark, German, Holland, Iran, Italy, Japan, Russian Federation and other countries. Most hydroponic systems operate automatically to control the amount of water, nutrients and photoperiod based on the requirements of different plants (Swain *et al.*, 2021).

Dr. W. F. Gericke in 1936 from the University of California was the first person to carry out large-scale commercial experiments in which he grew tomatoes, lettuce and other vegetables (Swain *et al.*, 2021). Gericke used hydroponics to grow tomatoes, beets, carrots, potatoes, fruits, flowers and more plant are grown in rows or on trellises, just like in a traditional garden, but they have their roots in water rather than in dirt. While working at the University of California at Berkeley he began to popularize the idea that plants could be grown in a nutrient solution and water instead of soil. Naturally, the general public, as well as William's colleagues, questioned this claim. They were quickly shown to be wrong when they saw the growth of very tall tomato plants using only water and nutrients. He decided to call this growing method of hydroponics. The shocking results of Gericke's experiment with tomatoes prompted additional research in the field. Then there were further investigations by the University of California in which scientists discovered a lot of benefits related to the cultivation of soilless plants. (Sonawane, 2018) Hydroponics was also applied during World War II between 1939 and 1945 in order to provide plants for the troops (in arid soils and in Greenland). Currently NASA uses the hydroponics technique to provide food for space travelers. Later on, during 1960s and 70s, commercial hydroponics farms were developed in Abu Dhabi, Arizona, Belgium, California, Denmark, German, Holland, Iran, Italy, Japan, Russian Federation and other countries. During 1980s, many automated and computerized hydroponics farms were established around the world. Home hydroponics kits became popular during 1990s (Sonawane, 2018).

Soil-less Media Used in Hydroponics

Nutrient Media

For vegetative and reproductive growth, plants require 17 essential components. Carbon, hydrogen, and oxygen are the initial three. Other 14 include macronutrients like as nitrogen, phosphorus, potassium, calcium, magnesium, and sulphur, the first three of which are categorised as primary nutrients and the other three as secondary nutrients.

- ✰ Iron, manganese, copper, zinc, boron, chlorine, molybdenum, and nickel are examples of microelements.
- ✰ Nutrient control is simple with hydroponic systems.
- ✰ Nitrogen is typically beneficial to the plant's vegetative development.
- ✰ Phosphorus and potassium aid in the plant's blooming and reproductive development.

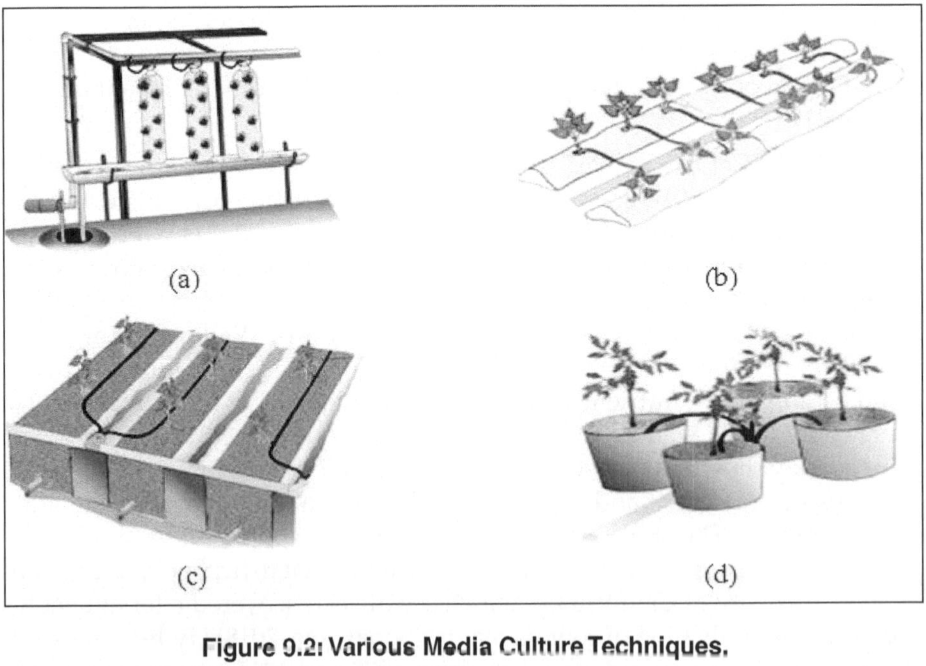

(a) (b) (c) (d)

Figure 9.2: Various Media Culture Techniques.

(a) Hanging bag technique, (b) Grow bag technique, (c) Trench or trough technique, (d) Pot technique (*Source:* Technical Bulletin on Hydroponics, Department of Agriculture, Ministry of Agriculture, Govt. of India).

In medium culture method a solid medium is used or the anchoring of roots. The culture is usually named after the inert medium. The media used can be abiotic like sand, gravel, rock or biotic like rice husk, wool or coir. (Sonawane, 2018) There are two main variations for each medium, sub-irrigation and top-irrigation. However, it is classified as follows:

(a) Hanging bag technique
(b) Grow bag technique
(c) Trench or trough technique
(d) Pot technique

Table 9.1: Various Types of Growing Media Used in Hydroponics

	Growing Media	*Description*	*Image*
A.	**Organic Growing Media**		
1.	**Coco Peat**	§ By-product of coconut husk. § Used for production of a wide range of soil-less crops like tomato, eggplant, cucumber, capsicum, *etc.* § High water holding capacity. § Provides a buffer in high temperatures	
2.	**Rice Hull**	§ By-product of the rice milling industry. § Extremely light in weight. § Very effective in improving drainage. § They break down very slowly like coco coir, making them suitable as a growing media for hydroponics § Referred to as either fresh, aged, composted and parboiled or carbonized. § Fresh rice hulls should be avoided as a growing media due to high probability of presence of the contaminants such as rice, fungal spores, bacteria, decaying bugs and weed seeds.	

	Growing Media	Description	Image
B.	**Inorganic Growing Media**		
1.	Perlite	§ Grey white silicate material of volcanic origin. § Help in holding the moisture and nutrients and make them available to plant roots. § Provide optimum aeration and drainage. § Perlite is sterile, it is free of diseases, seeds, and insects.	
2.	Sand	§ Sand is also commonly mixed with Vermiculite, perlite or coco coir. § It helps retaining moisture as well as helps in aerating the mix for the roots.	
3.	Vermiculite	§ Vermiculite has excellent exchange and buffering capacities as well as good ability to supply potassium and magnesium. § Vermiculite is considered as less durable than sand and perlite, its chemical and physical properties are very desirable for container media.	
4.	Rockwool	§ It retains moisture well, so there is less chances of underwatering. § It retains oxygen well, so there is less chance of overwatering.	

Source: Swain *et al.*, 2021

Table 9.2: List of Vegetable Crops for Profitable Purpose using Hydroponics Approach

Sl.No.	*Vegetable Classification*	*Vegetables*
1.	**Solanaceous**	Tomato, Chilli, Brinjal, Bell pepper
2.	**Legumes**	Green bean, Winged bean
3.	**Leafy Vegetables**	Lettuce, Kang Kong, Spinach, Celery, Swiss chard, Atriplex, Coriander leaves, Methi, Parsley
4.	**Cucurbits**	Cucumbers, Melons
5.	**Root Crops**	Radish, Beet,
6.	**Crucifers**	Cabbage, Cauliflower

Source: Sharma *et al.*, 2019.

Factors for Selection of Suitable Techniques for Hydroponics

Numerous aspects must be taken into consideration when selecting an efficient means for applying hydroponics. Despite the numerous alternative hydroponics cultivation techniques available, it is imperative to take these aspects into consideration before adopting them.

a) Availability and Cost of space and other related resources.
b) Availability of suitable growing medium.
c) Expected yield for profitability.
d) Expected productivity from the current method.
e) Quality expected from the application of hydroponics including crop size, colour and appearance.

Hydroponics-Structures and Classifications

Hydroponic system is customised and modified according to recycling and reuse of nutrient solution and supporting media. Commonly used systems are wick, drip, ebb-flow, deep water culture and nutrient film technique (NFT) which are described below:

A. Circulating Method/Continuous Flow Solution Culture

A continuous nutritional environment for roots may be offered by flowing solution culture systems. They are quite adaptable to automated control, but if the flow of solution ceases for any reason, they are vulnerable to fast plant desiccation. As a result, constant care is needed.

1. Deep Water Culture System

In deep water culture, air is delivered directly to the plant roots by an air stone while the plant's roots are suspended in nutrient-rich water. The typical example of this technology is the hydroponic bucket system. The roots of the plants are suspended in nutritional solution and placed in net pots, where they develop rapidly and in a mass. Algae and moulds may grow quickly in the reservoir, therefore it is essential to monitor the oxygen and nutrient concentrations, salinity, and pH (Domingues *et al.*, 2012). This technique is effective for bigger fruit-producing plants, particularly cucumber and tomato, which thrive in this arrangement. Vegetables like cucumber and tomato can be grown very successfully by adopting this method (Swain *et al.*, 2021).

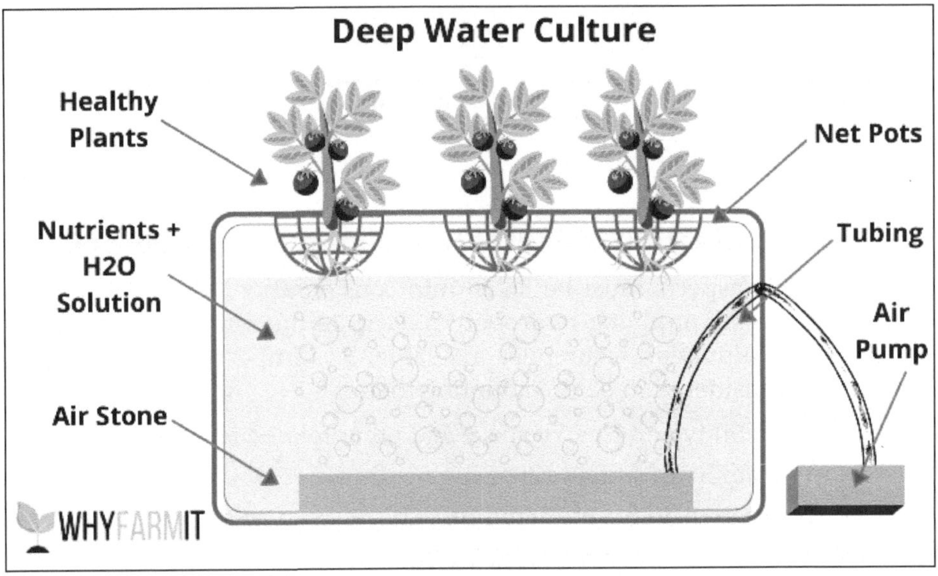

Figure 9.3
(*Source*: https: //whyfarmit.com/deep-water-culture/).

2. Nutrient Film Technique (NFT) System

To address the drawbacks of the ebb and flow approach, Dr. Alen Cooper created NFT in England in the middle of the 1960s. In this method, a water pump without a timer enters the growing tray and circulates water or a nutritional solution throughout the whole system (Domingues *et al.*, 2012). Due to the system's slight inclination, nutritional solution flows through the roots and back into a reservoir. With their roots hanging in a hydroponic solution, plants are put in a channel or tube. However, since roots are continually submerged in water or nutrients, they are prone to fungal infestation (Sharma *et al.*, 2018) This technique makes it simple to cultivate a variety of leafy greens, which are commercially most commonly employed to produce lettuce. Spinach could be grown year round using NFT in greenhouse if proper cultivars and appropriate solution temperature can be choosen (Swain *et al.*, 2021)

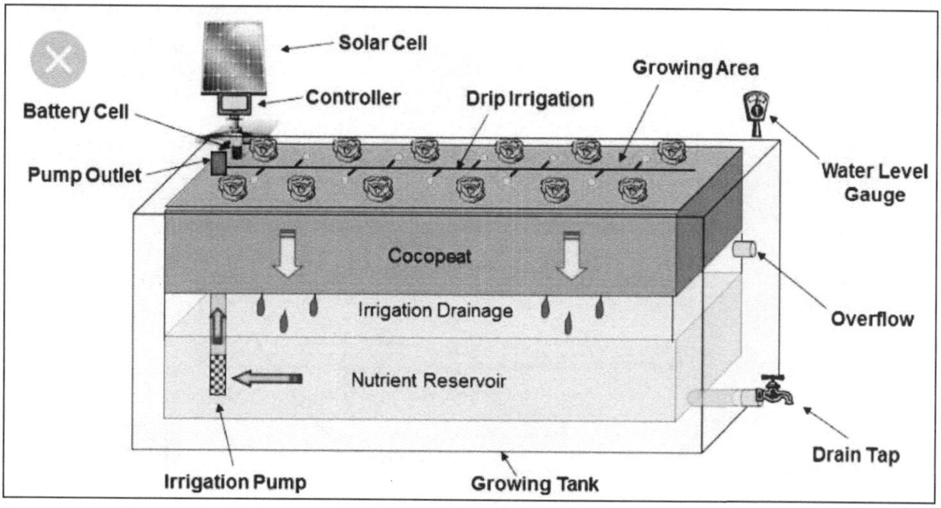

Figure 9.4

B. Non-Circulating Method (Open Systems)/Static Solution Culture

3. Wick System

It doesn't need highly developed apparatus or equipment, such motors and pumps, or extraordinary engineering feats. As there is no need for an electric pump or aerators, it is a passive sort of system. The roots of the plants are connected to a reservoir of nutritional solution by a nylon wick, which is put in an absorbent media such as coco coir, vermiculite, or perlite. Plants acquire water or nutritional solutions through capillary action. This method is effective for small plants, herbs, and spices but ineffective for plants that require a lot of water (Shrestha and Dunn, 2013). Capillary action delivers the nutrient solution to the plant roots in a hydroponic wick system. The method used by paper towels and sponges to absorb liquid from a surface called capillary action. Wick hydroponic systems are frequently used in conjunction with other growing systems that can support larger volumes of plant growth. These systems work best for growing smaller, non-fruiting plants like lettuce and herbs.

4. Ebb and Flow System

This is the first commercial hydroponic system that utilizes the flood and drain technique. In a bid to provide the plants with nutrients and moisture, nutrient solution and water from a reservoir are flooded by a water pump together into grow bed until they reach a predetermined level and sustain there for a certain period of time. This nutrient solution outflow from the growth media attracts air into the rooted bed, supplying O_2 (Swain *et al.*, 2021). Plants can acquire water and nutritional components from the wet rooting media.

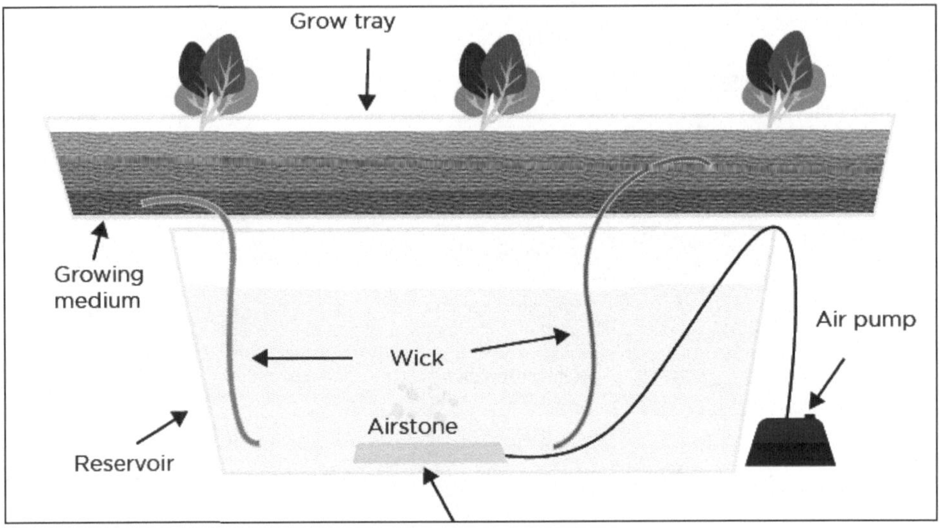

Figure 9.5: Wick Systems

(*Source*: https: //www.trees.com/gardening-and-landscaping/ hydroponic-wick-systems).

Again, in such a nutrient solution delivery system, the roots are exposed to a changing environment, which may not be appropriate for optimal plant growth and development, however plant performance is normally good with this hydroponic approach. (Jones, 2014) Likewise, it is feasible to grow numerous crops, however root rot, algae, and mould problems occur frequent (Nielsen *et al.*, 2006), thus a modified system with a filtering unit is needed (Sardare and Admane, 2019).

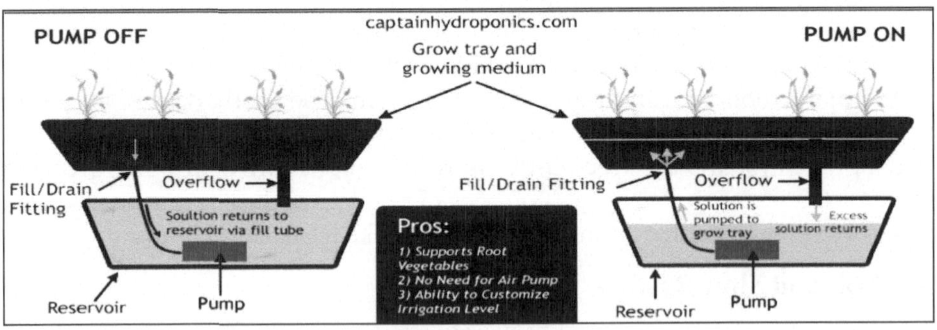

Figure 9.6: Ebb and Flow System

(*Source:* https: //visual.ly/community/Infographics/environment/ ebb-and-flow-hydroponic-system).

5. Drip System

Both business and residential growers frequently employ the drip hydroponic technique. With the aid of a pump, each plant root receives the appropriate amount of water or nutrition solution from the reservoir (Rouphael and Colla, 2005). In order to make the nutrient solution drip slowly, plants are generally planted in growth media that are only moderately absorbent. With increased water conservation, a variety of crops may be cultivated. This strategy may be used to raise a variety of vegetable crops.

pH management

The majority of the plant nutrients utilised in hydroponics are inorganic and ionic forms that are dissolved in water. Different chemical formulations have been used to supply all 17 elements necessary for plant development. The most prevalent nutrient solution for hydroponic systems is Hoagland's solution. For the plant to thrive at its best and for the nutrients in the solution to be available and absorbable, there is a certain pH range. However, the plant itself causes some changes to the solution's pH as it develops (Sonawane, 2018). Therefore, external pH monitoring and management are essential to the system's productivity. However, fluctuations in the solution's pH up to 0.1 units often are perceptible. The optimal range of pH for a variety of plants lies between 5.5 and 6.5 units. This is sufficient for most common species and certain more tolerant species can survive and even thrive at much higher or lower pH. In fact there is a significant variability of pH requirement in various species (Sonneveld, 2000). Cooper's 1988 and Imai's 1987 nutrient solutions were also used for growing leafy vegetables, tomatoes and cucumber. For the optimum plant performance, the nutrient solution's pH and EC must be maintained properly. Most crops can be grown hydroponically with an ideal EC range of 1.5 to 2.5 dS m^{-1}. Osmotic pressure caused by higher EC will prevent nutrients from being absorbed, while lower levels will adversely effect plant health and yield. So, appropriate management of EC in hydroponics technique can give effective tool for improving vegetable yield and quality (Gruda, 2009). As an example, yield of tomato under hydroponic system increased as EC of nutrient solution increased from 0 to 3 dSm^{-1} and decreased as the EC increased from 3 to 5 dS m^{-1} due to increase of water stress (Zhang *et al.*, 2016). Level of EC @1.5, 2 and 3 dS m^{-1} at vegetative, middle vegetative and generative phase, respectively had increased crop height, fruit number and pepper fresh weight. In a nutrient solution, pH determines the availability of essential plant elements. Optimum pH range of nutrient solution for development of plants is 5.5 to 6.5 (Trejo-Tellez and Gomez, 2012) for most species but some can differ from this range. Once the plant has grown, it will alter the composition of the nutrient solution by excreting either acidity or alkalinity to change the pH, depleting specific nutrients more quickly than others, and eliminating water from the solution. For maintaining an ideal solution pH of 5.5 to 6.5, Wang *et al.* (2017) revealed that a mixture of three acids (HNO_3, H_3PO_4, and H_2SO_4) was substantially more effective than a single acid. A pH change may lead to nutrient imbalance, and as a result, a plant may exhibit signs of toxicity or deficiency. Therefore, it is crucial to take precautions to maintain the ideal pH, EC, and nutrient levels in hydroponic solution.

Table 9.3: pH Requirement of Varied Vegetables grown under Hydroponics

Vegetable Crops	*Required pH Range*	*Vegetable Crops*	*Required pH Range*
Artichoke	6.5-7.5	Onions	6.0-6.7
Potato	5.0-6.0	Sweet Corn	6.0
Tomato	6.0-6.5	Turnip	6.0-6.5
Cucumber	5.5	Carrots	6.3
Okra	6.5	Pea	6.0-7.0
Cabbage	6.5-7.0		

Source: Hydroponic Food Production by Howard M. Resh Woodbridge Press, 1987.

Control of Contaminants

Under a highly sterile root-zone environment, contaminants must be eradicated. This is crucial for healthy plant vigour in a soil-free culture (Sonawane, 2018) This is the most challenging step since creating and maintaining a specific environment in the root zone is quite challenging. Nevertheless, it is crucial to reduce the population of plant pathogens like Pythium and Phytophthora, which destroy all and save main roots, Fusarium and Verticillum, which induce wilt. Although no safe fungicides for use in hydroponics have been found, sterilisation of the solution with thorough heat treatment (at around 20–22°C) has been found to be successful (Savvas and Gizas, 2002 and Sardare and Admane, 2019).

Benefits of Hydroponics

i. In hydroponics, healthiest crops are produced with high yields per unit area. Using nutrient solutions, artificial lights, heaters and other pieces of equipment, plants can be made to develop faster, produce larger yields and grow all year round.

ii. The conservation of water is the biggest advantage of hydroponics compared to the common crops. Water can be recirculated in hydroponics allowing unused water to be recycled back into a reservoir, ready for use in the future. Plants only use about 10 per cent of the water compared to the uses by a normal farm crop (Sharma *et al.*, 2018)

iii. The amount of nutrients that are to be given to plants can be controlled thereby reducing the cost of nutrition. As it is a closed system, nutrients are not allowed to flow out with water. An efficient hydroponic farm uses only 25 per cent of the fertiliser compared to a traditional open agricultural system. This also prevents eutrophication due to growth of aquatic plants (like algae) in excess on rivers and lakes caused by accumulation of fertilizer and plant nutrients leached into river water.

iv. Hydroponic crops are least affected by the environment and plants can therefore, be grown closer to consumers, reducing transport emissions and providing people with fresher goods. (Manzocco *et al.*, 2011).

v. Less occurrence of pests and diseases in hydroponic system of planting and it is much easier to treat them. There is no chance of soil-borne insect pest, disease attack and weed infestation. Soil pests are non-existent, and in enclosed greenhouses natural predators can be used to control insect pests and no pesticide is required.

vi. Nutrients are fed directly to the roots; as a result, plants grow faster with smaller roots.

vii. Crop growing is clean and extremely easy. Hydroponically grown crops are more palatable and better for consumption.

viii. Production of 'off-season' vegetable is possible when market prices are highest.

ix. Plants are grown locally hence carbon mileage can be reduced.

x. Less space is required (higher density planting) for growing plants in hydroponics.

xi. Farmers do not need to worry about exhausting their fields of certain nutrients through growing the same crop over and over – there is no need for crop rotation, so in-demand crops can be focused on (Resh, 2013).

xii. More complete control of the environment is possible by using the system (*i.e.* timely nutrient feeding, irrigation and root environment) and different greenhouse type operations, the light, temperature, humidity and composition of the air can be manipulated very easily.

xiii. Water which is carrying high soluble salts can be used if done with extreme care. If the soluble salt concentrations in the water supply are over 500 ppm, an open system of hydroponics can be used if thorough care is given to frequent leaching of the growth media in order to reduce the salt accumulations (Sharma *et al.*, 2018).

xiv. Plant grown in hydroponic system can be protected from UV radiations just like they are within a protected structure.

Limitations of Hydroponics

Technical knowledge and higher initial cost is fundamental requirement for commercial scale cultivation (Resh, 2013). Plant in a hydroponics system is sharing the exact same nutrient, and water borne diseases can easily spread from one plant to another (Ikeda *et al.*, 2002). Hot weather and limited oxygenation may limit production and can result in loss of crops. Maintenance of pH, EC and proper concentration of the nutrient solution is of prime importance. Finally, light and energy supply is required to run the system under protected structure.

i. Despite of many advantages, hydroponics has some limitations.
ii. Technical know-how for growing crops in hydroponics on commercial scale is required which is either not easily available or costlier.
iii. It requires high initial investment.
iv. Due to high cost, only high value crops are grown in hydroponics for high net returns.
v. High energy inputs are necessary to run the system.

Performance of Vegetables under Hydroponics

By using a hydroponics system, a lot of plants, crops, or vegetables can flourish. In general, organic soil-based production produces food with superior quality, flavour, and nutritional content. Leafy greens (lettuce, spinach, parsley, celery, *etc.*) can be grown successfully and simply in hydroponic systems, according to a number of experimental findings. Due to their higher growth and nutrient uptake capacities, lettuce and spinach are the most promising species to grow in integrated hydroponics and aquaculture system.

1. Leafy Vegetables

1.1 Lettuce

Life cycle of hydroponic lettuce is very short compared to traditionally grown lettuce. Hydroponic lettuce can be harvested after 35 to 40 days of production. Lettuce can be successfully grown in NFT system and more than 8 crops per year can be grown efficiently in this system. Horizontal and vertical hydroponic system was also evaluated with different nutrient solutions for yield optimization of lettuce (Touliatos *et al.*, 2016). Growing of lettuce in recirculating hydroponic system at spacing of 50 plants m^{-2} significantly increased yield and yield components (Maboko and Plooy, 2009). Frezza *et al.* (2005) found that there is significant difference in productivity and nitrate content of lettuce in both soil less (floating system and substrate culture) and soil culture however, other traits like leaf area, dry weight and ascorbic acid content were remain unaffected. In non circulated and non-aerated system, air space between nutrient solution and tank cover also determines optimum lettuce yield. Another study observed that marketable yield, shoot biomass and leaf area index of lettuce grown in floating system was not affected by nutrient solution composition (Fallovo *et al.*, 2009). In other experiment, it was observed that both the hydroponic and organic system perform equal in terms of lettuce yield, quality and nitrate content, whereas, delayed harvesting not only increased yield but lower down nitrate level and reduced health hazards (Sardare and Admane, 2019).

Figure 9.7: Different Vegetables grown under Hydroponic Technique.

1.2 Spinach

According to Ranawade *et al.* (2017), spinach yields in hydroponic, aquaponic, and conventional systems that employed sphagnum moss and perlite to support the plants were compared. The yield of spinach grown in aquaponic systems was marginally higher than that of spinach grown in hydroponic systems. According to Mwazi *et al*(2010).'s findings, salt has a detrimental effect on vegetative development, however spinach can tolerate salinity up to 5 ppt. Since spinach is a short-duration crop, the lack of aeration and hypoxia were not severe enough to effect yield and yield components, but quality was nevertheless impacted when spinach was grown in a floating system (Lenzi *et al.*, 2011).

1.3 Other Leaf Vegetables

Swiss chard produced hydroponically using the gravel film technique with a plant density of 40 plants per square metre and a harvest interval of 14 days increased crop production, leaf area, biomass, and leaf fresh weight (Maboko and Plooy, 2013). In contrast, lettuce, swiss chard, and sweet basil grown hydroponically have higher mineral contents, higher root-to-shoot ratios, and lower nitrate levels than those produced in soil cultures, but their yield and uptake of nutrients were lower (Bulgari *et al.*, 2016). In the NFT system, the efficacy of rice husk biochar alone and in conjunction with perlite as substrates for the growth of crops such cabbage, red lettuce, dill, and mallow was also assessed (Awad *et al.*, 2017).

2. Solanaceous Vegetables

2.1 Tomato

Tomatoes may be grown in a variety of hydroponic systems, however NFT and deep flow technique (DFT) are the most often used methods for effective tomato production. Tomato growth, productivity, and mineral composition were all increased in the NFT system when nutrient solutions were recycled regularly, while yield was decreased when nutrient solutions were recycled for an extended period of time. (Zekki *et al.*, 1996). Open and closed hydroponic systems were evaluated for performance of various cultivars of tomato and in closed system higher marketable yield was obtained as because of fruit cracking, yield was reduced in open system (Maboko *et al.*, 2011). Schmautz *et al.* (2016) compared yield, quality and overall tomato plant vitality in three different systems of hydroponics (NFT, drip system and floating raft) system (Sardare and Admane, 2019).

2.2 Pepper

On the production and quality of pepper produced hydroponically in different systems, researchers also examine the impacts of plant population, pruning, and plant growth regulators. Peat + perlite had the most significant impact on the green pepper's growing characteristics and yield when the effectiveness of three different substrates (vermiculite + sand, Peat + perlite, and rockwool) was compared (Majdi *et al.*, 2012).

Water Conservation by Adopting Hydroponics Technique

The use of hydroponics and other water-saving methods for crop production is absolutely required now and is expected to become more widespread over time as water becomes more scarce and valuable as a resource. Compared to soil gardening, hydroponics requires a lot less water. In soil farming, the majority of the water we give plants gets absorbed deeply into the soil and is inaccessible to the plant roots, whereas in hydroponics, plant roots are either completely submerged in water or constantly covered by a film of nutrients mixed with water, keeping the root zone hydrated and fed. Water is not wasted in this process, as it gets recovered, filtered, replenished and recycled. Waste nutrient solution can be used as an alternate water resource for crop cultivation under hydroponic system (Choi *et al.*, 2012). By recycling the run-off water, NFT-based hydroponics may cut irrigation water use by 70 per cent to 90 per cent. Under regulated hydroponic conditions, it is feasible to produce high-quality, high-value crops while consuming 85 to 90 per cent less water than conventional soil-based farming. Salinity, dissolved solids, and pathogens are common components of groundwater and dam/river water sources that can affect plant condition and yield. While some of these elements may help crops, others must be minimised (Sharma *et al.*, 2018).

CONCLUSION

Hydroponic culture is the most intense form of crop production in today's agriculture industry, and it is predominantly utilized in developed and developing countries to produce food in limited area. It is extremely productive, conserves water, is environmentally friendly, and can be done on a minimal piece of land. Hydroponics enables for up to 50 per cent quick and efficient growth than soil because it provides consistent and easily available nourishment. It also produces a higher yield than the traditional approach. In a short period of time, the concept of hydroponics has grown tremendously, leading to a growth in experimentation and research in the area of indoor and outdoor hydroponic agriculture.

Future Scope of this Technology

Hydroponics is the fastest growing sector of agriculture, and it could very well dominate food production in the future. As population increases and arable land declines due to poor land management, people will turn to new technologies like hydroponics and aeroponics to create additional channels of crop production (Maharana and Koul 2011). To get a glimpse of the future of hydroponics, we need only to examine some of the early adopters of this science (Singh and Singh, 2012).

Hydroponics also will be important to the future of the space program. NASA has extensive hydroponics research plans in place, which will benefit current space exploration, as well as future, long-term colonization of Mars or the Moon (Van Os *et al.*, 2002). As we haven't yet found soil that can support life in space, and the logistics of transporting soil via the space shuttles seems impractical, hydroponics could be key to the future of space exploration (Van Os *et al.*, 2002). The benefits of hydroponics in space are two-fold: It offers the potential for a larger variety of food, and it provides a biological aspect, called a bio-regenerative life support system (Dholwani *et al.*, 2018). This simply means that as the plants grow, they will absorb carbon-di-oxide and stale air and provide renewed oxygen through the plant's natural growing process. This is important for long-range habitation of both the space stations and other planets (Singh and Singh, 2012).

The possibility of using hydroponics in third-world nations with scarce water supply has already generated a significant amount of attention in the scientific community (Sardare and Admane, 2019). Although the initial capital costs of installing hydroponics systems are now a barrier, costs will eventually fall as with other technology, making this option far more practical. In regions of Africa and Asia where both water and food are in short supply, hydroponics has the potential to feed millions of people.

Significant agricultural innovation has also been applied to desert regions, such as those in Israel. The nation has been employing hydroponics to grow berries and bananas in shipping containers due to its desert climate and overall lack of water. Although these fruits cannot be cultivated in that climate, they can nonetheless produce 1,000 times more than they can in a normal climate. Even some very large companies have seen the benefits of hydroponic systems. In the spring of 2017, hydroponic gardens were erected at a few locations of the global retail chain Target as part of a series of trials. With little water use, these gardens can offer consumers incredibly fresh veggies and herbs.

REFERENCES

Awad, Y. M., Lee, S. E, Ahmed, M. B. M., Vu, N. T., Farooq, M., Kim, S., Kim, H. S., Vithanage, M., Usman, A. R. A., Wabel, M., Meers, E., Kwon, E. E. and Yong, S. O. (2017). Biochar, a potential hydroponic growth substrate, enhances the nutritional status and growth of leafy vegetables. *Journal of Cleaner Production*. 156: 581-588.

Bulgari, R., Baldi, A., Ferrante, A. and Lenzi, A. (2016). Yield and quality of basil, swiss chard, and rocket micro greens grown in a hydroponic system. *New Zealand Journal of Crop and Horticultural Science*. 45 (2): 119-129.

Butler, J. D. and Oebker, N. F. (2006). Hydroponics as a Hobby– Growing Plants Without Soil. Circular 844. Information Office, College of Agriculture, University of Illinois, Urbana, IL 61801.

Choi, B., Lee, S. S. and Sik Ok, Y. (2012). Effects of waste nutrient solution on growth of Chinese cabbage (*Brassica campestris* L.) in Korea. *Korean Journal of Environmental Agriculture*. 30 (2): 125-131.

De Kreij, C., Voogt, W. and Baas, R. (1999). Nutrient solutions and water quality for soilless cultures. Research Station for Floriculture and Glasshouse Vegetables (PBG), Naaldwijk, The Netherlands, Brochure 196.

Domingues, D. S., Takahashi, H. W., Camara, C. A. P. and Nixdorf, S. L. (2012). Automated system developed to control pH and concentration of nutrient solution evaluated in hydroponic lettuce production. Computers and Electronics in Agriculture 84: 53-61.

Fallovo, C., Rouphael, Y., Cardarelli, M., Rea, E., Battistelli, A. and Colla, G. (2009). Yield and quality of leafy lettuce in response to nutrient solution composition and growing season. *Journal of Food, Agriculture and Environment*. 7(2): 456-462.

Frezza, D., Leon, A, Logegaray, V. and Chiesa, V. Desimone, M. and Diaz, L. (2005). Soilless culture technology for high quality lettuce. Proc. IS on Soilless Culture and Hydroponics. Acta Horticulture 697: 43-47.

Gruda, N. (2009). Does soil-less culture systems have an influence on product quality of vegetables. *Journal of Applied Botany and Food Quality*. 82(2): 141-147.

Ikeda, H., Koohakan, P. and Jaenaksorn, T. (2002). Problems and counter measures in there use of the nutrient solution in soilless production. Acta Horticulturae. 578: 213-219.

Jan, S., Rashid, Z., Ahngar, T. A., Iqbal, S., Naikoo, N. A. and Majeed, S. (2020). Hydroponics: A review. *Int. J Curr. Microbiol. App. Sci;* 9: 1779-1787.

Jones, J. B. (Jr.). (2014). Complete Guide for Growing Plants Hydroponically. CRC Press, Taylor and Francis Group, Boca Raton, Florida, USA; 1-206. 6.

Lenzi, A., Baldi, A. and Tesi, R. (2011). Growing spinach in a floating system with different volumes of aerated or non aerated nutrient solution. Advance Horticulture Science 25(1): 21-25.

Maboko, M. M. and Plooy, C. P. (2009). Effect of plant spacing on growth and yield of lettuce (*Lactuca sativa* L.) in a soilless production system. *South African Journal of Plant and soil.* 26(3): 195-198.

Maboko, M. M. and Plooy, C. P. (2013). Effect of plant spacing and harvesting frequency on the yield of Swiss chard cultivars (*Beta vulgaris* L.) in a closed hydroponic system. *African Journal of Agricultural Research.* 8(10): 936-942.

Maboko, M. M., Plooy, C. P. and Bertling, I. (2011). Comparative performance of tomato cultivars cultivated in two hydroponic production systems. *South African Journal of Plant and Soil.* 28(2): 97-102.

Maharana, L. and Koul, D. N. (2011). The emergence of Hydroponics. Yojana (June). 55: 39-40.

Majdi, Y., Ahmandizadeh, M. and Ebrahimi, R. 2012. Effect of different substrate on growth indices and yield of green pepper at hydroponic cultivate. *Current Research Journal of Biological Science.* 4(4): 496-499.

Manzocco, L., Foschia, M., Tomasi, N., Maifreni, M., Costa, L.D., Marino, M., Cortella, G. and Cesco, S. (2011). Influence of hydroponic and soil cultivation on quality and shelf life of ready-to-eat lamb's lettuce (*Valerianella locusta* L. Laterr). *Journal of the Science Food and Agriculture.* 91(8): 1373-1380.

Mwazi, F. N., Amoonga, S. and Mubiana, F. S. (2010). Evaluation of the effect of salinity on spinach (*Beta vulgaris* var cicla) grown in hydroponic system along the coast of Namibia. Agricola 14-17.

Nielsen, C. J., Ferrin, D. M. and Stanghellini, M. E. (2006). Efficacy of biosurfactants in the management of Phytophthora capsici on pepper in recirculating hydroponic systems. *Canadian Journal of Plant Pathology.* 28(3): 450-460.

Ranawade, P. S., Tidke, S. D. and Kate, A. K. (2017). Comparative cultivation and biochemical analysis of Spinacia oleraceae grown in aquaponics, hydroponics and field conditions. *International Journal of Current Microbiology and Applied Science.* 6(4): 1007-1013.

Resh, H. M. (2013). Hydroponic Food Production: A Definitive Guidebook for the Advanced Home Gardener and the Commercial Hydroponic Grower. CRC Press, Taylor and Francis Group, Boca Raton, Florida, USA, 1-560.

Rouphael, Y. and Colla, G. (2005). Growth, yield, fruit quality and nutrient uptake of hydroponically cultivated zucchini squash as affected by irrigation systems and growing seasons. *Scientia Horticulturae.* 105 (2): 177- 195.

Sardare M and Admane S. (2019). A Review on Plant Without Soil -Hydroponics. *International Journal of Research in Engineering and Technology.* 2(3): 299-304.

Savvas, D. and Gizas, G. (2002). Response of hydroponically grown gerbera to nutrient solution recycling and different nutrient cation ratios. *Scientia Horticulturae,* 96(1-4): 267-280.

Schmautz, Z., Loeu, F., Liebisch, F., Graber, A., Mathis, A., Bulc, G. T. and Junge R. (2016). Tomato productivity and quality in aquaponics: comparison of three hydroponic methods. Water 8: 1-21.

Sharma, N, Acharya, S., Kumar, K., Singh, N. and Chaurasia, O. (2019) . Hydroponics as an advanced technique for vegetable production: An overview. *Journal of Soil and Water Conservation.* 17. 364-371.

Shrestha, A. and Dunn, B. 2013. Hydroponics. Oklahoma Cooperative Extension Services HLA-6442. Touliatos, D., Dodd, I.C. and McAinsh, M. (2016). Vertical farming increases lettuce yield per unit area compared to conventional horizontal hydroponics. *Food and Energy Security* 5(3): 184-191.

Singh, S. and Singh, B. S. (2012). Hydroponics: A technique for cultivation of vegetables and medicinal plants. In. Proceedings of 4th Global conference on ¯Horticulture for Food, Nutrition and Livelihood Options N"è%?" Bhubaneshwar, Odisha, India. p.220.

Sonawane, M. S. (2018). Hydroponics: An upcoming and innovative way of future farming. *International Archive of Applied Sciences and Technology,* 9(1): 69-74.

Sonneveld, C. (2000). Effects of salinity on substrate grown vegetables and ornamentals in greenhouse horticulture. Ph. D. Thesis, University of Wageningen, The Netherlands.

Swain, Chatterjee, A., Subhrajyoti and Viswanath, M. (2021). Hydroponics in vegetable crops: A review. *The Pharma Innovation Journal.* 10(6): 629-634.

Touliatos, D., Dodd, I. C. and McAinsh, M. (2016). Vertical farming increases lettuce yield per unit area compared to conventional horizontal hydroponics. *Food and Energy Security,* 5(3): 184-191.

Trejo-Tellez, L. I. and Gomez, M. F. C. (2012). Nutrient Solutions for Hydroponics Systems, Hydroponics - A Standard Methodology for Plant Biological Researches, Dr. Toshiki Asao (eds). ISBN 978-953-51-0386-8.

Van Os, E. A., Gieling, H. and Ruijs, M. N. A. (2002). Equipment for hydroponic installations. In: Hydroponic Production of Vegetables and Ornamentals (Savvas D; Passam H C, eds), pp 103–141. Embryo Publications, Athens, Greece.

Wang, L., Chen, X., Guo, W., Li, Y., Yan, H. and Xue, X. (2017). Yield and nutritional quality of water spinach (Ipomoea Aquatica) as influenced by hydroponic nutrient solutions with different pH adjustments. International Journal of Agriculture and Biology 19: 635-642.

Zekki, H., Gauthier, L. and Gosselin, A. (1996). Growth, productivity, and mineral composition of hydroponically cultivated greenhouse tomatoes, with or without nutrient solution recycling. Journal of American Society of Horticulture Science. 121(6): 1082- 1088.

Zhang, P., Senge, P. and Dai, Y. (2016). Effects of salinity stress on growth, yield, fruit quality and water use efficiency of tomato under hydroponics system. Reviews in Agricultural Science 4: 46-55.

Chapter 10

Importance and Scope of Medicinal Plant for Ecosystem and their Scientific Validation

Nishant[1]*, *Ankita Sharma*[2], *Abhinav*[3], *Reshav Naik*[1] and *Rehan*[2]

[1]*Ph.D. Research Scholar, Division of Vegetable Science, ICAR-Indian Agriculture Research Institute, New Delhi – 110012*

[2]*Ph.D. Research Scholar, Department of Horticulture, College of Agriculture, Jawaharlal Nehru Krishi Vishwa Vidhyalaya, Jabalpur, Madhya Pradesh*

[3]*Department of Fruit Science, Dr. YSP University of Horticulture and Forestry, Nauni, Solan, Himachal Pradesh.*

**e-mail: nishant1998katoch@gmail.com*

INTRODUCTION

The existence of humans without plants on earth is inconceivable. Since the beginning of human civilization, mankind has indeed been reliant on vegetation. Medicine is frequently obtained from medicinal plants (Aslam and Ahmed, 2016). According to a widely held conventional belief, factors including the genetic makeup of the relevant plants, their biological habitats and the nutrients in the soil all have a profound impact on the performance and quantity of crude pharmaceuticals derived from medicinal plants (Dai *et al.*, 2003; Jia *et al.*, 2016; Sherameti *et al.*, 2005). But in past few years, it has become increasingly apparent that endophytic fungi or endophytes had quite a profound impact on the performance and quantity of crude drugs through a specific fungus-host engagement, indicating that greater comprehension of the assessment of the relationship among endophytic fungi and medicinal plants is needed to promote the production of crude drugs (Faeth and Fagan, 2002; Jia *et al.*, 2016).

Due to the fact that natural medicines are regarded as "green medicine" and are therefore always expected to be safe, there is currently a resurgence of interest in drugs with a natural origin. Medicinal plants have a broad spectrum of medicinal qualities, including anti-inflammatory, antifungal and anticarcinogenic both analgesic and anti-malarial properties. In present scenario from several of the biggest challenges facing global health is cancer. As the major cause of mortality across all epidemic diseases, cancer emerges as a threat to health globally. According to statistics, lung cancer affects men more frequently than women, accounting for 17 per cent of all cancer cases in men, whereas breast cancer affects approximately 23 per cent of women's cancer cases (Jemal *et al.*, 2011; Raina *et al.*, 2014).

But it was not until the 1950s that plant-based chemicals came into utilisation to treat cancer. Vinca alkaloids, vinblastine, vincristine, and cytotoxic podophyllotoxins were among the very first plant-based anticancer medications. According to records, clinical studies have been conducted so far on 16 plant-derived anticancer medications (Belayachi *et al.*, 2013; Raina *et al.*, 2014). A significant component of such clinical trials is flavopiridol, which was isolated out from Indian tree *Dysoxylum binectariferum*, and meisoindigo which was isolated by the Chinese plant Indigofera tinctoria. Both of these compounds have been shown to be less toxic than traditional chemotherapeutic anticancer drugs (Saklani and Kutty, 2008; Raina *et al.*, 2014).

If the fungus-host connections and their metabolic pathways under cultural settings are well established, an alternate technique to directly create desired medications could be established using the specified medicinal herbs and endophytic fungi (Kumaran *et al.*, 2008; Jia *et al.*, 2016). Isabgol is a medicinal and therapeutic crop grown in India and the country's most important export of medicinal plants. The environment that is semi-arid and arid is best for the produce. Due to its therapeutic characteristics, Isabgol and its value-added products are becoming more and more prominent on the domestic and worldwide markets. Given the enormous potential for value addition, Indian Isabgol is getting more and more appreciated. Isabgol's husk contains mucilage, which is utilized across several medical practices (Jain, 2014; Rathore and Mathur, 2019).

Because of their therapeutic properties, herbal medicines are in higher demand in the beauty and pharmaceutical and food industries. The purpose of the current assessment is to collect relevant information about the various herbal medicines with anticancer characteristics that have a tremendous possibility of treating cancer and to know about the different advantages of herbal extracts on human health.

Current Scenario of Medicinal Plants on Global Scale

In 2001, it was estimated that pure medicinal, aromatic, and plant components were imported globally for more over $1 billion USD. Approximately 13 per cent of global market share came from India, which was the second-largest supplier of raw medicinal plant products after China. About Rs. 4000 crores worth of goods or services related to medicinal plants are sold domestically. The commercialization of the Indian medicinal vegetation economy now amounts to Rs. 5200 crores, with exports at a value of Rs. 1200 crores. Government of India organisations including the Department of Biotechnology, Council of Scientific and Industrial Research (CSIR), Department of AYUSH, Ministry of Health and Family Welfare, and others have started to make efforts in the biological studies of therapeutic and aromatic plants. A cluster of 19 CSIR facilities as well as other authorities and campuses working in the subject of folk medicine have joined forces with CSIR to launch an unified initiative on pharmaceutical research. One of the most cutting-edge biomedical field initiatives, the New Millennium Indian Technology Leadership Initiative (NMITLI), was backed by the Indian Planning Commission. (Subramoniam, 2014). India is the second-largest supplier of MAPS in Asia, accounting for 8.75 per cent of MAPs trade, trailing China, which holds the top position with a 1.48 per cent share (Karki, 2017; Riaz *et al.*, 2021). About 12,000 tonnes of natural medicinal species materials worth US $4.5 million are prevalent in Bangladesh, but the country additionally imports US$ 8 million worth of material comprising more than 5000 tonnes. Bangladesh's domestic natural medicinal investment is worth US $60 million. Similar to that, Nepal trades medicinal and aromatic herbs for an estimated USD 33 billion each year (Merry and Shahjahan, 2013; Pyakurel *et al.*, 2019; Riaz *et al.*, 2021).

Personal hygiene products and beauty products composed of natural ingredients have suddenly become an extremely mainstream trend on global commerce. The physiologically active ingredients in these cosmetics have the potential to treat skin specific conditions and have the desired bioactivities, smoothing, relaxing and conditioning effects on face and eyes. Due to this reason, this industry has significant economic effect on the uplift of a country's economy as a result of the development of improved technology in cosmetics, skin care, and hair care goods. In contrast, Aloe Vera, which has excellent sun protection and calming attributes, is widely used in the cosmetic sector. For the therapy of issues relating to the complexion, Ginkgo, Ginseng, Ashwagandha and Saw palmetto are utilised in place of synthetic chemicals. The use, manufacturing and export of related medicinal plants have expanded significantly as a result of the natural compounds found in many age-defying, hair volumizing stimulants, sun protection moisturizers and gels, cosmetic scrubs and protective eyewear, as well as other goods (Kim, 2016; Dorni *et al.*, 2017; Rao *et al.*, 2017; Riaz *et al.*, 2021).

Vegetation are transforming their ranges as a result of climatic changes. Since an attempt to reclaim suitable growing zones, their ranges have started to move northward and upward. Numerous indigenous species might go eradicated worldwide due to ecological loss and migratory issues caused by global warming. Severe weather including cyclones, periods of drought and storm surges have reportedly grown extremely frequent and more severe all-around world in recent decades, as per investigations, statistics and documentation. (Sharma *et al.*, 2020). Evidences suggest that unusual weather conditions can have a consequence on landowners' and collectors' efficiency to cultivate and gather various medicinally important plant species. Extremely harsh climate across all of European continent is having an impact on the sustainability of medicinal plants from germination to maturity, like chamomile in both Germany and Poland (Pompe *et al.*, 2008; Sharma *et al.*, 2020).

Table 10.1: Medicinal and Aromatic Plants Uses and Propagation Methods

Medicinal Plants	*Active Ingredients*	*Uses*	*Propagation*
Ashwagandha	Withanine	Rheumatism	Seeds
Aloe	Aloin	Stomach disorder	Suckers
Belladona	Tropane	Gout	
Cinchona	Quinine	Malaria	Cuttings
Datura	Atropine	Sedative	Seeds
Guggul	Guggalsterone	Obesity	Cuttings
Isabgol	Plantiose	Gastro-intestinal disorder	Seeds
Liquorice	Saponin	Cough	Seeds
Opium poppy	Morphine	Sedative	Seeds
Priwinkle	Vinblastine	Cancer therapy	Seeds
Senna	Sennoside	Constipation	Seeds
Safed Musil	Saponins	General debility	Tubers
Aromatic Plants	Active ingredients	Uses	
Basil	Linalool	Flavouring food stuff	Seeds
Citronella	Citronellol	Soap	Slips
Geranium	Geranial	Perfumes	Cuttings
Jasmine	Benzyl benzoate	Cosmetics	Cuttings
Lavender	Linalyl acetate	Perfumes	Seeds

Due to the adverse circumstances in Bulgaria throughout the springtime, fennel (Foeniculum vulgare, Apiaceae) reportedly reported to have produced almost negligible output during the first year. There have already been instances where it has been impracticable for harvesters to undertake second cuttings of the aerial sections of cultivated herbs like peppermint because of Serbia's long and dry summers, which are often accompanied by other inclement weather like intense showers and thunderstorms (Pal *et al.*, 2004; Schar *et al.*, 2004; Sharma *et al.*, 2020). For India, where the yearly monsoon dominates the country's climate, it appears to be getting worse and more irregular. Recent studies have revealed that although the amount of rainfall during the monsoon season has generally managed to remain constant over the last period in central India, the frequency of modest rainfalls has decreased dramatically and the intensity of heavy rainfall events has substantially enhanced since early 1980s (Bhardwaj *et al.*, 2007; Sharma *et al.*, 2020). In 2006, 2007 and 2008 regions such as Gujarat and Rajasthan, saw thunderstorms and rain at occasions when these instances often did not occur over the preceding fifty years. Psyllium, wheat and cumin harvests have also been negatively affected in the region by extreme weather and storms.

According to estimates, 33 plant species among India's 1500 threatened plant species have reportedly disappeared, 157 are threatened with extinction, 114 are vulnerable, and 246 are rare and over 70 per cent of the therapeutic plants that are in danger are traded actively. 235 species (or around 38 per cent) of the 625 species of therapeutic plants traded today are herbs, and their gathering is typically deleterious, because of how severe and significant the loss of biodiversity is, action must be taken quickly and urgently to minimize the impact. However, in order to meet the demands of the millions and millions of people living across the world, conservation measures for medicinal and aromatic plants must take into account both the continued existence of the species and a steady supply of raw materials. For medicinal plants, good agricultural practises (GAP) have been established to regulate productivity, guarantee quality, and allow the standardisation of herbal medicine (Rasool *et al.*, 2020).

The demand for plant-based medicines is certain to rise given the upsurge of interest in using natural remedies for treating a variety of persistent disorders. This situation necessitates a thorough examination into the accessibility and spontaneous selection processes of high-value Himalayan herbal drugs, which are widely used for biopharmaceuticals in a variety of healthcare disciplines. Establishing a measure to fight the anticipated expanding imbalance between supply chain (availability of raw materials) will be made easier with the aid of such a study (Dhar *et al.*, 2002). Natural products potentially herald a new era of medicinal plants for the treatment of human ailments over the next several generations. The majority of people in the globe (around 80 per cent) get their primary healthcare services from conventional medicine. The exploration of therapeutic plants as a potential source of possible herbal medicines has regained popularity over the past 10 years. With the use of enhanced screening techniques from medicinal and aromatic plants sources, investigations for the preparation and characterization of novel natural medications must be advanced. As science progressed, it was able to apply AYUSH to address the fresh problems facing the contemporary health service (Shakya, 2016).

Many herbs can be grown extensively for regional economic production of pharmaceuticals and herbal dietary supplements. It is absolutely imperative that whatever is known is put to good use for both monetary gain and the betterment of our person's quality of life. One must acquire the essential skills for indigenous medical advancement, and purposeful attempts should be made to promote domestic industrialization of medicinal herbs so that cultivation will be feasible and thus contribute to the redistribution of wealth (Rasool *et al.*, 2020).

Figure 10.1: Medicinal Plants for Treating Human Ailments Medicinal Plants as Best Option for Treatment of Human Ailments.

The principal antioxidant characteristics of herbal plant extracts chemical constituents play a significant vital role in their pharmacological properties. The primary groups of bioactive substances essential for the antioxidant effect include flavones, isoflavones, flavonoids, anthocyanins, tannins, lignans, catechins, and isocatechins (Nema *et al.*, 2013; Raina *et al.*, 2014). Garlic is scientifically referred as *Allium sativum*, is a member of the Alliaceae family. It usually contains minerals like selenium, several enzymes, 17 amino acids, and sulphur containing substances. Selenium, an antioxidant that may also have chemoprotective properties, is abundant in garlic (Ip and Lisk, 1996; Raina *et al.*, 2014). Known as "Noni" domestically, *Morinda citrifolia* is a species belonging to Rubiaceae family. In terms of pharmacology, it can be used to treat conditions like malaria, jaundice, hypertension, boils, carbuncles, gastrointestinal disorders, stomach-aches, cracks, insulin resistance, indigestion, urogenital problems, cramps, fistulae and human vitamin A insufficiency. The juice of the noni fruit involves two distinctive glycosides that have been isolated. These glycosides, 6-O-(D-glucopyranosyl)-1-O-octanoyl-D-glucopyranose and asperulosidic acid, were found to be extremely effective at preventing TPA- or EGF-induced cell transformation as well as related AP-1 activity in the mouse epidermal JB6 cell line (Liu *et al.*, 2001; Raina *et al.*, 2014). Ashwagandha is the popular name for *Withania somnifera*, a small wooded shrub in the Solanaceae family. The key phytochemical component are steroidal alkaloids and steroidal lactones, which are collectively referred to as withanolides. This plant has adaptogenic, aphrodisiac, anti-inflammatory, deobstruent, antibiotic, diuretic, sedative, abortifacient, immune stimulating, astringent and the anti-tumour ability is governed by the chemical constituent withaferin (Raina *et al.*, 2014).

In Tamil Nadu, there are several medicinal plants which are widely used to treat various diseases and disorders *i.e.* the leaves of Koyya is used to treat dysentery, Nochi leaves for the treatment of viral infection such as cold, fruit juice of Alangimaram is used for the treatment of eye specified disorder (Munuswamy *et al.*, 2013). There are many herbs whose biochemical and antibacterial capabilities have indeed been catalogued. The queen of spices (Chotti elaichi) *Elettaria cardamomum*, *Trachyspermum ammi* (Omum), *Anethum graveolens* (Dill), *Foeniculum vulgare* (Fennel), and *Syzygium aromaticum* (Clove) have all been used to treat induced gastric disorders. *Viola odorata* has also been utilized to treat respiratory illnesses. There have been reports of these plants natural herbal extracts having both pharmacological, antibacterial and antifungal properties (Arora and Kaur, 2007). Various portions of a particular plant can be utilized to treat a variety of medical conditions, such as *Citrullus colocynthis*, a creeper that has the ability to treat leucoderma rheumatism, paronychia, abnormal menstrual periods, and bowel problems. Identical to this, the plant *Zizyphus nummularia* is used for treating a wide range of ailments by using its roots, leaves, and bark (Parveen *et al.*, 2007). Flowers when collected for eating purposes should not be sprayed with pesticides or untreated manure because they can be toxic or cause illness due to the presence of pathogens. Some floral parts such as the pedicel, sepals, pistils and stamens of most flowers are removed prior to use (Sharma and Nair, 2020) Pollens of some flowers can detract flower's flavour and cause hay fever, asthma, allergies. The sepals are sour except that of Violas, Pancies and hence should be removed. In Flowers of rose, lavender, tulip, calendula, and chrysanthemum only the petals are edible. The flowers should be cooled immediately after harvest to maintain freshness. Flowers with white and bitter petal bases such as chrysanthemum, dianthus, marigold, and rose should be removed as it is bitter to taste (Sharma and Nair, 2020).

Geographically inside the Indo-Burma Biodiversity Hotspot, the North-East of India is a wonderful source of genetically diverse species. Ever since heritage, almost every tribal group in Manipur has developed its very own distinct knowledge of how to use medicines derived from plants. They also have developed its own traditional medical expertise, which is frequently distinct from that of the other tribes. The current inhabitants of this area have been treating cases of stones with a variety of therapeutic plants. It was additionally discovered that many ethnic populations employ various herbs for renal stone treatment, indicating that local cultural heritage is particular to peers. For example, in the case, it was discovered that only the Khoibu group used *Anneslea fragrans*, while the Thadou tribe used *Saccharum officinarum* to treat urinary infections, proving that these traditional practises were specific to the communities that held that kind of expertise and then were down from generation to generation greatly from one person to the next. In contrast, the majority of species utilized by the Muslims and Meeitei residents of Manipur are similar, with only minimal distinctions in how they are used. This is likely because these communities speak a similar language and are more closely clustered together in their colonies than the tribal groups, who occupied the majority of the hilly terrain with a more dispersed pattern of territories and different languages and culture of the various tribes (Mikawlrawng *et al.*, 2014).

Numerous herbs utilised by the locals in Churu district are mentioned in historical articles and are also employed in other medical systems, including Ayurveda and Unani. In contrast, Citrullus colocynthis, that is commonly used by local residents to reduce inflammation and skin disorders, are included in the Ayurvedic medicine as a treatment for the identical illnesses. These are some of the primary components of the ancient herbal remedies "Narayan Churna" and "Sukhvirechani vati." This is recommended in the Unani health care system to cure "kala-azar" (Guha Bakshi *et al.*, 2000; Parveen *et al.*, 2007). Likewise, *Momordica charantia*, one that the natives use to address piles and blood disorders, is also utilised in Ayurvedic system of medicine to treat various diseases such as ulceration and chest infections. In Unani, it is used to treat rheumatoid arthritis and syphilis (Kritikar and Basu, 1981; Parveen *et al.*, 2007).

Herbs are progressively being used as medication. Consumers nowadays are reconnecting to natural order in wealthy nations as well. (Parveen *et al.*, 2007) Throughout almost every developing economy, the use of conventional healers is the core component of universal health coverage. The contributing factors for the widespread use of conventional healers include

a) Deep relations between people and their native vegetation and their faith in cultural education regarding herbs and plants.
b) Ease with which indigenous medicinal plants can be obtained.
c) Restricted availability to and exorbitant prices of synthetic drugs medications.
d) Socioeconomically, disadvantaged designation of the indigenous populace.

CONCLUSION

The dialectical interaction between indigenous knowledge and practice shapes ecology, which has an influence on plant populations. By merging indigenous knowledge and scientific research, new ideas for sustainable resource conservation can be established. Because of shifting perspectives of local people and the ever-increasing effect of globalization and socioeconomic development, indigenous knowledge of plant resource usage is continually declining. The number of precious plant resources is dwindling at an alarming pace due to a lack of regulated scientific and sustainable monitoring cultivation and harvesting, correct management practices, and awareness of societal issues.

Furthermore, indigenous knowledge of lesser-known plant applications is fast disappearing. Recent research has shown the bioactivities of medicinal plants and phytoextracts, indicating their potential therapeutic application in the treatment of a variety of illnesses. Local communities lack suitable understanding of wild plant populations, marketing and selling, insufficient regulation and legal protection, and restricted access to appropriate agricultural plantation and harvesting technology. Local communities, too, deserve aid and encouragement in order to protect their knowledge and resources. Because the current generation is less interested in traditional knowledge of medicinal plants as a result of modernity, there is an urgent need to document traditional knowledge of medicinal plants in the study site before it is lost to society. Plant-based treatments may become more popular as a result of the many negative effects of modern allopathic drugs, and traditional knowledge of plants and folk remedies may be preserved.

REFERENCES

Arora, D. S. and Kaur, G. J. (2007). Antibacterial activity of some Indian medicinal plants. *Journal of Natural Medicines*, 61(3), 313-317.

Aslam, M. S. and Ahmad, M. S. (2016). Worldwide importance of medicinal plants: Current and historical perspectives. *Recent Adv Biol Med*, 2(2016), p.909.

Belayachi, L., Aceves-Luquero, C., Merghoub, N., Bakri, Y., de Mattos, S. F., Amzazi, S. and Villalonga, P. (2013). Screening of North African medicinal plant extracts for cytotoxic activity against tumor cell lines. *Eur J Med Plants*, 3, 310-32.

Bhardwaj, J., Singh, S. and Singh, D. (2007, September). Hailstorm induced crop losses in India: some case studies. In Abstract for presentation at 4th European Conference on Severe Storms in Trieste, Italy (pp. 10-14).

Dai, C. C., Yu, B. Y., Xu, Z. L. and Yuan, S. (2003). Effect of environmental factors on growth and fatty acid composition of five endophytic fungi from *Sapium sebiferum*. *J. Appl. Ecol.* 14, 1525–1528.

Dhar, U., Manjkhola, S., Joshi, M., Bhatt, A., Bisht, A. K. and Joshi, M. (2002). Current status and future strategy for development of medicinal plants sector in Uttaranchal, India. *Current Science*, 83(8), 956-964.

Dorni, A. C., Amalraj, A., Gopi, S., Varma, K. and Anjana, S. N. (2017). Novel cosmeceuticals from plants. An industry guided review. *Journal of Applied Research on Medicinal and Aromatic Plants* 7: 1-26. https: / / doi.org / 10.1016 / j. jarmap.2017.05.003

Faeth, S. H., and Fagan, W. F. (2002). Fungal endophytes: common host plant symbionts but uncommon mutualists. *Integr. Comp. Biol.* 42, 360–368. doi: 10.1093 / icb / 42.2.360

Guha Bakshi, D. M., Sensarma, P. and Pal, D. C. (2000). A lexicon of medicinal plants of India 1, 446–448.

Ip, C. and Lisk, D. J. (1996). The attributes of selenium-enriched garlic in cancer prevention. *Dietary Phytochemicals in Cancer Prevention and Treatment*, 179-187.

Jain, L. K. (2014). Economics and gap analysis in Isabgol cultivation through frontline demonstrations in western Rajasthan. *International Journal of Agricultural Extension*, 2(2), 109-114.

Jemal, A., Bray, F., Center, M. M., Ferlay, J., Ward, E. and Forman, D. (2011). Global Cancer Statistics. CA Cancer J Clin 61: 69–90.

Jia, M., Chen, L., Xin, H. L., Zheng, C. J., Rahman, K., Han, T. and Qin, L. P. (2016). A friendly relationship between endophytic fungi and medicinal plants: a systematic review. *Frontiers in microbiology, 7*, pp.906.

Karki, M. B. (2017). Challenges, opportunities and trade-offs in commercialization of medicinal and aromatic plants in South Asia Region. In *Invited paper presented at the workshop on current challenges and recommendations*. New Delhi: Government of India.

Kim, S. K. (2016). Marine cosmeceuticals: Trends and prospects. *CRC Press*; 2016 Apr 19. https: //doi.org/10.1201/b10120.

Kritikar, K. R. and Basu, B. D. (1981). Indian Medicinal Plant, vols. 1–11. Oriental Enterprises, Uttaranchal.

Kumaran, R. S., Muthumary, J. and Hur, B. K. (2008). Taxol from *Phyllosticta Citricarpa*, a Leaf Spot Fungus of the Angiosperm citrus medica. J. Biosci. Bioeng. 106, 103–106. doi: 10.1263/jbb.106.103

Liu, G., Bode, A., Ma, W. Y., Sang, S., Ho, C. T. and Dong, Z. (2001). Two novel glycosides from the fruits of *Morinda citrifolia* (noni) inhibit AP-1 transactivation and cell transformation in the mouse epidermal JB6 cell line. *Cancer Research*, 61(15), 5749-5756.

Merry, S. R. and Shahjahan, M. (2013). Country status report on medicinal and aromatic plants in Bangladesh. Expert consultation on promotion of medicinal and aromatic plants in the Asia-Pacific Region: Proceedings, 75.

Mikawlrawng, K., Kumar, S. and Vandana, R. (2014). Current scenario of urolithiasis and the use of medicinal plants as antiurolithiatic agents in Manipur (North East India): a review. *International Journal of Herbal Medicine*, 2(1), 1-12.

Munuswamy, H., Thirunavukkarasu, T., Rajamani, S., Elumalai, E. K. and Ernest, D. (2013). A review on antimicrobial efficacy of some traditional medicinal plants in Tamilnadu. *Journal of Acute Disease*, 2(2), 99-105.

Nema, R., Khare, S., Jain, P., Pradhan, A., Gupta, A. and Singh, D. (2013). Natural products potential and scope for modern cancer research. *American Journal of Plant Sciences* 4: 1270–1277.

Pal, J. S., Giorgi, F. and Bi, X. (2004). Consistency of recent European summer precipitation trends and extremes with future regional climate projections. *Geophysical Research Letters*, 31(13).

Parveen, Upadhyay, B., Roy, S. and Kumar, A. (2007). Traditional uses of medicinal plants among the rural communities of Churu district in the Thar Desert, India. *Journal of Ethnopharmacology*, 113(3), 387-399.

Pompe, S., Hanspach, J., Badeck, F., Klotz, S., Thuiller, W. and Kuhn, I. (2008). Climate and land use change impacts on plant distributions in Germany. *Biology Letters*, 4(5), 564-567.

Pyakurel, D., Smith-Hall, C., Bhattarai-Sharma, I. and Ghimire, S. K. (2019). Trade and conservation of Nepalese medicinal plants, fungi, and lichen. *Economic Botany*, 73(4), 505-521.

Raina, H., Soni, G., Jauhari, N., Sharma, N. and Bharadvaja, N. (2014). Phytochemical importance of medicinal plants as potential sources of anticancer agents. *Turkish Journal of Botany, 38*(6), 1027-1035.

Rao, S., Hegde, S. K., Baliga-Rao, M. P., Palatty, P. L., George, T. and Baliga, M. S. (2017). An aloe vera-based cosmeceutical cream delays and mitigates ionizing radiation-induced dermatitis in head and neck cancer patients undergoing curative radiotherapy: a clinical study. *Medicines*, 4(3), 44.

Rasool, A., Bhat, K. M., Sheikh, A. A., Jan, A. and Hassan, S. (2020). Medicinal plants: Role, distribution and future. *Journal of Pharmacognosy and Phytochemistry*, 9(2), 2111-2114.

Rathore, R. and Mathur, A. (2019). Scope of cultivation and value chain perspectives of medicinal herbs in India: A case study on aloe Vera and Isabgol. *J Pharmacogn Phytochem, 8*, 243-246.

Riaz, U., Iqbal, S., Sohail, M. I., Samreen, T., Ashraf, M., Akmal, F., Siddiqui, A., Ahmad, I., Naveed, M., Khan N. I. and Akhter, R. M. (2021). A comprehensive review on emerging importance and economical potential of medicinal and aromatic plants (MAPs) in current scenario. *Pakistan Journal of Agricultural Research*, 34(2), 381-392.

Saklani, A. and Kutty, S. K. (2008). Plant-derived compounds in clinical Trials. *Drug Discov Today* 13: 161–171.

Schar, C., Vidale, P. L., Luthi, D., Frei, C., Haberli, C., Liniger, M. A. and Appenzeller, C. (2004). The role of increasing temperature variability in European summer heatwaves. *Nature*, 427(6972), 332-336.

Sharma, A. and Nair, R. (2020). Edible Flowers: An Extreme and Elegant Source of Minerals. In Pandey S.K., Nair R and Bhan K, Horticultural Crops- Emrging Issues and Techniques (pp 193-204). Biotech Books.

Sharma, M., Thakur, R., Sharma, M., Sharma, A. K. and Sharma, A. K. (2020). Changing Scenario of Medicinal Plants Diversity in Relation to Climate Changes: A Review. *Plant Archives*, 20(2), 4389-4400.

Sherameti, I., Shahollari, B., Venus, Y., Altschmied, L., Varma, A. and Oelmuller, R. (2005). The endophytic fungus Piriformospora indica stimulates the Expression of nitrate reductase and the starch-degrading enzyme glucan-water dikinase in tobacco and Arabidopsis roots through a homeodomain Transcription factor that binds to a conserved motif in their promoters. *J. Biol. Chem.* 280, 26241–26247. doi: 10.1074/jbc.M500447200

Subramoniam, A. (2014). Present scenario, challenges and future perspectives in plant-based medicine development. *Annals of Phytomedicine*, 3(1), 31-36.

Chapter 11

Irradiation of Fresh Fruits and Vegetables

Shalini Singh[1], Vibhu Pandey[1], Khursheed Alam[2] and Amit Kumar[1]*

[1]Ph.D. Research Scholar, Department of Fruit Science,

[2]Ph.D. Research Scholar, Department of Vegetable Science,

Sardar Vallabhbhai Patel University of Agriculture and Technology, Meerut – 250 110, Uttar Pradesh

**e-mail: shalinisingh1523@gmail.com*

INTRODUCTION

The practice of centrally processing fresh produce and appropriately packaging it for distribution and marketing is growing in both developed and many developing nations. The rise in demand for freshly cut or minimally processed fruits and vegetables has been attributed to changes in demography, lifestyles and eating patterns among other factors. Freshly cut produce continues to be quite popular (Acinf, 1996). Because of this, food-borne diseases linked to produce are of major interest (Akamine and Moy, 1983).

Low dosage irradiation is one of the most promising new methods for ensuring the safety of fresh and freshly cut fruits and vegetables (Anonymous, 1987).

Food irradiation is a physical method of food processing that involves exposing pre-packaged or bulk foods to gamma rays, X-rays, or electrons. Foods are typically exposed to gamma radiation from a radioisotope source, electrons, or X-rays produced by an electron accelerator, or a combination of these radiations (Barbosa-Canovas *et al.*, 1998). The non-residual property of ionizing radiation is a significant benefit due to the great need to limit the usage of pesticides applied to fruits and vegetables. Food irradiation is a 'cold' process and after more than 50 years of study and development, food irradiation has proven to be a secure and reliable method of food processing and preservation (Bender, 1986; Bord, 1991; Cast, 1989; Crawford and Ruffeh, 1996; Defesche, 1983; Diehl, 1990, 1993; Ehlermann, 1983; Elias, 1987; FAO/IAEA/WHO, 1993, 1999).

The scientific community has long agreed that gamma irradiation of food is safe, but despite a number of recent reports showing growing interest in food irradiation from government regulators and industry and declining public apprehension, the general public continues to be wary of this technique (Bender, 1986; Feenstra and Scholten, 1991; Fellows, 2001; Ghosh, 1989). Any technology for processing food can only be employed if it outperforms alternatives in terms of efficiency, public health, cost, convenience, and so on.

The deliberate exposure of food to ionizing radiation is referred to as "food irradiation". Irradiation of food is not a recent innovation. Two British scientists filed the first patent for food preservation in 1905. The biomedical literature from the 1920s and 1930s has a large number of studies on the impact of X-rays on food and food components. When irradiation was deemed a way to provide safe food and man-made radioisotopes were made available in the 1940s and 1950s, research in this field was intensified (Goldblith, 1966; Hasegawa and Moy, 1973; Hoppe *et al.*, 1983; IAEA, 2006; IFT, 1983; Josephson, 1983). Food irradiation was authorized in the United States in 1963 for the control of insects in wheat and wheat flour (Kadar, 1986) and in the European Economic Community in 1987 for the irradiation of some specific foods. More than 40 nations have now approved the consumption of more than 100 food items that have undergone food irradiation (Barbosa-Canovas *et al.*, 1998).

Radiation

Any energy that moves through space as waves or particles is known as radiation. The term "radiation" encompasses the entire electromagnetic spectrum as well as all high-energy atomic and subatomic particles. Depending on its energy, it can be categorized as ionizing and non-ionizing radiation.

Compared to non-ionizing energy, ionizing energy has shorter wavelengths but a higher frequency and energy. Infrared, radio, microwave and radio waves have enough energy to cause molecular vibrations and excitations but not ionization. Far UV rays X-rays and γ-rays, in contrast, have higher energies that can eject electrons from atoms, resulting in the ionization of molecules. Additionally, ionizing radiation can disrupt the chemical connections that hold molecules together, altering how cells normally function.

Electromagnetic radiation having a wavelength of at least 1.0 nm is referred to as non-ionizing radiation. The rest of the electromagnetic spectrum and all high-energy atomic and subatomic particles fall under the category of ionizing radiation.

Action of Ionizing Radiation

The ability of radiation to ionize the atoms and molecules of the substance it passes through is referred to as ionization. Depending on the nature of the matter and the radiation's properties, ionizing radiation penetrates matter to varied degrees. The penetration strengths of all atomic and subatomic particle types are inferior to those of gamma and X-rays. However, the ability of radiation to remove electrons from the atoms of the substance exposed to it determines how effectively it produces radiation effects. Gamma rays typically have a greater ability than beta particles to ionize materials. Positive and negative ions are produced when ionizing radiation interacts with the atoms and molecules of the material being irradiated by transferring energy to electrons (Kilcast, 1995).

There are direct and indirect actions that radiation can have on biological materials. In direct action, chemical reactions take place as a direct result of the radiation's energy being transferred to the target molecule; however, in indirect action, intermediates like the hydroxyl radical (OH-), hydrated electron (e aq), hydrogen atom, or hydrogen peroxide formed from the radiolysis of water are involved in the energy transfer and consequently the occurrence of chemical reactions (Loaharanu, 1995; Loaharanu and Murrell, 1994).

The hydrogen radical is a strong reducing agent, whereas hydrogen peroxide and the OH- radical are strong oxidizing agents. The mechanisms by which microorganisms, enzymes and food components are altered in irradiated material include the formation of ion pairs, free radicals that are components of molecules, groups of atoms, or single atoms that possess an unpaired electron, the reaction of free radicals with other molecules, the recombination of free radicals and related physical and chemical phenomena.

Sources of Ionizing Radiation

Food is irradiated using ionizing radiation of electromagnetic and particle origin. It includes X-rays produced by machine sources operating at or below 5 MeV, gamma rays produced by naturally occurring radioactive elements and artificially induced radioactive isotopes created in nuclear reactors, typically ^{60}Co, less frequently ^{137}Cs, and electrons produced by accelerators operating at or below a level of 10 MeV (Matsuyama and Umeda, 1983; Maxie and Abdel-Kader, 1966).

After the disintegration of the ^{60}Co isotope, two gamma rays with average energy of 1.17 and 1.33 MeV are generated. The electron and X-ray energy used in food irradiation applications are restricted to 10 and 5 MeV, respectively, in order to reduce the amount of nuclear reaction that could result in radioactivity in the product. Additionally below this threshold is the mean energy of gamma radiation from ^{60}Co (1.25 MeV).

A beam of high-energy photons can be created by the accelerated electrons interacting with a suitable target or by using the electrons directly. These photons are typically referred to as X-rays. Electromagnetic radiation includes gamma and X-rays, which are high-energy photons with the same physical characteristics as visible light but much higher energy. A machine can be turned off when not in use, but radioactive decay cannot be arrested, which is a key distinction between radioactive and machine sources. The different ionizing radiation causes the same chemical changes even though it can be produced using various pieces of equipment. The only real distinctions in the evaluation of radiation used in food processing are the radiation's penetration strength and, consequently, the size and density of the food product being irradiated (May, 1983; Mcmurray, 1990).

The majority of industrial gamma irradiation plants employ the radioisotope ^{60}Co. Another radioisotope source, though not currently used extensively, is nuclear fuel, which produces ^{137}Cs (Mitchell, 1994).

Doses Employed in Radiation Processing

Ionizing radiation's ability to cause biological effects is influenced by both the intensity of incident radiation and the quantity of radiant energy that is transferred to the unit mass or volume of the target material. In irradiation processing, the energy that the target material absorbs is referred to as the absorbed dose or simply dose. A gray (Gy) unit, which represents the energy absorption of one joule per kilogram, is used to measure dose. Kilograys are used to signify large doses (kGy). In the past, the rad (1 rad = 0.01 Gy) was also commonly used.

Depending on the chemical make-up and physical makeup of the food, irradiation can have a wide range of effects on the organoleptic quality of food. As a result, the amount of radiation given to a food relies on the resistance of the organisms present and the treatment's objective. The maximum dose advised for food processing is 15 kGy, with an average exposure of not more than 10 kGy (Monk *et al.*, 1995; Moseley, 1989; Moy, 1993).

Depending on the purpose of the processing and the dose used, food irradiation methods can be divided into different categories. The dose required to prevent sprouting differs by a factor of a thousand from the dose required to eradicate all bacteria. Potato, onion, and garlic sprouting is inhibited by very low doses of radiation (less than 0.1 kGy) (Murray, 1990; Niemira and Fan, 2006). Low doses (below 1 kGy) appear to give an adequate technical solution for the use of pesticides, which are now prohibited or limited in many countries (Potter and Hotchkiss, 1995). Low doses are effective for disinfestation and also extend the shelf life by delaying ripening. At these radiation dosages, microbial and insect contamination of fruits and vegetables which are very prone to infestation can also be decreased. Low dosage irradiation can be used to postpone the deterioration of fresh produce (Proctor and Goldblith, 1951).

If they are exposed to low amounts of radiation, mould growth on fruits like strawberries and the ripening of fruits and vegetables can be prevented. By destroying or lowering populations of microorganisms in foods at still higher levels (medium doses) using irradiation, food-borne illnesses and food deterioration can be greatly reduced. These doses are referred to as pasteurizing doses.

Highly perishable meats and fish can delay food spoilage by being pasteurized using radiation and refrigeration. One of the most significant uses of irradiation is anticipated to be the elimination or reduction of microorganisms that cause food poisoning. High radiation dosages are used for sterilization, also known as radappertization (25–50 kGy). It is made to get rid of the majority of germs in food, if not all of them. The outcomes are comparable to those obtained when low acid food is canned.

Scope of Irradiation

Food preservation is essential to life. Traditional preservation techniques include salting, smoking, and drying meat, fish, fruits, and herbs. They also include protective storage of grains or seeds in jars, granaries, or sloes (vegetables, such as whole green beans, olives, some meats and fish) (Proctor and Goldblith, 1951). In addition to the conventional preservation procedures, there are now techniques for heat sterilization, refrigeration, controlled atmospheric storage of fruits, freeze-drying and vacuum packaging. At the start of the 20th century, food preservation through irradiation was added to this list of methods. The potential applications of irradiation can be summarised as: control of insect pests, reduction in microbial contamination, prolonging shelf-life and improvement of quality of product (Rahman, 1999; Resureccion *et al.*, 1995).

1. Control of Insect Pests

Tropical fruits and grains may contain insect and larval infestations. They need to be managed. Typically, chemicals are employed for this purpose. Irradiation can replace chemicals like ethylene dibromide and methyl bromide, which are used for disinfestations, because low doses, below 3 kGy, are effective in destroying insects, at various stages of development, which may be present on the food. It has been noted that irradiation can harm an insect's capacity to reproduce or grow to adulthood (Resureccion *et al.*, 1995).

The insects are considered to be killed by doses in the range of 1-3 kGy as currently requested by law. The most potential use of food irradiation processing is the radiation-based prevention of pest infestation. It is commonly employed as an efficient fruit fly quarantine treatment.

2. Reduction in Microbial Contamination

The main advantages of sterilization by radiation are lowering the microbial contamination and reducing the common health risks brought on by foodborne illnesses. In terms of biological function, nucleic acids and membrane lipids are the main targets of gamma rays. By altering the structure of nucleic acids and cell membranes and preventing the activity of metabolic enzymes and cell division, the reactive intermediates created after irradiating food harm or kill bacteria. The most prevalent and problematic pathogenic bacteria are susceptible to irradiation and can be effectively rendered inactive by doses as little as 10 kGy.

3. Prolonging Shelf-life

Some foods, like potatoes, yams, and garlic, can have their shelf lives prolonged by applying relatively low radiation doses to prevent sprouting (0.05–0.15 kGy). Another technique used to increase shelf life is to delay the ripening and senescence of specific fruits. By irradiating some tropical fruits at 0.12-0.75 kGy, this approach is widely used to extend the shelf life of fruits such bananas, litchis, avocados, papayas, and mangoes (Rohlman *et al.*, 1994). When strawberries and tomatoes are irradiated at low radiation doses, their shelf life can be increased by around two to three times. However, these items should be mature before irradiation since irradiation inhibits hormone synthesis and disrupts the biochemical processes of cell division and growth, which prevent fruits and vegetables from ripening and maturing (Sahasrabudhe, 1990).

4. Improvement of Quality of Product

One of the most contentious issues surrounding radiation-processed food is the nutritional value of irradiated foods. Irradiation alters the nutrient makeup of the treated food, just like other methods of food processing. The main variables influencing the nature and scope of these changes are composition, radiation dose, processing temperature, air presence or absence, storage conditions, packaging, and so on. Irradiated foods have similar levels of nutrient loss to those from cooking, freezing, and even simple storage. The vitamins that are most vulnerable to irradiation are thiamine, ascorbic acid, and vitamins A and E. It has been demonstrated that irradiating fruits at a dose level of few kGy improves product recovery by increasing juice production.

Advantages and Limitations of Irradiation

Irradiating food has advantages that are obvious as well as drawbacks, much like any other food preservation technique. It is neither as fantastic as some supporters assert nor as awful as detractors allege. Five major advantages of irradiation processing are reduced food waste, increased global trade, improvement in public health, a substitute to fumigation and greater energy savings (Thayer and Rajkowski, 1999).

The postharvest losses brought on by insect infestation in grains, pulses and fruits can be managed and reduced with the help of irradiation's advances in disinfestation and shelf-life extension. For underdeveloped nations where postharvest losses are significant, this argument is more crucial (Urbain, 1983).

According to one estimate, food irradiation might prevent 30 per cent of all food storage losses. Food might be transported without suffering significant losses from infestations from the region where it is produced to the location where it is needed (Hasegawa and Moy, 1973).

Numerous foods include parasites or pathogenic bacteria. Food irradiation effectively rids food of dangerous bacteria, yeasts, moulds and insects while also reducing microorganisms that cause foodborne deterioration. This disinfection can enhance the food's hygienic quality, eliminate any health risks and significantly improve public health problems.

In most subtropical nations, losses from fresh food production due to pest infestation, spoilage or short shelf life range from 10 to 30 per cent, and in certain circumstances even up to 50 per cent (Vas, 1979). This hampers the growth of the global market for fresh foods grown in tropical and subtropical regions. By offering an efficient quarantine technique for infected or infested foods or by extending shelf life, irradiation can be utilized to promote or improve international trade of fresh food (Potter and Hotchkiss, 1995; Proctor and Goldblith, 1951; Resureccion *et al.*, 1995; Webb and Lang, 1988).

In contrast to fumigants like ethylene dibromide (EDB), methyl bromide (MB) and ethylene oxide, which are either forbidden or subject to increasing restrictions in the majority of developed nations due to concerns about public health, the environment, or workplace safety, irradiation within the dose range of 0.2–0.7 kGy has been shown to be an effective substitute. Many foods and food products may come into less contact with substances intended to stop the growth of invading germs after being exposed to radiation. The usage of food additives, fumigants, and preservatives, which pose risks to both consumers and workers in the processing industry, can be completely replaced by irradiation or greatly reduced (Barbosa-Canovas *et al.*, 1998; WHO, 1977).

Compared to the energy needed for canning, refrigeration or frozen storage, the energy required for irradiation preservation of food is minimal. Wherever possible, irradiation in conjunction with refrigeration can be utilized as a replacement for frozen storage for the distribution of raw food with a significant savings in energy and cost.

Even though the irradiation process uses little energy and has minimal running costs, it has substantial startup costs and needs a certain minimum capacity and output to be profitable (WHO, 1981, 1994). There is no preservation technique that can be employed for all foods, just as the food irradiation process is not appropriate for all foods. Above the threshold radiation levels advised for treating food, organoleptic changes and the development of off flavours occur and removal of all microorganisms or their toxins is not feasible at low doses. When the food is exposed to radiation when it is frozen, some of the organoleptic changes can be avoided (Willemot *et al.*, 1996).

It is challenging to standardise the radiation processing due to the variability of the effects. Vitamin losses are another frequent point of criticism for irradiation. In actuality, compared to other heating procedures, irradiation processing often results in smaller and less significant nutritional losses.[53] Lack of enzyme inhibition in foodstuffs, even at the high levels of radiation doses needed for commercial sterilization, is a glaring drawback of irradiation in comparison to thermal processing. Utilizing irradiation in conjunction with other food preservation techniques is one option to get around these drawbacks and restrictions. When used in conjunction with other methods, irradiation has its highest potential (Feenstra and Scholten, 1991). The only preservation technique that is capable of inactivating dangerous germs like Salmonella in frozen foods is irradiation. It can be used on foods in tiny packages or in bulk, in a frozen condition or at room temperature (Vas, 1979; Willoneti *et al.*, 1996).

Irradiating food has its name as its worst drawback. Unpleasant connections with radioactivity, nuclear threats, advanced technology, genetic mutation, and cancer are brought on by ionizing radiation (Barbosa-Canovas *et al.*, 1998; Willoneti *et al.*, 1996).

Future Prospects of Irradiation in Fruits and Vegetables

Reducing agricultural waste after harvest is a significant issue in most nations, particularly in emerging and less developed nations. It is well recognized that a significant portion of the food produced around the world, including fruits and vegetables, is wasted due to bacterial, mould, insect and other pest damage. Irradiation offers a long-term, eco friendly and less energy-intensive approach to control insect pests, organisms that cause spoilage and physiological processes, notably in the case of perishable crops.

Food irradiation is taking centre stage as a preservation method alongside other technologies following decades of study, development and discussion in the media. As a recognized sanitary, phytosanitary and preservation procedure, irradiation of fruits and vegetables is emerging to play a significant role in enhancing food safety and security and promoting expanded trade. The near future will see this job grow in significance. However, it is frequently difficult to standardize the irradiation treatment due to the heterogeneity in effects. For a given commodity and cultivar, a successful and reliable treatment requires the control of a wide range of variables, including radiation dose, dose rate, degree of maturity, physiological status of fruits, temperature and atmosphere during and after treatment, pre- and postharvest treatments, and handling susceptibility of the microorganisms to the radiation used. In order to implement a therapy that is efficient and effective, more research will be required, particularly in the case of fresh fruits and vegetables. On the contrary, in order to solve a technical or food trade issue, a nation's infrastructure must be present and appropriately developed.

The commercial viability and economic benefit of irradiation as a postharvest treatment for particular commodities of tropical origin have been demonstrated by a number of fruitful pilot-scale storage studies and market trials conducted in various nations across the world. The successful outcomes of these studies inspired some nations to build industrial facilities for the care of bulb and tuber crops. If scientific advancements and the trend toward increased regulatory acceptance of fruit and vegetable irradiation continue, it is anticipated that the usage of this technique will rise even higher and play a significant part in future international food exports and imports. To accomplish these objectives, additional international co-trials and discussions are required to address the challenges of developing consistent regulations and laws relevant to the trade in irradiated products.

The prohibition of numerous fumigants that were previously used for disinfestation is an additional aspect encouraging the application of irradiation to the preservation of fruits and vegetables. Radiation will increasingly be used to decontaminate food when fumigant use is reduced and banned. After the ban, several farmers attempted to utilise the hot water dip method to rid mangoes and papayas of pests, however it was found that this technique led to physiological problems with the fruits, most notably browning and discoloration of the pulp. Irradiation's ability to extend storage life makes it especially beneficial for fruits, which are frequently infested and need a long shelf life to reach consumer markets in satisfactory quality.

The development of straightforward and accurate detection techniques, the harmonisation of legislation, the commitment of the food industry, and consumer attitudes are thought to be the main elements influencing the future of food irradiation. Therefore, a standardised method of detecting irradiated foods and irradiation dosages, consumer education and consultation, and well-defined standards for irradiated food at national and international levels will be the key components for future advancements in the field of food irradiation, particularly in the field of fruit and vegetable irradiation.

CONCLUSION

Fresh fruit and vegetable quality and safety can be enhanced and maintained using low-dose irradiation. Higher doses may result in worse quality effects. Irradiation has advantages over other methods now in use, notably as it preserves the fresh product's natural rawness. It is more environmentally friendly than other technologies and can even be used on the finished product. Despite these benefits, low dose irradiation as a postharvest method is underutilized mostly because of labeling restrictions and a lack of knowledge about how ionizing energy functions. Its application as a phytosanitary treatment has grown dramatically in the last ten years. Low-irradiation dosage effects on fresh plant foods are quite diverse and depends on a wide range of elements such as variety, maturity, packing, and storage and distribution conditions. Research must be done to comprehend the mechanisms by which irradiation modifies the physicochemical properties of plant meals at permitted dosages in order to guarantee the highest quality.

REFERENCES

Acinf (1996). Report on the safety and wholesomeness of irradiated foods, London, Department of Health and Social Security, 52.

Akamine, E. A. and Moy, J. H. (1983). Delay of postharvest ripening and senescence of fruits, In Preservation of Food by Ionizing Radiation Vol. III, Josehson E. S. and Peterson M. S. (eds), Boca Raton, F. L., CRC Press, 129–58.

Anonymous (1987). 'Perspective on food irradiation'. *Food Technology*. 41, 100–1.

Barbosa-Canovas, G. V., Pothakamury, U. R., and Palou, E. (1998). Nonthermal Preservation of Foods. Marcel Dekker, New York.

Bender, A. E. (1986), 'Food irradiation', *Journal* of the *Royal Society* of *Health,* 3, 80–1.

Bord, R. (1991). Consumer acceptance of irradiated foods in the United States, In Food Irradiation, Thorne S. (ed), New York, *Elsevier Applied Science*. 61-86.

Cast (1989). 'Ionizing energy in food processing and pest control II', Applications Report No. 115, Ames, Iowa.

Crawford, L. M. and Ruffeh (1996). A review of the safety of cold pasteurization through irradiation. *Food Control*, 7, 87.

Defesche, F. (1983). 'Consumer attitudes towards irradiation of food and some suggestions on how to achieve acceptance of irradiated food', In Marketing and Consumer Acceptance of Irradiated Foods, Joint FAO/IAEA, Vienna, IAEA, 47-54.

Diehl J. F. (1993). Will irradiation enhance or reduce food safety. *Food Policy*, 18, 143–51.

Diehl, J. F. (1990). Safety of Irradiated Foods, New York, Marcel Dekker, 217-53.

Ehlermann, D. A. E. (1983). 'Future prospects for irradiation processing of food', In Recent Advances in Food Irradiation, Elias P. S. and Cohen A. J. (eds), Amsterdam, Elsevier Biomedical Press, 331-54.

Elias, P. S. (1987). Food irradiation, In Toxicological Aspects of Food. Miller K. (ed), New York, *Elsevier Applied Science*. 295–346.

FAO/IAEA/WHO (1993). International Symposium on Cost/Benefit Aspects of Food Irradiation Processing, Aix-en-Provence, Report of the working group.

FAO/IAEA/WHO (1999). High-dose irradiation: wholesomeness of food irradiated with dose above 10 kGy, WHO, Geneva, Technical Report Series, 890.

Feenstra, M. H. and Scholten, A. H. (1991). 'Consumer acceptance of irradiated foods', In Food Irradiation, Thorne S (ed), New York, *Elsevier Applied Science*, 97-128.

Fellows, P. J. (2001). Food Processing Technology, Second edition, Chambridge, CRC Press, Woodhead Publishing, 199.

Ghosh P. (1989). A taste for irradiated food, New Scientist, 123, 40–1.

Goldblith, S. A. (1966). 'Historical development of food irradiation', In Food Irradiation. Proceedings of the International Symposium on Food Irradiation, STI/PUB 127, Vienna, IAEA, 3-21.

Hasegawa, Y. and Moy, J. H. (1973). Reducing oligosaccharides in soybeans by gamma-radiation-controlled germination. Joint FAO/IAEA Proc. Symp. Radiation Preservation of Foods, STI/PUB/317, 89–103.

Hoppe, W., Lohmann, W., Markl, H. and Zeiger, H. (1983), Biophysics, Berlin, Germany, Springer-Verlag.

IAEA. (2006). Use of Irradiation to Ensure the Hygienic Quality of Fresh, Pre-Cut Fruits and Vegetables and Other Minimally Processed Food of Plant Origin. Proceedings of a final research coordination meeting organized by the Joint FAO/IAEA Programme of Nuclear Techniques in Food and Agriculture and held in Islamabad, Pakistan.

IFT (1983). Radiation preservation of foods. Expert panel on food safety and nutrition, Food Technology, 37-55.

Josephson, E. S. (1983). An historical review of food irradiation. Journal of Food Safety. 5, 161-90.

Kadar, A. A. (1986). Potential applications of ionizing radiation in postharvest handling of fresh fruits and vegetables. *Food Technology*, 40, 117–21.

Kilcast, D. (1995). Food Irradiation: current problems and future potentials. *International Biodeterioration* and *Biodegradation*. 36, 279.

Loaharanu, P. (1995). Food irradiation: current status and future prospects, In New Methods of Food Preservation. Gould G. W. (ed), Glasgow, Blackie Academic and Professional. 90–109.

Loaharanu, P. and Murrell, D. (1994), 'A role for irradiation in the control of food borne parasites', Trends Food Science Nutrition, 5, 190-5.

Matsuyama A. and Umeda K. (1983). Sprout inhibition in tubers and bulbs. In Preservation of Food by Ionizing Radiation Vol. III, Josehson E. S. and Peterson M. S. (eds), Boca Raton, FL, CRC Press, 159–213.

Maxie, E. C. and Abdel-Kader, A. S. (1966). Food irradiation-physiology of fruits as related to the feasibility of the technology. *Advance Food Research*. 15, 105-45.

May, J. H. (1983). 'Radurization and radicidation: Fruit and vegetables', Preservation Of Food by Ionizing Radiation Vol. III, Josephson E. S. and Peterson M. S. (eds), Boca Raton, FL, CRC Press, 83–103.

Mcmurray, C. H. (1990). Food irradiation: the challenge', In Food Irradiation and Chemist, Johnston D. E. and Stevenson M. H. (eds), Cambridge, UK, Royal Society of Chemistry, 1-12.

Mitchell, G. E. (1994). 'Irradiation preservation of meats', *Food Australia*, 46, 512–17.

Monk, J. D., Beuchat, L. R. and Doyle, M. P. (1995). Irradiation inactivation of food borne microorganisms. *Journal on Food Protection*. 58, 197-208.

Moseley B. E. B. (1989). 'Ionizing radiation: action and repair', In Mechanisms of Action of Food Preservation Procedures, Gould G. W. (ed), London, UK, *Elsevier Applied Science*, 43.

Moy, J. H. (1993). Food irradiation-lessons and prospects for world food preservation and trade, In Development of Food Science and Technology in South East Asia, Liang O. B., Buchanan A. and Fardiaz D. (eds), Bogar, IPB Press, 86.

Murray, D. R. (1990). Biology of Food Irradiation, Taunto, GB, Research Studies Press, 25-32.

Niemira, B. A. and Fan, X. (2006). Low-dose irradiation of fresh-cut produce: Safety, sensory, and shelf life. *Food irradiation research and technology*. Editors: Sommers, H. S. and Fan, X. Blackwell Publishing, U.S.A. pp.169-184.

Potter, N. N. and Hotchkiss, J. H. (1995), Food Sciences, fifth edition, London, UK, Chapman and Hall, 245.

Proctor, B. E. and Goldblith S. A. (1951). Food processing with ionizing radiation. *Food Technology*. 5, 376-80.

Rahman, M. S. (1999). Irradiation preservation of foods. In Handbook of Food Preservation, New York, Marcel Dekker, Chap. 13, 397.

Resureccion, A. V. A, Galvez, F. C. F., Fletcher, S. M. and Misra, S. K. (1995). 'Consumer attitudes toward irradiated food: results of a new study'. *Journal of Food Protection*. 58, 193-6.

Rohlman, A. J., Wood O. B. and Mason A. C. (1994). 'Influence of audiovisuals and food samples on consumer acceptance of food irradiation'. *Food Technology*. 48, 46-9.

Sahasrabudhe, M. R. (1990). Food irradiation: current status, concerns, limitations and future propects. *Journal of the Canadian Dietetic Association*, 51, 329–34.

Thayer, D. W. and Rajkowski, K. T. (1999). Developments in irradiation of fresh fruits and vegetables. *Food Technology*. 53(11): 62-65.

Urbain W. M. (1983). Food Irradiation, New York, Academic Press.

Vas K. (1979). Food preservation by irradiation. In Developments in Food Science II, Chiba H, Fujimaki M., Iwai K., Mitsuda H. and Morita Y. (eds). Proceeding of a congress, Tokyo, Kodansha Ltd., 205–11.

Webb, T. and Lang, T. (1988). 'Food irradiation', In Irradiated Foods, Darrington H., London, UK, Morgan-Grampian, 40-54.

WHO (1977). 'Wholesomeness of Irradiated Food', Report of the Joint FAO/IAEA/WHO Expert Committee, WHO Technical Report Series No. 604, HMSO, London, UK.

WHO (1981), 'Wholesomeness of Irradiated Food', Report of the Joint FAO/IAEA/WHO Expert Committee, WHO Technical Report Series No. 659, HMSO, London, UK.

WHO (1994),'Review of the safety and nutritional adequacy of irradiated food', Report of a WHO Consultation, WHO, Geneva.

Willemot, C., Marcotte, M. and Deshenes, L. (1996). 'Ionizing radiation for preservation of fruits', *Processing Fruits: science and technology*, Somogy L. P., Ramaswany H. S. and Hui Y. H. (eds), Vol. 1, Lancaster, Technomic Publishing, 221-60.

Willoneti C., Marcotte M. and Deshenes L. (1996). Ionizing radiation for preservation of fruits. In Processing Fruits: Science and Technology Vol. I., Somogyi L. P., Ramaswamy H. S. and Hui Y. H. (eds), Lancaster P. A., Technomic Publishing, 221–69.

Chapter 12

Microgreens: A Unique Food Product

Vibhu Pandey[1], Vyankatesh Bagul[2], Anuj Pal[3] and Kaushlendra Pratap Singh[3]

[1]Ph.D. Research Scholar, Department of Fruit Science,

[2]Ph.D. Research Scholar, Department of Agricultural Biotechnology,

[3]Ph.D. Research Scholar, Department of Horticulture,

Sardar Vallabhbhai Patel University of Agriculture and Technology, Meerut – 250 110, Uttar Pradesh

**e-mail: Pandey.vibhu007@gmail.com*

A "microgreen" is a term that refers to an edible plant in its early stages of development. A microgreen is distinguished from a "sprout" by the fact that it has been harvested (removed) from the plant's seed and root system before to consumption. Microgreens, frequently called 'vegetable confetti', are a new class of speciality crop, defined as tender immature greens produced from the seeds of vegetables, herbs, or grains, including wild species **(Xiao, Lester, Luo, and Wang, 2012).** A microgreen is typically comprised of the plant stem, cotyledons (first leaves contained in the seed), and the first "true leaves" of that plant, depending on the plant species. Depending on species and growing conditions, microgreens are generally harvested at the soil level, *i.e.* at the base of hypocotyls, upon appearance of the first pair of true leaves, when cotyledons are fully expanded and still turgid, usually within 7-21 days from seed germination depending on the species **(Sun *et al.*, 2013).** Microgreens can be harvested before or directly after the first true leaves have appeared, depending on the grower's and customer's preferences. Microgreens, which range in size from 5 to 7.5 cm, are prized for a variety of qualities, including nutrition, flavour, colour, form and texture. Microgreens have much stronger

Figure 12.1

flavor enhancing properties than sprouts, and a broad range of leaf color, variety and shape **(Ebert, 2012).** Microgreens have received a lot of attention in recent years due to their potential profitability, which is based on a quick production cycle, low input costs, and a small footprint (do not take large amounts of space or resources to produce). However, profitability is ultimately determined by the capacity to sell such a product, thus the market you intend to sell into should be carefully considered. Producing the product, as with other crops, is one thing, but selling it is quite another. To ensure that their crops can be sold, producers must establish relationships with target markets and clients early in the company development process.

Varieties

Many microgreen varieties can be cultivated, but the most important consideration for growers is that the plant is appropriate and safe for human consumption. Over a hundred distinct plant species are routinely grown and sold as microgreens in the context of this essential factor (*e.g.* mustard). The hundreds of cultivar variants of these plants are not included in this statistic (*e.g.* Golden Frills mustard, Ruby Streaks mustard, or Green Wave Mustard).

Although microgreens have a short crop cycle, different species and types grow at different rates. Vegetables produced as microgreens are ready to harvest in 7 to 14 days, whereas herbs cultivated as microgreens take longer to mature, taking 16 to 25 days on average.

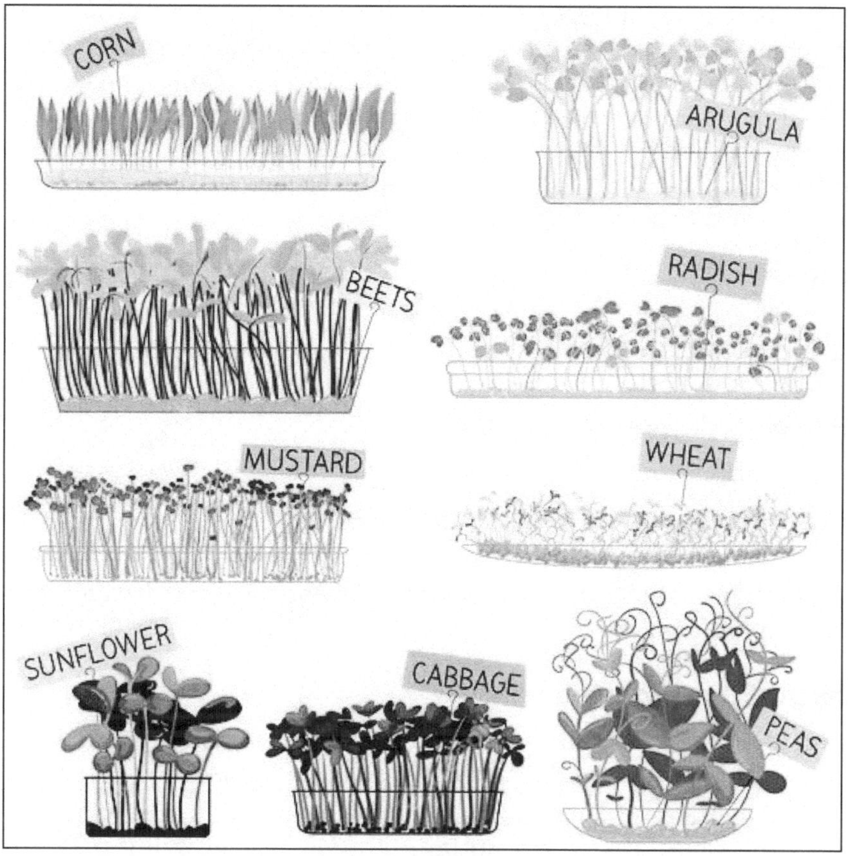

Figure 12.2

See the list below with common species grouped by their crop cycles.

Common Species

- ☆ Fast growing vegetables (7 to 14 days)
  Cabbage, Corn, Kale, Mustard, Radish
- ☆ Slow growing vegetables (15 to 25 days)
  Amaranth, Beet, Carrot, Swiss chard
- ☆ Slow growing herbs (15 to 30 days)
  Anise, Basil, Dill, Fennel, Parsley, Saltwort, Sorrel

Methods of Production

Seeds are demanded in large quantity and represent a major cost for the production of quality microgreens **(Di Gioia, Mininni, and Santamaria, 2015).** Preliminary germination test per seed lot is advisable for adjusting sowing rate **(Di Gioia *et al.*, 2015)**. Many species germinate easily and grow promptly while others are slow and may require pre-sowing treatments to improve, standardize and shorten the production cycle **(Lee, Pill, Cobb, and Olszewski, 2004)**. Growers must determine how they want to harvest their crops. Each kind of production has its own set of advantages, challenges and maintenance requirements. On a commercial basis, the following methods of manufacturing have been used to generate microgreens:

Soil-based

- Garden beds
- Raised planters

Soil-less/Hydroponics

- Various forms of hydroponic crop production.
    - Substrate culture where the roots grow within a pH neutral media.
    - Solution culture where the roots grow within nutrient solutions.
- The roots are grown in a continual flow of nutrient rich water (nutrient solution) in the

Nutrient Film Technique (NFT)

- Aeroponics is a method of growing plants in which the root systems are suspended and a nutritional solution is misted over them.
- Regardless of the facility used, the steps of production will be largely the same. It's worth noting, though, that growing microgreens necessitates a food-safe facility that complies with local requirements.
- Sources of seeds when compared to standard crop development, microgreens require a lot more seeds. Seed costs can contribute a significant amount to the overall cost of manufacturing and in some situations will determine whether a product is profitable or not. It's critical to find multiple, trustworthy, and cost-effective bulk seed suppliers. Having multiple main suppliers decreases the danger of shortages, limited diversity, food safety and traceability difficulties.
- A producer should seek out suppliers who can provide traceability, food safety assurance, and/or organic certifications for their products.
- Some seed species require soaking in water or in an acidic solution to germinate (scarification- used to break the hard seed coat of certain species). Because each species has varied germination requirements, it's vital that growers understand their species and how to propagate it.
- Soaking solutions, such as diluted hydrogen peroxide, can also be employed as disinfectants, reducing the danger of pathogen and bacterial infection.

Figure 12.3

- Similarly, disinfection may be aided by a hot water treatment. Some seeds, such as arugula, basil, chia and flax, fall into the mucilaginous seed category.
- When this type of seed comes into touch with water, a thick covering of a gluey/gel like substance forms on the seed's exterior.
- This layer aids in the germination of the seed by retaining moisture. To ensure optimal germination, extra steps and care may be required when growing these seeds.
- Before planting, mucilaginous seeds should not be moistened. Seeds are shaken or mechanically seeded, depending on the planting medium, to ensure that they are equally distributed at a density that corresponds to the type of seed and intended completed result.
- Growers apply a thin coating of fine grade vermiculite, single ply paper towel, or a thin layer of soilless growing media on top of the seeds to preserve moisture after planting.

Common seeding densities are shown here.

- 100 g to 120 g – Sunflower, Peas, Corn
- 60 g to 70 g – Mustard, Broccoli, Radish, Swiss chard, Red cabbage
- 50 g to 60 g – Dill, Basil, Arugula

Germination and Growth

In order to produce microgreens, a grower must master recreating the environment and necessities that seeds require to germinate. This method may include the use of heating mats, increased lighting or complete darkness and less or more watering, depending on the species.

Environment

Microgreens are produced in a variety of environments (open air, protected environment, indoor) and growing systems (soil, soilless), depending on the scale of production. Containerized production, adaptable both to micro-scale urban and large scale commercial operations, allows for commercialization of the product while growing on the media, to be harvested directly by the end user. This approach bypasses harvesting and many postharvest handling issues, and may ensure freshness and high quality **(Di Gioia *et al.*, 2015).** Although the ideal environment for microgreen development varies per variety, most plants will thrive in a temperature range of 18°C to 24°C and a relative humidity of 40 per cent to 60 per cent. The likelihood of insect and disease issues may increase when the temperature and humidity rise. Air circulation (mixing, exchanging and circulation) is essential in any growing environment. The use of air circulation helps to maintain a consistent temperature and humidity throughout the growth area. To mix and exchange air within the growing space, growers must employ horizontal air flow fans, as well as forced air or natural air vents. If plants are grown in a vertical rack system, air movement between layers must be monitored; otherwise, air trapped between layers can soon become stagnant, increasing disease risk. Multiple smaller fans fastened at each level of expansion on the rack systems can govern air flow for these systems.

Lighting

Many plants react differently to artificial light than they do to natural light. Keep an eye on the length of the light, its intensity and its distance from the crop. Stretching or "legginess" may occur if the light source is too far away from the crop, for example; when the light intensity is insufficient, the crop colour may appear washed out. Some light systems, such as fluorescent lights, produce heat while in use. When a facility contains more than 150 fluorescent lights, this effect can have a significant impact on the air temperature. Because of the lower heat emissions and flexibility to adapt the light spectrum to certain crops in production, light emitting diodes (LED) light fixtures have become more common in enclosed production facilities (such as warehouses) in recent years.

Optimal management of light intensity may enhance photosynthetic activity and phytochemical content in vegetables, whereas excessive irradiance can provoke photo damage with detrimental effects on plant growth and product quality **(Bian *et al.*, 2015).** The effects of five LED irradiation levels (545, 440, 330, 220 and 110 μmol m-2 s-1 323) on nutritional quality of Brassica microgreens (kohlrabi, mustard, red pak choi and tatsoi) were investigated by **Samuolienë *et al.* (2013)**

In a greenhouse environment, using lights for 12 to 18 hours per day to replace natural sunlight up to 4 to 6 hours per day for supplemental lighting will significantly raise the cost of microgreens production. LED lighting systems have been proven to use significantly less energy than their fluorescent and halogen counterparts. Despite significant cost reductions in recent years, the initial cost of LED lighting systems might still be prohibitively high. Growers and potential growers should have a clear awareness of the estimated capital and operating costs of cultivating microgreens before producing to ensure a profitable enterprise.

Watering

Plant irrigation must be consistent and calculated when growing microgreens. To avoid displacing the seeds, misting nozzles should be used for the initial seed watering during germination. Shower head nozzles can be used to meet the increasing water demands of the plants during the growth stage of watering. To deal with any extra water from crop watering, effective floor drainage should be in place regardless of the facility. Standing water on the floor can serve as a breeding ground for insects and diseases, as well as a possible concern to workers' safety.

Harvesting

You may be needed to get a food handling permit and operate within an approved food facility if you sell your microgreens as living trays of plants or harvest them yourself. Prior to manufacturing, double-check your local and provincial legislation and standards. Microgreens come in a variety of finished product forms, but the most prevalent are as follows:

Fresh cut produce/harvested microgreens

- ☆ Living tray microgreens/live plant sales
- ☆ Mixed microgreen combinations
- ☆ Value-added/processed products that include microgreens

They require careful, often tedious harvesting, and quick cooling to remove vital heat and suppress the rate of respiration, spoilage and senescence. Harvesting microgreens is labor intensive and can have a direct impact on the cost of production, especially when production is implemented in trays that require harvesting with scissors. Use of loose substrates in trays slows down the harvesting process, whereas seeding on synthetic fiber, food-grade plastic or burlap-type mats can facilitate easier handling, and faster harvesting and cooling of the product **(Treadwell *et al.*, 2010).** Microgreen harvesting is a crucial phase in the production process, which can take a long time and need a lot of effort. Harvesting these "micro" plants can be challenging due to their small size, and should be done with caution. To ensure postharvest freshness, crop quality, and appropriate food safety measures, refrigeration is required and recommended as soon as possible. Microgreens are typically cut as close to the stem base as possible without contaminating the final product with growing medium or seeds. Keep in mind that the simpler your harvesting tools are, the easier it will be to sanitize them properly. The method of microgreens cutting varies by the scale of production, but the following methods are common:

- Pair of scissors
- Electric knife
- Very sharp chef's knife
- Hand-held harvesting mechanism
- Automated harvesting conveyor system

You can trim the crop right into the final packaging, although most microgreens that are collected are additionally washed before being packaged. If you're washing the microgreens, make sure you're transferring them to clean storage containers.

Washing and Drying

Washing the microgreens is a very important step that also adds value to your product. The procedure informs the buyer that your product has been thoroughly cleansed and is safe to eat. Furthermore, because microgreens are frequently consumed without further processing or heating, microgreen items leaving a facility should be presumed to be consumed as-is. Microgreens are already highly clean, although seeds, seed husks, root systems and soil medium may be found in the harvested product after harvesting. Multiple washings and draining of the microgreens will help to remove these pollutants. Washing the microgreens with cold water reduces the crop temperature, reducing the potential for microbial growth. To limit and lessen the possibility of microbial development and contamination, it is critical to thoroughly dry the microgreens. There should be no extra delays between harvesting, washing, drying, and packaging to maximize shelf life.

Packaging

Packaging might differ depending on the market and the target audience. Product quantity, pricing, and marketing decisions will all be aided by the packaging. The packaging should preserve the finished product from contamination and damage while also complying with product labelling regulations.

Here are the most common forms of packaging:

- Various sizes of clear plastic bags
- Small plastic cups with lids
- Various shapes and sizes of plastic clamshells

Growers often have differing opinions on what types of packaging maintain or extend microgreens shelf life. Proper storage and transportation conditions are the most essential variables in preserving predicted shelf life. Harvested microgreens should be kept as cold as possible (at or below 4°C) to limit the danger of microbial growth while preserving freshness. Microgreens should be packaged in shallow containers because they allow faster cooling and reduce the risk of pathogen contamination.

Labeling

To decrease the danger of misidentification and contamination, a proper labelling system should be in place throughout the production cycle. All food labels must conform with federal legislation for sales and marketing, regardless of where items are distributed, according to the Canadian Food Inspection Agency (CFIA). Additional labelling, such as lot identification (traceability codes), is required in the event of a product recall and also aids in effective stock rotation.

Specific core labelling requirements must be included on the microgreens label:

- Common name
- Allergens
- Net quantity
- Nutrition labelling (unless exempt)
- Dealer name and address
- Durable life date
- List of ingredients
- Bilingual labelling (unless exempt)

Storage and Shipping

- Microgreens should be kept as cool as possible after harvesting to avoid harming the product.
- A need should be to keep the product at room temperature for as little period as possible.
- To avoid microbial growth, microgreens should be stored in a clean, refrigerated environment at 4°C or less. In a temperature-controlled setting, a thermometer should be mounted and checked on a regular basis.
- Microgreens should also be kept in a cool, dry place to avoid spoilage. Stacking too many clamshell packets or bags of goods on top of each other might damage the product at the bottom of the stacks, increasing the danger of the final product deteriorating faster.

- While shelf life will vary from species to species and variety to variety, most microgreens will have a shelf life of approximately 5 to 10 days with proper handling and storage.
- The product's high level of care, attention, and correct storage must be transmitted to the distribution stages.
- To avoid contamination and microbial growth, microgreens must be carried in refrigerated rooms. Although it is responsibility of the end user to observe basic food safety guidelines, growers should instruct customers on how to handle and store the product when the opportunity arises.
- Using the best practices and proper crop management discussed in this factsheet can give growers the best results for their business goals.

REFERENCES

Bian, Z. H., Yang, Q. C. and Liu, W. K. (2015). Effects of light quality on the accumulation of phytochemicals in vegetables produced in controlled environments: a review. *Journal of the Science of Food and Agriculture*, 95, 869–877.

Di Gioia, F. and Santamaria, P. (2015). Microgreens: Microgreens: Novel fresh and functional food to explore all the value of biodiversity (pp 118). ECO-logica srl Bari, Italy.

Di Gioia, F., Mininni, C. and Santamaria, P. (2015). How to grow microgreens. In F. Di Gioia and P. Santamaria (Eds.), Microgreens: Microgreens: Novel fresh and functional food to explore all the value of biodiversity (pp. 51-79). ECO-logica srl Bari, Italy.

Lee, J. S., Pill, W. G., Cobb, B. B. and Olszewski, M. (2004). Seed treatments to advance greenhouse establishment of beet and chard microgreens. *Journal of Horticultural Science and Biotechnology*, 79(49), 565–570.

Samuolienë, G., Brazaitytë, A., Jankauskienë, J., Virðilë, A., Sirtautas, R. and Novièkovas, A. (2013). LED irradiance level affects growth and nutritional quality of Brassica microgreens. *Central European Journal of Biology*, 8(12), 1241-1249.

Sun, J., Xiao, Z., Lin, L.-Z., Lester, G. E., Wang, Q. and Harnly, J. M. (2013). Profiling polyphenols in five Brassica species microgreens by UHPLC-PDA-ESI/HRMSn. *Journal of Agricultural and Food Chemistry*, 61, 10960–10970.

Treadwell, D., Hochmuth, R., Landrum, L. and Laughlin, W. (2010). Microgreens: A new specialty crop. University of Florida, IFAS Extension, HS1164.

Xiao, Z., Lester, G.E., Luo, Y. and Wang, Q. (2012). Assessment of vitamin and carotenoid concentrations of emerging food products: Edible microgreens. *Journal of Agricultural and Food Chemistry*, 60, 7644-7651.

Chapter 13

Micropropagation in Horticultural Crops

Ravi Kumar[1] *, *Alamgir*[1], *Vishwajeet Yadav*[1], *Devendra Pal*[2] *and Rishabh Shukla*[2]

Ph.D. Research Scholar, Department of Agriculture Biotechnology,

Ph.D. Research Scholar, Department of Horticulture,

Sardar Vallabhbhai Patel University of Agriculture and Technology, Meerut – 250 110, Uttar Pradesh

**e-mail: ravikm4@gmail.com*

INTRODUCTION

Micropropagation, often known as clonal propagation, is a modern commercial use of plant tissue culture methods. The approach is chosen over conventional clonal propagation methods. Plant micropropagation is a typical horticultural procedure for preserving favorable traits of chosen germplasm or cultivars. Plants are propagated in vitro by cultivating stems, leaves, apical and axillary buds, immature seeds, embryos, and cotyledons. Because of its advantages as a multiplication technique, micropropagation has become an essential aspect of the industrial propagation of many plants (Iliev *et al.*, 2010). The most significant approach for propagating many commercial plants is the creation of micro-propagated plants in plant-tissue-culture facilities and nurseries. Micropropagation based on the tissue-culture technique entails large-scale propagation since it allows for the multiplication of a large number of true-to-type explants in a very short period and in a very restricted space, as well as all year round, regardless of climate Abdalla *et al.* (2022). Because stringent aseptic conditions must be maintained in the development of in vitro cultivated plants, cuttings and plants are cultured and transferred in germ-free laminar flow cabinets, and the containers in which plants are cultured are stored airtight. The laminar-flow cabinets are intended to generate sterile air. Instruments needed for cutting and transferring plant parts, such as scalpels, scissors, and forceps, are sterilized using an alcohol dip and flame or dry heat. Similarly, the media are sterilized. The purpose of this chapter is to describe and highlight current improvements in clonal propagation technologies for horticulture crop enhancement.

Gottlieb Haberlandt (1902) employed totipotency and individual cell culture in one of the earliest studies on the subject of plant tissue culture. This was followed by further attempts to cultivate isolated root tips in an aseptic environment. In 1934, Philip White discovered that removed root tips grown in liquid culture conditions with inorganic salts, sugar, and yeast extract could develop indefinitely. Tissue culture was seen as a viable alternative strategy for quick clonal plant replication in the early 1960s when totipotency of plant cells was established and reproduction of many plants from tiny somatic tissue parts was made possible (Murashige 1974).

Definitions

The process of micropropagation "ensures true-to-type, rapid, and mass propagation of cloned copies of individual plants in a short time and lack of space, resulting in regeneration of old varieties and quick re-generation of different varieties resistant to biotic and abiotic stresses"

"Micropropagation is a technique that can produce a lot of clonal explants in a short period of time".

In vitro clonal multiplication of significant horticultural, silvicultural, and medicinal plants from tiny plant parts is achieved via micropropagation, one of the fastest in vitro commercial uses of plant tissue culture. Additionally, the rate of multiplying is frequently much higher in clonal propagation, and the micro-propagated plants may develop more desirable new qualities, such as an increase in the number of runners in strawberries and a bushier habit in ornamental plants (Kane 2011; Bhojwani and Dantu 2013a).

Important Developments in the History of Plant Tissue Culture and Micropropagation

a)	Father of horticulture/Totipotency	G. Haberlandt (1902)
b)	First attempt embryo culture	Hanning (1904)
c)	Develop root culture in Tomato	White (1934)
d)	*In-vitro* adventitious root formation in tobacco	Skoog (1944)
e)	Single-cell culture	Muir *et al.* (1953)
f)	Regulation of organogenesis by auxin and cytokinin ratio	Skoog and Miller (1957)

g)	Discovery of somatic embryogenesis	Steward *et al.* (1958)
h)	Clonal propagation (Production of virus-free plants)	Morel and Martin (1960)
i)	Regeneration of the first plant from a protoplast	Takebe *et al.* (1971)
j)	First somatic hybrid	Carlson *et al.* (1972)

Sources: Instant biotechnology, A competitive approach.

Basic Micropropagation Method Steps

Micropropagation is a step-by-step procedure. The four stages involved in the intricate process of clonal propagation are outlined. Murashige identified the four steps of overall micropropagation in 1978. Stages I and III were studied in vitro, and stage IV was studied in the greenhouse. Debergh and Mane introduced Stage 0 into the micropropagation system in 1981.

Step 0: Mother Plant Selection and Maintenance

The initial stage in micropropagation is to identify mother plants with appropriate properties for large-scale duplication. After choosing the mother plants, they are kept in a controlled environment with low humidity, watering, and no systemic microbial infection for three months to allow culture initiation.

Step I: Culture Development and Establishment

Explants chosen in Phase 0 are prepared for inoculation in this step after being sterilized and grown on a good culture medium. Surface sterilization of the axillary bud, shoot tip, and meristem is accomplished with different chemicals such as 5 per cent sodium hypochlorite, 0.1 per cent mercuric chloride, or 70 per cent alcohol using varying contact durations in combination or alone depending on the extent of surface contamination. Following surface sterilization, the plantlets are inoculated onto MS medium (or any other acceptable basal medium) with different growth regulators, vitamins, and sugar. A growing medium containing cytokinins (1-3 mg/L 6-benzyl amino purine) and less auxin (indole-3-acetic acid) was utilized for clonal propagation. The auxin 2,4-dichlorophenoxyacetic acid is beneficial because it inhibits organogenesis while stimulating callus development. The cultures are maintained at 3000-10,000 flux light intensity for 16 hours.

Step II: The Explants' Multiplication

Explant multiplication takes a long period due to the rejuvenation of shoots from explants. Only single shoots emerge from the apical shoots in certain circumstances, and these shoots are employed to excise a number of explants. To isolate nodal explants for shoot proliferation growth, a cytokinin-rich medium is employed. During this procedure, several shoots may be formed through

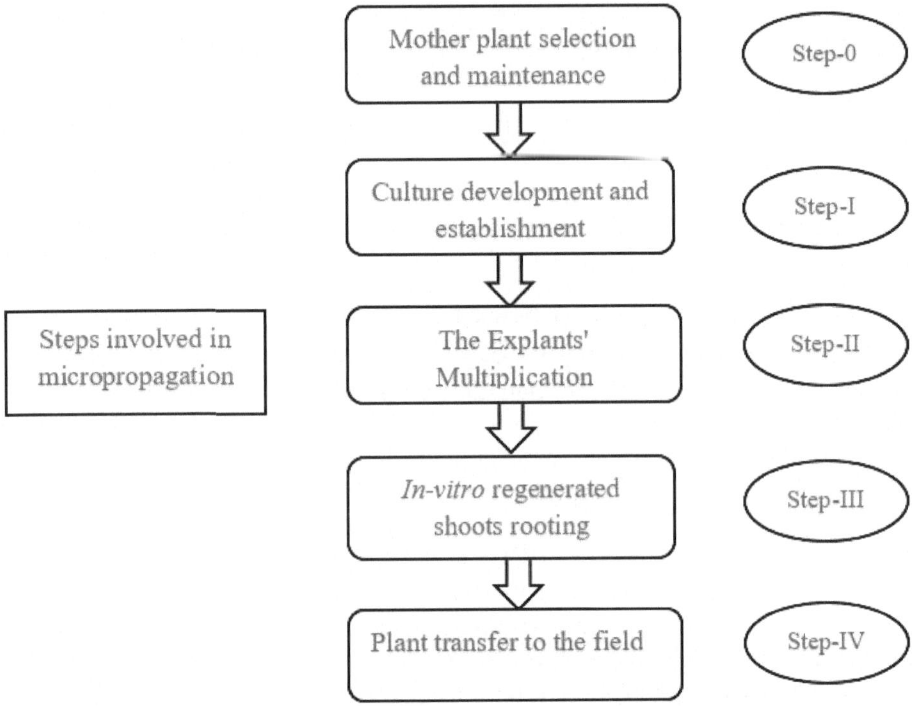

Figure 13.1: Various Steps of Micropropagation.

organogenesis or somatic embryogenesis directly on explants. Each explant yields five to six shots in four to five weeks. Single explants would generate 5^{10}-6^{12} plants in a year if all of the plants survived.

Step III: *In-vitro* Regenerated Shoots Rooting

Multiple shoots developed during stage II stimulate rooting in fresh media. Individually divided shoots are inoculated into the rooting media (containing auxins), and in other situations, rooting is directly stimulated in the soil under high moisture conditions. The planted plants are highly susceptible and sensitive to dampness. Germination of somatic embryos is allowed to produce plantlets before they are introduced to the soil. Following the hardening process, the plantlets are gradually transplanted into the soil. The hardening medium should be pearlite, peat, or vermiculite, capable of retaining significant moisture and remaining at high humidity levels until being transferred to the soil.

Step IV: Plant Transfer to the Field

The explant of stage III is ready for the outside soil properties from their in-vitro circumstances in this step. This includes the plants being resistant to stress, wetness, and disease, resulting in autotrophic plants from heterotrophic in vitro cultivation settings. Plantlets must be shielded from direct sunshine, and relative humidity must be reduced gradually over time. During this acclimation stage, the plantlets generate well-developed roots and cuticular wax is created in the floral tissues. Following this, the plantlets are ready to be transferred to outdoor areas (Gaspar T., 1991).

Some species or branches become brittle, glossy, and water soaked under in vitro conditions, indicating vitrification. Vitrification is caused by aberrant stomatal function, poorly formed vascular bundles, and low wax quality, which results in plantlet loss. This situation can be remedied by using growth inhibitors, bottom chilling of culture tubes, adding a heavy amount of agar (1 per cent), and adding hydrolysate chemicals (Ozdemir *et al.*, 2014). The high cost of plantlets developed by micropropagation is the primary limitation on their usage and development in large-scale vegetative propagation Singh *et al.* (2011). Thus, cost reduction is the most significant advancement of this approach in the twenty-first century. The following are the primary benefits and drawbacks of micropropagation:

Mains Advantages of Micropropagation

a) From a single individual, a huge number of plants can be grown.

b) Aseptic cultures require very short parts of plant tissue to be initiated.

c) Because cultures can more efficiently manage nutrition and growth regulator levels, temperature, and light, the rate of multiplication in vitro is generally much faster than in vivo techniques of vegetative propagation.

d) It applies to many of these genotypes where in vivo plant multiplication is difficult or even impossible.

e) Throughout the entire year.

f) If the plants are developed from virus-free stock, they are immune to re-infection, as well as the micro-propagated plants could be transferred without quarantine.

g) Microorganism infection is negligible in plants.

h) In vitro development can be better prepared by keeping cultures at moderate temperatures during low market demand seasons.

i) Micro-propagated plants may invent different attractive characteristics, like the branching habit of ornamental plants and an increase in the number of runners in strawberries.

Disadvantages of Micropropagation

a) It is essential to have expensive and complex facilities, skilled employees, and specific processes.

b) High production costs are caused by costly facilities and high labor inputs
c) Contamination or pest and diseases can result in significant losses in a short period of time.
d) Somatic variation at a higher level.
e) Plant growth in the field was poor.

Use of Micropropagation in Horticulture Crops

Herbaceous/ornamentals:

a) Aechmea-Cueva et al. (2006)
b) Gardenia-Hatzilazarou et al. (2006)
c) Philodendron-Jámbor-Benazir and Márta-Riffer (1990)
d) Pulmonaria-Stimart et al. (1998)
e) Tillandsia-Cueva et al. (2006)

Fruits:

a) Citrus-
b) Malus (apple)-Zimmerman (1984)
c) Olea-Leva (2011)
d) Passiflora-Faria and Segura (1997)
e) Prunus-
f) Rubus-
g) Vaccinium-Grout and Read (1986)
h) Vitis-Gago et al. (2009)

Nut crops:

a) Castanea-Yang et al. (1986)
b) Corylus-Nas and Read (2001)
c) Juglans-Leslie et al. (2010)
d) Pistachia-Benmahioul et al. (2012)

Woody ornamentals

a) Acanthopanax-Yang and Read (1997)
b) Acer-Preece et al. (1991ab)
c) Berberis-Uno and Preece (1987)
d) Cercis (redbud)-
e) Hibiscus-Airò et al. (2009)

Medicinal and herbs

a) Allium-Mohamed-Yasseen et al. (1993)
b) Balanites-Gour et al. (2007)
c) Hypericum-Karppinen et al. (2006)
d) Rosemarinus-Tawfik et al. (1992)
e) Satureja-Arrebola et al. (1997)

Since Murashige's (1974) listing of more than 100 species and more than 50 genera that had been micro-propagated to that point, a maximum number of plant species and genotypes have been successfully micro-propagated. Numerous decorative, fruit, nut, and medicinal plants are now routinely micro propagated thanks to the contribution of micropropagation to horticulture, including the following:

Based on this knowledge and the fact that plants propagated sexually or asexually by other methods cost significantly less per unit than plants propagated by micropropagation (Georget *et al.*, 2017; Singh *et al.*, 2011); vitro propagation as a method was found to be of restricted use and could only be applied to some significant horticultural and forest species and circumstances, as described below:

a) Plants with significant dormancy limits (strawberries), those with regulations on genetic material and amount from the mother plant (*e.g.*, Anthurium), and plants with prolonged juvenile phases as a result of sexual reproduction (most fruit trees), whether or not they are associated with high heterozygotes where maintenance of the primary characteristics of I is possible (Anthurium, eucalyptus, orchids, pineapple).

b) Species with few or restricted additional techniques of vegetative or asexual development, such as bananas or pineapples. The mother plant's developmental time is extended in these plants, a restricted number of new shoots are developed through typical propagation of plants, and new shoot roots occur straight from mature plants. Micropropagation, on the other hand, results in the more efficient and fast mass spread of new shoots.

c) When there are a limited number of mother plants available to begin propagation: mini-cuttings are the primary propagation technique employed by the industry during eucalyptus clonal propagation. However, after fresh clones are generated from seeds (only when one or a few parent plants are available), the business mostly uses micropropagation to swiftly mass multiply and raise the number of planting materials.

d) The eradication of major diseases from old mother plants grown asexually typically causes significant cultural loss, including in potato tubers. Due to the existence of a large number of germs as well as other fungus and bacteria illnesses after multiple rounds of propagation, the mass spread of crops is almost impossible. Virus and bacterium elimination may necessitate further treatment using antibiotics, antiviral medicines, or heat therapy, in addition, to shooting meristem culture.

e) A comparison of the outcomes of research aimed at increasing carnation micropropagation utilizing a novel bioreactor based on Temporary Immersion Bioreactor (TIB) and solid culture. Ahmadian and colleagues (2017).

f) Micropropagation is the most common method for clonally propagating gerbera in vitro, yielding millions of plants grown each year. For gerbera, several forms of explants and micropropagation procedures have been developed and employed. Gerbera is a popular cut flower throughout the world, according to Cardoso and Teixeira da Silva (2013)

g) Clonal propagation is used to grow plants of specified sex, such as papaya (Carica papaya), which requires only intersex and female plants to produce fruit. Micropropagation methods are important tools for current plant development programs for introducing new features into specific plants, propagating superior selections, and developing optimal cultivars in the shortest amount of time (Taji *et al.*, 2002). Clonal propagation of certain strawberry plants, on the other hand, is critical since transplanting the mother plant is prohibitively expensive.

CONCLUSION

One *in-vitro* approach that has been adopted as an industrial technology is micropropagation, which is utilised all over the world for quick clonal replication of significant horticultural, medicinal, and silvicultural plants. Furthermore, the only method for clonally propagating orchids that is economically viable is tissue culture. For species that are valuable or difficult to reproduce using traditional cloning techniques, micropropagation is very helpful. It enables year-round, disease-free, fast clonal plant reproduction under controllable environmental conditions. It contributes to the continued good health of virus-free plants. Future scholars or students may find the knowledge or material in this chapter to be helpful.

REFERENCES

Abdalla, N., El-Ramady, H., Seliem, M. K., El-Mahrouk, M. E., Taha, N., Bayoumi, Y., Shalaby, T. A. and Dobránszki, J. (2022). An Academic and Technical Overview on Plant Micropropagation Challenges. *Horticulturae*, 8, 677.

Bhojwani, S. S. and Dantu, P. K. (2013a). Micropropagation. Plant tissue culture: an introductory text. *Springer*, New Delhi, pp 245–274.

Cardoso, J. C. and Teixeira da Silva, J. A. (2013). Gerbera micropropagation, *j. biotechadv*.05.008.

Georget, F., Courtel, P. and Garcia, E. M. (2017). Somatic embryogenesis-derived coffee plantlets can be efficiently propagated by horticultural rooted mini-cuttings: a boost for somatic embryogenesis. *Sci Hortic*, 216: 177–185.

Haberlandt, G. (1902). Culturver suche mit isolierten Pflanzenzellen. Sitz-Ber. Mat Nat Kl Kais Akad Wiss Wien 111: 69–92 2.

IAEA-TECDOC, (2004). Low-cost options for tissue culture technology in developing countries. IAEA-TECDOC-1384.

Iliev, I., Gajdošová, A., Libiaková, G. and Jain, S. M. (2010). Plant Micropropagation. In: Davey M.R., Anthony P. (eds). *Plant Cell Culture*. John Wiley and Sons. Ltd. Publications. pp. 1–24.

Kane, M. (2011). Propagation by shoot culture. In: Trigiano RN, Gray DJ (eds) Plant tissue culture, development and biotechnology. CRC Press, LLC, Boca Raton, pp 181–191.

Singh, H. P., Uma, S., Selvarajan, R. and Karihaloo, J. L. (2011). Micropropagation for production of quality banana planting material in AsiaPacific. *Asia-Pacific Consortium on Agricultural Biotechnology* (*APCoAB*), New Delhi, India.

Taji, A., Kumar, P. P. and Lakshmanan, P. (2002). In Vitro Plant Breeding, *Food Pro- ducts Press,* New York, 167 pp

White, P. R. (1934). Potentially unlimited growth of excised tomato root tips in a liquid medium. *Plant Physiol* 9: 585–600.

Chapter 14

Mulching Techniques in Horticulture

Avneesh Rathour

Ph.D. Research Scholar, Institute of Agricultural Sciences, Banaras Hindu University, Varanasi – 221 005, Uttar Pradesh

e-mail: avneeshrathour47@gmail.com

INTRODUCTION

Applying a layer of plant debris, other natural or synthetic materials on the soil surface is known as mulching. In other words, it is the practice of enclosing plants with a layer of protection, such as bark chips, straw, or plastic sheeting, on the ground in order to control weed development, hold moisture in the soil, or avoid the freezing injury of the roots. A mulch is a substance or any material spread on soil to cover the surface of soil. It reduces weed development, enhances soil fertility and improves soil health. Mulches, however, are mostly employed in agriculture for the purposes of water conservation and preventing weed germinations and soil erosion in arid regions (Nithisha *et al.*, 2022).

In 1960, Waggoner introduced the term "mulch." Probably from the German term "molsch" which meaning soft to decay, came the word mulch. A soil cover made of leaves and straw by gardeners has been mentioned (Kasirajan and Ngouajio, 2012).

Objectives of Mulching

1. To conserve soil moisture
2. To moderate soil temperature
3. To suppress the weed seeds germination
4. To improve soil properties
5. To minimize soil erosion caused by water and wind

6. To reduce excess evaporation loss of soil moisture
7. To ensure economic use of irrigation water
8. Enhancement of crop growth, yield and quality produce.

Importance of Mulching in Horticultural Crops

Mulching is a process in which a protective layer of straw, hay, organic manure, or polythene mulch is applied to the soil surface surrounding the plants. This helps to produce an environment that is beneficial for plant growth. Mulches deter some insects, function as a barrier against soil diseases and can slow the growth of weeds. Mulch is said to be a wonderful substitute for chemical herbicides and has a negative effect on weed growth. It serves as a layer of protection that keeps sunlight from directly hitting the soil. This regulates soil temperature, speeds up seed germination and increases crop yield. Reduces soil temperature, which will conserve water and make the soil healthier. It also increases soil organic matter, which promotes soil health and avoid letting plants come into touch with dirt; this will help them avoid pests and diseases. It has a significant impact on the earliness, yield and quality of crop. However, it is favoured in nurseries, vegetable farms, flower crops, fruit orchards and plantation crops (Bisen and Pardhi, 2021).

Benefits of Mulching (Kaur and Bons, 2017)

1. **Keeps soil moisture in check:** Mulching helps to preserve soil moisture by reducing the amount of water that evaporates from the soil. By lowering watering frequency, it eventually conserves your precious irrigation water.
2. **Minimises weed growth:** Mulch slows the development of weeds by preventing sunlight from directly hitting the soil.
3. **Keeps the root zone at a comfortable temperature:** The soil's coating serves as an insulator, preserving the ideal root temperature. So, controlling temperature aids in boosting soil microbial activity. In actuality, it gently lowers the temperature on hot days and slightly raises it on cold days.
4. **Minimises soil erosion:** The protective covering also lessens soil erosion by preventing the damaging effects of direct rainfall on bare soil.
5. **Enhances soil structure:** Mulch enhances soil structure by minimising unwelcome soil compaction brought on by weed growth and other physical pressures. Furthermore, organic mulch increases the amount of organic matter in the soil, which improves soil aeration and water-holding capacity.

6. **Prevents infections that are transmitted through soil:** The mulch layer prevents some soil-borne illnesses from spreading by preventing fruits and leaves from coming into direct touch with the soil. Low-growing fruits and vegetables like tomato, brinjal and others benefit from this strategy.
7. **Reduces infiltration of water:** The amount of crop residue or organic mulch material on the soil directly affects how much rainwater infiltrates in the soil and evaporates. Mulch cover slows down surface runoff and keeps precipitation on the soil surface longer, allowing it to soak into the ground.
8. **Add organic matter into the soil:** After decomposing, organic mulches improve the physical, chemical and biological characteristics of the soil, reintroducing organic matter and plant nutrients that in turn boost crop. The loose, friable soil beneath the mulch continues to provide a favourable environment for root penetration. The addition of organic matter to the organic mulches increases the soil's nutrient content while also preserving the soil's moisture (Bhardwaj, 2013).

Classification of Mulch Types

Mulching materials come in a wide variety; however, they may be broadly categorised into two groups (Prem *et al.*, 2020):

1. Live Mulch

Legume crops are typically the best option for live mulch. In plantation and fruit crops, it is used. You may also plant intercrops between the rows of a productive crop that are effective in smothering weeds. Cover crop, inter crop, green crop, velvet bean, subabool and barnyard are a few examples of living much.

2. Dead Mulch

Dead mulch includes synthetic, soil/dust, inorganic and organic mulch.

- ☆ **Residue mulch:** This kind of mulch is termite-attacked and when it decomposes, it produces compounds that might be phytotoxic to crops. Because of this, it is preferable to remove the mulch from the field 35 to 40 days after the crop was sown, once the weed-control function has been achieved. Crop wastes are an excellent source of recycled materials. These have an impact on the biological, chemical and physical characteristics of soil and water. For instance, dry plant or crop leftovers, straw, sawdust and banana pseudo stems.
- ☆ **Organic matter mulch:** Organic materials are utilised as mulch; the thickness should be 3″. Too much mulching lowers soil oxygen levels and interferes with a crop's ability to form roots. Perennial weeds may be able to break through the mulch layer, however annual weeds can

be effectively controlled by this sort of mulch. In the mulch layer, annual weeds may also sprout and flourish. Compost, FYM, sludge, *etc.* are all included.

- ☆ **Inorganic mulches:** Typically, polythene mulches with varying colours may change the temperature below and above the mulch by absorbing, transmitting and reflecting solar radiation. This has an impact on the plant's immediate microenvironment. The amount of soil warming is also influenced by the amount of mulch contact with the ground. The effectiveness of the mulch's warming capabilities increases with the degree of soil contact. Today, plastic films made of LDPE and LLDPE are often utilised for mulching. The most preferred mulch film is LLDPE black, which has the dual advantages of downgauging and improved puncture availability.

Mulches can be again Classified Based on Material

1. Organic Mulches

The practise of using plant and animal remains to safeguard the soil's surface and provide an ideal habitat for plant development. Organic mulches break down over time, adding nutrients back to the soil. straw, leaves, manure, or sawdust are examples of organic mulches.

Various Organic Mulches: Dry leaf, paper mulch, Bamboo mulch, sawdust, dried-up coconut husk, sparse grass Mulch made up of water hyacinth *etc.*

- ☆ Paper mulch: Both whole and torn up papers are used. To hold it in place, apply two or three layers of paper at a time and cover with something living, such as leaf mulch or blades of grass.
- ☆ Dry leaf mulch: Before using them as mulch, shred the leaves if possible. Consider the leaves of sugarcane, bananas, or other plants. affordable and easily accessible.
- ☆ Mulch made of sawdust can disintegrate quickly and compress into itself, but it requires more nitrogen to do it. After being treated for termites, sawdust is applied.
- ☆ Mulch made of straw and hay is effective for protecting against the winter elements. A soil amendment with more nitrogen should be made if used in the summer. Example: paddy straw.

2. Inorganic Mulches

Inorganic mulches are typically used to inhibit weed growth. Plastic films and other inorganic mulch, such as rocks or gravel, do not disintegrate easily. According to the type of material utilised, there are basically two sorts of mulches. They are listed below (Ray and Biswas, 2016):

a. *Plastic Mulch*

It is possible to regulate the soil environment by choosing the right plastic mulch composition, colour and thickness.

(i) **Black plastic mulch:** This mulch prevents sunlight from penetrating the ground. The absence of sunshine and the presence of a dark layer prevent photosynthesis from taking place in the soil. Stop the growth of weeds as a result. A black plastic cover aids with moisture preservation and weed management. But it can raise the temperature of the soil. For 7–14 days, it offers a head start on harvest. In particular for green crops, it helps to enhance yields.

(ii) **Transparent and clear plastic:** This mulch allows sunlight to flow through and encourages the growth of weeds; however, by applying a herbicide coating to the inner side, weed growth can be inhibited. It works well as a solar-reinforcing film to eradicate soil-borne pathogens and other weeds by destroying germs in the soil. This is an application for creating a nursery by establishing sun beds before planting seeds to grow a nursery that offers nearly 100 per cent germination and a disease-free nursery.

(iii) **Invisible Film**: Film helps raise soil temperature in steep places during the chilly winter months.

(iv) **Plastic mulch with two colours on one side:** Also known as wavelength selective or photo selective films, these materials are made to absorb a certain range of solar radiation, altering the spectrum of light that passes through them or is reflected back into the canopy of the plants. The growth and development of plants were significantly impacted. The impacts include increased soil temperature and a reduction in weed growth. Examples are red and black, white and black and yellow and black.

- ☆ Red/black mulch: Partially translucent, allowing radiation to penetrate and warm the soil but also reflecting radiation back into the plant canopy, changing the ratio of R: FR light, leading to changes in plant vegetative, flower development and metabolism, leading to early fruiting and increased yield. This mulch is particularly applied on tomato crops.
- ☆ Yellow and black: these colours draw specific insects and serve as traps for them, preventing sickness.
- ☆ White or black: Let the dirt cool.
- ☆ Black and silver: discourages thrips and aphids.

b. *Mulches that degrade naturally*

(i) **Photo-degradable mulch**: Under ultraviolet light, these mulches disintegrate, the iron and copper compounds in the mulch are light-sensitive and only gradually disintegrate when exposed to light. compostable mulch gets broken down in a natural environment and mingle with the soil after a time of mulching. Mulches constructed of a starch-and-biodegradable-plastic blend.

(ii) **Aluminium foil or aluminium-painted mulch:** This mulch may deter aphids and thrips by reflecting light off the undersides of leaves. The spread of viruses is also decreased when vector insects are driven away.

(iii) **Rock mulch:** Stones don't hold onto moisture, but they do help soils retain heat. It is primarily utilised in kitchen gardens and homes.

Mulch Film Laying

Tractor mounted mulch laying equipment has been created and is in use, a variety of coloured mulch film is decided upon based on a particular goal, such as weed management, soil temperature control, soil cooling, disease control, increased plant growth, *etc.* The most common film sizes are 90 cm and 120 cm.

When should Use Mulch?

The timing of application depends on the goals we have for mulching. Mulches help to regulate soil temperature by creating an insulating barrier between the soil and the air. As a result, a mulched soil will be cooler in the summer than an adjacent unmulched soil and it may not freeze as deeply in the winter. However, as mulch serves as an insulating layer, mulched soils typically warm up more slowly in the spring and cool down more slowly in the fall than unmulched soils. It is recommended to apply mulches to vegetable or flower gardens when the soil has warmed up in the spring. In general, cool, damp soils tend to inhibit seed germination and accelerate seed and seedling deterioration. Wait until the soil is thoroughly warmed up before adding more layers of mulch to existing perennial beds. The optimum times to apply mulch are in the middle to end of spring and in the fall, when the soil is warm and moist. Mulching should be avoided throughout the winter and early spring when the soil is too cold and during the dry summer months. They can be used on established beds and specimen plants as well as around new plantings (SK *et al.*, 2020).

Choice of Mulch

The first step in choosing a particular type of mulch material, whether it be organic or non-organic, is to consider its durability, suitability/compatibility, desired above- and below-ground effects and effectiveness with the surrounding environment as well as the type of crop (along with any advantages and disadvantages). Plastic mulches are the most often used inorganic mulch applications, particularly for growing commercial crops. Because plastic mulching modifies both the light environment around the plant and the hydrothermal regimes of the soil, this may have an impact on the morphogenesis and growth of the plant. In many parts of the world, extrapolations of the optical characteristics of various plastic kinds of varying thicknesses with varied crops have been made. Additionally, it has been asserted that, in general, 50–75 micron thick black and white polymers are the best synthetic mulches for use in winter and summer, respectively, in dry and semi-arid environments. By combining knowledge of crop biology and surface energy balance, one may select the mulch and crops that work best together.

Second, when using organic mulch, care must be taken that it has certain essential qualities, including being easily decomposable, able to add a significant amount of organic matter to the soil, not releasing allelochemicals, leguminous in nature (especially for nitrogen-deficient soils), not having much economic value and not posing an excessive amount of acidic, basic, or toxic effects. When applying organic mulches, the ideal thickness of the mulch components should be prioritised over their type. Mulch depths should typically not go over 2 inches if you have shallow-rooted plants growing in clay soils, which drain poorly. On the other hand, if you have species with deeper roots that are growing in loams or sandy soils with greater drainage, your plants would benefit from 3–4 inches of mulch. To keep weeds at bay and avoid needless soil drying in the summer, you might just need a covering of 1-2 inches. Additionally, it is advised to use a variety of mulch particle sizes for a continuous decomposition of the mulch and release of nutrients. To lessen the consequences of acidic or basic soil reactions, microbial growth and other hazardous effects, using the same organic mulch at the same location for longer than four years must be avoided (Kader *et al.*, 2019).

Application of Mulch

Applying mulch should be done slightly away from the plants. Complete mulching of beds and borders is possible as long as low-growing plants are not smothered or woody plant stems are not pressed up against the mulch. A 2–3-inch layer of mulch is required to cover weeds and keep soil wet. Mulch that is less than 2 inches thick won't block enough light to prevent weed seeds from sprouting. The following precautions should be followed before spreading mulch:

- The mulch should be spread out on the bed while being tightly gripped and free of any grooves and it ought to be set up in a calm environment.
- Borders (1 cm) should be buried 7 to 10 cm deep in narrow furrows that are angled at 45° degrees.
- Mulch material should be spread out on the bed and pierced at the necessary intervals based on crop spacing. In the holes, the seeds or seedlings should be sowed or transferred.
- Direct use of sawdust, wood chips, or bark in horticultural gardens Wood takes a long time to break down, and sawdust can upset the nitrogen balance in the soil.

- Fields or tree canopies are covered with mulch.
- Field operations including tillage, weeding, adding farm yard manure (FYM), watering and hoeing can be finished before spreading mulch. The ground should also be flattened to lower the contact angle between the sheet and the soil for greater efficacy.
- When using plastic film to cover an area with fruit plants, place the plant in the middle of the film using the cut-out piece of the film. It is important to take care that the clipped piece of the film is not facing the wind. If not, air could get within the film, risking harm to the sheet.
- The holes in plastic sheets are also used for transplantation.

Horticultural Crops in which Mulching is Generally Practiced

Vegetable crops: Cabbage, Chilli, Okra, Brinjal, Tomato, Cauliflower, Capsicum *etc.*

Orchard crops: Pomegranate, Citrus, Banana, Peach, Guava, Papaya, Apricot, Grapes *etc.*

Mulch Material Utilised in Vegetable Crops

- Using grass clippings as a mulch
- Straw mulching
- Mulched with chopped leaves
- Mulching using peat moss
- Mulching using pine bark
- Using newspaper as mulch
- Shredded cardboard

Limitations/Drawbacks of Mulching (Singh, 2016)

1. The scarcity of mulch materials such as husk, hay, sawdust, *etc.*
2. Residue mulch is only accessible for a limited time.
3. Challenges in removing and discarding mulch, especially synthetic mulch.
4. Specialized equipment is needed for particular procedures, which increases the cost.
5. Mulch can prevent seedling emergence by generating an impenetrable barrier if it is applied too heavily.
6. It requires knowledge and expertise and is a labour-intensive process.
7. When straw, peat, or wood chips are utilised as mulching materials together with contaminated weed seeds, it might bring additional pests and illnesses into the main field (especially organic mulches).

8. Plastic mulching is not environmentally friendly or sustainable.
9. Clear plastic mulch encourages the growth of weeds, which deprives the crop of soil moisture and nutrients. In comparison to black plastic mulch, this mulch may be able to warm the soil more effectively.
10. Materials with a high carbon content put N- immobilisation at danger. The decomposing bacteria proliferate swiftly in the soil when organic matter is added. They require nutrients, particularly nitrogen, for growth, so when supplied plant material is insufficient, soil-dwelling microorganisms suck it up and temporarily fix it before releasing it. The crop may become malnourished at this time because the bacteria and plants are competing for nutrients. When straw or grain husks are used, this issue might also occur.

Figure 14.1: Mulching in Strawberry Field.

CONCLUSION

Mulches have a beneficial and a detrimental impact on crops and soil, just like every coin has two sides. Mulches, with preserving moisture, preventing weed development and increases agricultural productivity and there it also influences the pH of the soil. By regulating soil moisture, temperature and nutrients, as well as providing protection from weeds, insects, disease and other environmental conditions, mulching improves plant development and fruit production. Although broad mulching treatments responded favourably to plant growth, fruit excellent fruit quality and production. It is possible to draw the conclusion that the improved physical environment brought about by mulches promotes better nutrient utilisation of both native and applied nutrients in surface soil, moderates hydrothermal regimes of soil, adds organic matter to the soil, increases seed emergence, inhibits weed growth and ultimately results in higher yields and water use efficiency. In the end, we determined that mulching is a great horticultural method that enhances the growth, quality and production of fruit trees. Mulch must therefore be chosen and considered under factors in the soil kind, the region of climate area and crop to be grown (Tyagi, 2017).

REFERENCES

Bhardwaj, R. L. (2013). Effect of mulching on crop production under rainfed condition-a review. *Agricultural Reviews*, 34(3): 188-197.

Bisen, A. and Pardhi, S. (2021). Effect of Mulching on Horticultural Crop Production under Rainfed Condition. A Review. *International Journal of Current Microbiology and Applied Sciences*. 10(1): 693-700.

Kader, M. A., Singha, A., Begum, M. A., Jewel, A., Khan, F. H. and Khan, N. I. (2019). Mulching as water-saving technique in dryland agriculture. *Bulletin of the National Research Centre*. 43(1): 1-6.

Kasirajan, S. and Ngouajio, M. (2012). Polyethylene and biodegradable mulches for agricultural applications: A review. *Agronomy for sustainable development*. 32(2): 501-529.

Nithisha, A., Bokado, K. and Charitha, K. S. (2022). Mulches: Their impact on the crop production. *The Pharma Innovation Journal*. 11(7): 3597-3603.

Prem, M., Ranjan, P., Seth, N. and Patle, G. T. (2020). Mulching techniques to conserve the soil water and advance the crop production–A Review. *Current World Environment*. 15: 10-30.

Ray, M. and Biswas, S. (2016). Impact of mulching on crop production: A review. *Bioscience trends*. 9: 757-767.

SK, P. G. P., Debnath, S. and Maitra, S. (2020). Mulching: Materials, advantages and crop production. Protected cultivation and smart agriculture, eds. Maitra, S. *et al.*, New Delhi Publishers, New Delhi, India, 55-66.

Tyagi, S. (2017). Mulching for Commercial Fruit Production. Kalyani Publishers, Ludhiana, India, 292-307.

Chapter 15

Organic Vegetable Production

Prashant Gautam[1]*, Jagraj Singh[2], Rahul Kumar[3] and Imran Ali[4]

[1]Ph.D. Research Scholar, [4]M.Sc. Fruit Science, Department of Fruit Science,

[2]Ph.D. Research Scholar, [3]M.Sc. Vegetable Science, Department of Vegetable Science,

Sardar Vallabhbhai Patel University of Agriculture and Technology, Meerut – 250 110, Uttar Pradesh

**e-mail: prashantgautam151999@gmail.com*

Definition

The science and art of cultivating vegetables using organic farming practices, such as soil conservation, insect control, and soil building, are known as organic vegetable production.

As opposed to employing inputs that have negative impacts, organic vegetable production uses ecological processes and biodiversity cycles that are suited to local conditions to maintain soil health, a safe environment, and human health. Tradition, modernity, and science are all included in organic veggies to benefit the ecosystem as a whole, foster just relationships, and improve everyone's quality of life.

What is Organic Farming?

Organic farming is a method that, to the greatest extent practical, focuses on crop rotation, crop residues, animal manures, and off-farm organic waste and avoids or substantially excludes the use of synthetic inputs (such as fertilizers, pesticides, hormones, feed additives, *etc.*). Agro-ecosystem health, including biodiversity, biological cycles, and soil biological activity, are promoted and enhanced by this all-encompassing production management approach.

Why Organic?

New issues in agriculture include the growth of organic farming and the implementation of new laws or private standards. Sustainable development trends have an impact on agriculture as well, necessitating the reduction of cultural traditions' environmental impact **(Neeme, 2006)**. The increasing pressure on land and water resources is a result of population growth and rising consumer demand for agricultural products. The general consensus is that organically grown food meets the aforementioned criteria, has fewer negative effects on the environment, and has superior nutritional content.

Since the previous few decades, it has been scientifically recognised that intensive crop farming methods using high-analysis synthetic fertilisers have seriously harmed the quality of the air, water, and land resources. Therefore, everyone now believes that the only source of hope we have left is the adoption of organic farming techniques, not only to counteract the negative effects of overusing synthetic fertilisers, pesticides, and weedicides, but even to completely eliminate their use in agriculture production systems in order to conserve natural resources. More significantly, organic farming is a cultivation method Table 15.1 that traps soil organic carbon, which ultimately improves the quality of the environment. Increased soil carbon results in more soil organic matter, better soil water retention, resource conservation, and better crop yields.

Feature of Varieties for Organic Cultivation

1. They should be resistant/tolerant to stress conditions.
2. They should be resistant to diseases, insects, and pests.
3. Should be produced and evaluated under organic condition
4. They should be better adapted to local environment.
5. They should be responsive to organic sources of plant nutrients.

Table 15.1: Vegetable Crops Suitable for Organic Cultivation

Sl.No.	*Name of Vegetables*	*Botanical Name*
1.	Potato	*Solanum tuberosum* L.
2.	Tomato	*Lycopersicon esculentum* Mill.
3.	Brinjal	*Solanum melongena* L.
4.	Water melon	*Citrullus lanatus*
5.	Musk melon	*Cucumis melo*
6.	Cowpea	*Vigna unguiculata*
7.	Peas	*Pisum sativum* L.
8.	Cluster bean or guar	*Cyamopsis tetragonoloba*
9.	Chilli	*Capsicum annum* L.

Sl.No.	Name of Vegetables	Botanical Name
10.	Garlic	*Allium cepa* L.
11.	Radish	*Raphanus sativus* L.
12.	Bottle gourd	*Lagenaria siceraria* (Mol.)
13.	Coriander	*Coriandrum sativum* L.
14.	Fenugreek (Methi)	*Trigonella foenue-graecum*

Objectives of Organic Farming

i. To create a harmonious balance of crop production and animal husbandry.
ii. To minimize all forms of pollution.
iii. To use renewable resources in locally organized production systems.
iv. To encourage and enhance biological systems within the farming system involving microorganisms, soil flora and fauna, plant and animals.
v. To maintain the genetic diversity of the production system.
vi. To interact in a constructive and life-enhancing way with natural systems and cycles.
vii. To maintain and increase the long-term fertility of the soil.

Postharvest Management of Organic vegetables

Fresh veggies are extremely perishable and go through an active senescence process. Injuries, low air humidity, and excessive temperatures can all result in losses of fresh veggies. Such injury frequently happens as a result of improper handling, leading to internal bleeding, splitting, and skin breaches, which exacerbate water loss.

Causes

The harvest that was early, premature, or overmature. Faulty postharvest procedures, subpar sorting and grading procedures, subpar temperature and relative humidity management in storage, subpar packaging and washing, delay, and subpar transit to market, brought on by external factors including rotting and insect pests. Lack of understanding of postharvest methods. Water loss leads to weight loss and shrinkage. Stores with poor airflow. Small land holdings, sparse production pockets spread out far from market hubs, reckless handling leading to internal bruising, splitting, and fractures' inadequate current technologies and abilities.

Components/Inputs of Organic Vegetable Production

1. Organic Manure

Natural substances derived from plants and animals, known as organic manures, are used in enormous proportions and contain only trace levels of plant nutrients. When applied to soil, organic manures improve soil structure, create a better environment for food proliferation, increase water holding capacity, reduce soil erosion, increase nutrient uptake, and enhance soil microbial population, all of which lead to improved nitrogen fixation, phosphorus mobilization, and solubilization. All of these efforts result in increased vegetable growth, yield, and quality. Organic manures have had a positive impact on vegetable crops, resulting in increased production and quality.

2. Crop Residues Management

Non-economic plant pails left on the field after harvest, as well as residues from packing houses and processing facilities, are referred to as crop residues. Additionally, they contribute to the formation of a strong organic matter base by serving as a potential source of plant nutrients.

3. Green Manuring

An old technique for crop manuring is called green manuring. Crops used in cropping systems as "green manuring" are those that are cultivated to enhance the amount of organic matter in the soil. Green manuring is reported to boost yields by 15 to 20 percent while also enhancing vegetable quality. Sun-hemp, Dhaincha, black gram, cowpea, and guar are used as green manure crops.

4. Legumes in Rotation

Legume crops, including peas, cowpeas, guar, and others, have been demonstrated to increase yields by 30 to 35 percent while also enhancing soil fertility by fixing atmospheric nitrogen. In a vegetable rotation, growing legume crops can enhance both output and quality.

Sorghum – pea – okra

Tomato/Cabbage – cauliflower – pea and maize – garlic

Maize – cotton, chilies – onion and brinjal

5. Biofertilizers

Biofertilizers, which include both symbiotic and non-symbiotic nitrogen fixers, phosphorus solubilizers, and mobilizers, are substances containing living cells of various microorganisms that have the capacity to mobilize nutrients from non-usable forms through biological processes. Biofertilizers are an important part of organic farming because they have nutritional, stimulating, and medicinal properties that enhance soil fertility while also improving vegetable output and quality. *Azotobacter* and *Azospirillum* also showed a substantial impact on vegetable crops, leading to an increase in production of 5-25 per cent and a nitrogen economy of 20-25 per cent.

How Apply Biofertilizers?

Biofertilizers can be applied to seeds, seedlings and soil as follows:

1. **Seed:** Mix 200 g of carrier-based inoculums with 500 ml of water to create a thick slurry for seed treatment. The stickiness of the biofertilizer on the seed surface can be significantly increased by mixing a 10 per cent jaggery solution, a 5 per cent sugar solution, and 40 per cent boiling and cooled gum solution or rice-gruel water. The treated seeds must never be left out in the sun for an extended amount of time, as UV rays reduce the number of inoculation bacteria on the seed surface.
2. **Seedling:** Vegetable crops that are transplanted work best with this application approach. The roots of transplant-ready seedlings are submerged in a liquid biofertilizer slurry (500 g in 2.5 litres of water) for 20 minutes prior to transplanting in this technique.
3. **Soil:** All types of biofertilizers should be administered to the soil, with the exception of Rhizobium. After being combined 1: 2 with dry FYM, compost, or vermicompost, the biofertilizer is employed. 1-2 kg/ha is the recommended amount for crops that last six months. This can be increased to 2-4 kg/ha for crops that survive longer than six months. Perennial crops should have 10–25 g of biofertilizer administered to the root zone in the first year, followed by 25–50 g in succeeding years. This might happen when sowing seeds, transplanting plants, or cultivating crops in between.

6. Plant Protection

Organic producers should take all precautions possible to stop the spread of weeds, diseases, and pests. Choose naturally resistant varieties and rely on a large population of soil bacteria, insects, birds, and other organisms to keep the pest problem under control. Organic producers think that "prevention is better than treatment."

Pest Management

For the control of pests including pod borer, whitefly, aphids, mites, thrips, and others on vegetable crops, *Bt.*, N.P.V., Beauveria and Metarrhizium have been shown to be effective. The biological control of pests also includes the use of Trichogramma, Ladybird Beetles, Chrysopa, Spiders, and other helpful insects (predators and parasites).

Monitoring adult moths in the field is facilitated by the placement of a small number of pheromone traps (5/acre). In order to manage the early larval stage of the pod borers, it is helpful to spray with Nuclear Polyhedrosis Virus (NPV) 500 ml; ha 4-5 rounds ten days after spotting the moths in the traps. Neem seed kernel extract is effective against mites, aphids, and thrips (NSKE 3 percent). Release of *Chrysoperia carnea* larvae can be used to control thrips and mites.

Some other Pest Control Formulations

Many organic farmers and NGOs have developed a large number of innovative formulations which are effectively used for the control of various pests. Although none of these formulations have been subjected to scientific validation their wide acceptance by farmers speaks of their usefulness. Farmers can try these formulations, as they can be prepared on their own farms without the need for any purchases. Some of the popular formulations are listed below:

1. **Cow urine-** Cow urine diluted with water in a ratio of 1: 20 and used as a foliar spray is not only effective in the management of pathogens and insects, but also acts as an effective growth promoter for the crop.
2. **Fermented curd water-** In some parts of central India fermented curd water (buttermilk or Chhaachh) is also being used for the management of white fly, jassids aphids *etc.*
3. **Dashparni extract-** Take crushed neem leaves 5 kg, *Vitex negundo* leaves 2 kg, Aristolochia leaves 2 kg, papaya (*Carica Papaya*) 2 kg, *Tinospora cordifolia* leaves 2 kg, *Annona squamosa* (Custard apple) leaves 2 kg, *Pongamia pinnata* (Karanja) leaves 2 kg, *Ricinus communis* (Castor) leaves 2 kg, *Nerium indicum* 2 kg, *Calotropis procera* leaves 2 kg, Green chilli paste 2 kg, Garlic paste 250 gm, Cow dung 3 kg and Cow Urine 5 lit in 200 lit water ferment for one month. Shake regularly three times a day. Extract after crushing and filtering. The extract can be stored up to 6 months and is sufficient for one acre.
4. **Neem-Cow urine extract-** Crush 5 kg neem leaves in water, add 5lit cow urine and 2 kg cow dung, ferment for 24 hrs with intermittent stirring, filter squeeze the extract and dilute to 100 lit, use as foliar spray over one acre. Useful against sucking pests and mealy bugs.
5. **Mixed leaves extract-** Take Crushed 3 kg neem leaves in 10 lit cow urine. Crush 2 kg custard apple leaf, 2 kg papaya leaf, 2kg pomegranate leaves, 2 kg guava leaves in water. Mix the two and boil 5 times at some interval till it becomes half. Keep for 24 hrs, then filter squeeze the extract. This can be stored in bottles for 6 months. Dilute 2-2.5 lit of this extract to 100 lit for 1 acre. Useful against sucking pests, pod/fruit borers.
6. **Chilli-garlic extract-** Crush 1 kg Ipomea (Besharam) leaves, 500 gm hot chilli, 500 gm garlic and 5 kg neem leaves in 10 lit cow urine. Boil the suspension 5 times till it becomes half. Filter squeezes the extract. Store in glass or plastic bottles. 2-3 lit extract diluted to 100 lit is used for one acre. Useful against leaf roller, stem/fruit/pod borer

7. **Disease Management:** The principal illnesses of semi-arid environments are rot and dieback caused by Colletotrichum sp., as well as bacterial wilt. Diseases will be controlled by careful seed selection and the use of phytosanitary measures. The early removal of diseased plants will help to control disease spread and limit the spread of diseases. In the nursery, seed treatment with Trichoderma prevents seedling rot.

Some Important Formulations for Soil Enrichment

1. **Preparation of liquid manures:** Many variants of liquid manures are being used by farmers of different states. Few important and widely used formulations are given below:
2. **Sanjivak-** Mix 100 kg cow dung, 100 lit cow urine and 500 gm jaggary in 300 lit of water in a 500-lit closed drum. Ferment for 10 days. Dilute with 20 times water and sprinkle in one acre either as soil spray or along with irrigation water.
3. **Jivamrut-** Mix cow dung 10 kg, cow urine 10 lit, Jaggary 2 kg, any pulse grain flour 2 kg and Live Forest soil 1 kg in 200 litres of water. Ferment for 5 to 7 days. Stir the solution regularly three times a day. Use in one acre with irrigation water.
4. **Amritpani**
5. **Panchagavya**

Organic Production Systems Prevalent in India

- Biodynamic
- Rishi Krishi
- Panchagavya Krishi
- Homa organic Farming

Biodynamic

Under the current circumstances, biodynamic agriculture appears to be a viable option. Its foundation is the methodical and coordinated harnessing of the energy of the universe, the earth, plants, and cows **(Steiner, 1920).** Additionally, **Rupela (2003)** promoted the idea that BD preparations have demonstrated 3.24^{10} to 6.90^{10} g^{-1} of bacteria hostile to disease-causing fungus. A few of the preparations, such as BD-500-508, Cow Pat Pit, and others that are administered in very little amounts have extraordinary impacts on plant growth **(Carpenter, 2000)**, metabolism, crop output, and crop quality **(Pfeiffer, 1984).**

Panchagavya

It is a unique formulation prepared from five cow by-products and a few other ingredients that have been incubated for a set amount of time in an earthen or wide-mouth plastic container **(Ramasamy, 2004).** Mix 5 kg of fresh cow dung, 3 lit of cow urine, 2 lit each of milk and curd, and 1 kg of cow butter oil. Ferment for 7 days while stirring twice daily. Spray the soil with a solution of 3 lit of Panchagavya and 100 lit of water. Per acre, 20 lit Panchagavya and irrigation water are required for soil application.

Procedure for Preparation

Mix thoroughly fresh cow dung and cow ghee

Incubate for two days

↓

Add cow urine + 10 liters of water

↓

Stir properly in the morning and evening

↓

Incubate for 15 days

↓

Add sugarcane juice*

+

Add cow milk + Add cow curd + Add coconut water**

Toddy (Palms flowers extract) *** if available and 12 ripe bananas

↓

Incubate for 2 weeks

Recommended Dosages

For all uses, it has been discovered that a spray of a 3 percent (3 kg/100 litres of water) solution works well. It is advised to use a high-volume sprayer to apply the mixture after filtering it with muslin cloth. It can be applied to seeds, seedlings, and seeds before storage.

Rishi Krishi

In this system following aids are used to maintain soil fertility and crop productivity.

1. **Angara:** Bhoomi Sanskar is performed to make the soil fertile, wherein, 15 kg of rhizosphere soil of the banyan tree "Vatavriksh" (*Ficus benghalensis*) is broadcasted over an acre of land.
2. **Amrit Pani:** It is made with honey, cow ghee, and cow dung. 10 kg of cow dung and 500 g of honey should be properly combined to create a creamy mixture. 250 gm of cow desi ghee should be added, then quickly mix. Add 200 lit of water to dilute. Sprinkle this suspension over soil or irrigation water in an area of one acre. Apply a second dosage between each row of plants or through irrigation water after 30 days. This is used for a number of operations such as:
   - Bhoomi Sanskar is known as a Soil treatment.
   - Beej Sanskar is known as a Seed treatment.
   - Vanaspati Sanskar (Spray on Crop).

Precautions

- Use of agrochemicals needs to be avoided in Rishi Krishi
- Mulching particularly with organic wastes needs to be encouraged.
- As for as possible, dung from indigenous cows with humps should be preferred.

Homa Organic Farming

Different countries in Europe (Austria, Germany, Poland), as well as South America (Peru, Venezuela), that employ Homa therapy, have seen amazing results in terms of crop output, animal and human health, and the construction of pollution-free environments.

It is simply founded on the tenet that by healing the environment, you will also heal yourself. The most straightforward and commonly used is Agnihotra, which must be combined with other organic production systems to create the Jaivik production package.

Homa farming is a totally revealed science. Agnihotra was regularly performed to purify the atmosphere since ancient times. The basic process used in Homa farming is Agnihotra. A copper pyramid shape container is used in the process.

Materials Required

- Pyramid
- Rice
- Cow ghee
- Dried cow dung cake

Feeding Plants

Soil feedings and foliage sprays are recommended every alternate week. Agnihotra ash should be spread on the ground in close vicinity of the plants and sprayed on the leaves to protect the plant from insect and disease damage.

REFERENCES

Carpenter-Boggs, L., Reganold, J.P. and Kennedy, A. C. (2000). Effects of biodynamic preparations on compost development. *Biol. Agric. Hort,* 17: 313-328.

Nesme, T., Bellon, S., Lescousret, F. and Habib, R. (2006). Survey-based analysis of irrigation and N fertilization practices in apple orchards. *Agron. Sustain. Dev*, 26: 215-225.

Pfeiffer, E. (1984). Using the Biodynamic compost preparations and sprays in Garden, Orchard and Farm. Biodynamic Farming and Gardening Association, Inc, Kimberton, PA, 64 p.

Ramasamy, K., Vedivel, E. and Murugesan, R. (2004). Panchagavya as potentiator of living cell; Effects on crop plants and biochemistry that validates the effects. National Conference on Organic farming for Sustainable Production, NASC Complex, New Delhi, 15 p.

Rupela, O. P., Gopalakrishna, S., Krajewski, M. and Sriveni, M. (2003). A novel method for identification and enumeration of microorganisms' potential for suppressing fungal plant pathogens. *Biology and Fertility of soils,* 39(2): 131-134.

Steiner, R. (1920). Spiritual Science and Medicine, The Rudolf Steiner Publishing Co. London.

Singh, S., Sachan, K., Alam, K. and Kumar, A. (2022). Modern Concept in Agriculture. Edition: 1(1), Akinik Publications, Delhi, pp 17-34.

Chapter 16

Physiology of Pruning and Training on Fruit Trees

Vibhu Pandey[1], Shalini Singh[1], Amit Kumar[1] and Khursheed Alam[2]*

[1]Ph.D. Research Scholar, Department of Fruit Science,

[2]Ph.D. Research Scholar, Department of Vegetable Science,

Sardar Vallabhbhai Patel University of Agriculture and Technology, Meerut – 250 110, Uttar Pradesh

**e-mail: Pandey.vibhu007@gmail.com*

INTRODUCTION

Woody plants are pruned to maintain a particular size, shape, and growth pattern. Fruit trees are pruned to increase fruit quality by promoting an optimal balance between vegetative (wood) and reproductive (fruiting) growth, as opposed to ornamental plants, which are pruned to enhance the aesthetic quality of the plant. Fruit trees should always be pruned annually to increase fruit quality while decreasing yield. By removing extra flower buds and promoting the growth of new shoots with high-quality flower buds, pruning boosts fruit size. Light is necessary for the development of flower buds, fruit set and growth and the development of the red colour, and pruning increases light penetration into the canopy. Pruning also enhances pest control by allowing better spray penetration into the tree. Additionally, pruning increases air movement across the canopy, which improves drying conditions and lessens the severity of many diseases. All plants can benefit from the information, but fruit trees are the ones to which it is most relevant. Fruit tree pruning involves a certain amount of art that is based on knowledge of the physiology and growth of plants. In other words, by comprehending how plants develop and how they will react to various plant manipulations, we may modify vegetative development and fruiting to produce trees and fruit with desired traits. Pruning can only be understood if you have a fundamental understanding of several aspects of plant physiology. Only two methods exist for plants to grow. The height and width of the canopy are increased through primary growth, which is the lengthening of shoots and roots. Stems and roots undergo secondary growth, which thickens them. Cell division, cell expansion, and cell differentiation are prerequisites for both types of growth. Some of the cultivated fruit trees grow wild and do not give sufficient yield unless pruned or trained to a specific form. To enhance the functionality of the tree structure and correct or maintain it. It is done to boost the plant's productivity and fruit quality, lengthen the tree's life, shape it into a manageable size, and to get the most out of the orchard. When a branch is removed, less potential leaf surface is created in addition to the removal of stored carbohydrates. It boosts growth near the cut, and improves fruit size, and nitrogen per growing point. It is typically done to regulate the size of the plant, to regulate the form (structural make-up of the plant), which involves the number, placement, relative size, and angle of branches, to improve the quality of fruits by better light distribution, to remove diseased, crisscrossed, dried, and broken branches, to remove the non-productive parts in order to direct the energy into those parts that are capable of bearing fruits, to develop the right proportion of root-shoot ratio, to regulate the plant's. Some fruit trees does not require pruning at all and grows naturally *e.g.* Pineapple and papaya and in some fruit trees pruning varies *e.g.* Mango, sapota, *etc.* Most deciduous tree like apple, pear, almond, apricot, walnut *etc.* and grapes, ber, fig, citrus, pomegranate, guava *etc.* required pruning to train them for desired shape.

Pruning

Pruning is the practice of cutting out unwanted woody material from a tree. It is essential to develop a balanced tree in its formative years. Thereafter most trees need on-going pruning to produce quality fruit and to maintain structure and balance. It is done to divert a part of plant energy from one part to another part of plant. Pruning is a dwarfing process and can be used to maintain any desired tree size.

Jadhav *et al.* (2002) noticed that earliest emergence of vegetative bud sprout, shoot length, number of flowers per shoot and number of fruits per shoot, on severely pruned (60 cm) trees of guava was found to be significantly more than mild pruned (30 cm) trees and control. Rupankar Bhagawati *et al.* (2015) opined that severely pruned shoot of Guava has fewer number of shoots per pruned shoot. This might be due to a smaller number of vegetative buds in severely pruned shoot. While the new shoots per shoot was lowest in shoots without pruning. Meland (2009) stated that apple fruit weights and soluble solids contents values were highest with the lowest crop load and decreased with increasing crop load. Trees with the highest crop load had the lowest crop load in the following year. Fruit quality was generally high for all treatments

Objectives of Pruning

- ☆ To control flowering and fruiting
- ☆ To augment production in plants which bear on new shoots
- ☆ To obtain regular bearing
- ☆ To remove diseased, damaged, insect infested and week shoots
- ☆ To thin out flowers and fruits
- ☆ To ensure access to sunlight to bearing shoots
- ☆ To have a balance between vegetative and reproductive growth

Effect of Pruning

- ☆ It increases new vegetative growth
- ☆ In young trees flowering will be delayed
- ☆ In old trees there will be new vigorous vegetative growth which bears fruit
- ☆ It reduces bearing surface are as a result tree remain dwarf which in compensated by accommodating more number of dwarf trees (because pruning is a dwarfing process)
- ☆ Improvement in size, colour and quality of fruits

Principles of Pruning

- ☆ Young trees are pruned to train it to acquire a desired shape.
- ☆ In old trees light heading back is done to stimulate the flowering.
- ☆ In bearing trees light pruning is done to stimulate fresh growth. It bearing flower buds on fresh growth.
- ☆ In old trees heavy pruning is done to restore a vigorous.
- ☆ All the diseased, weak, dead or shading branches must be removed.

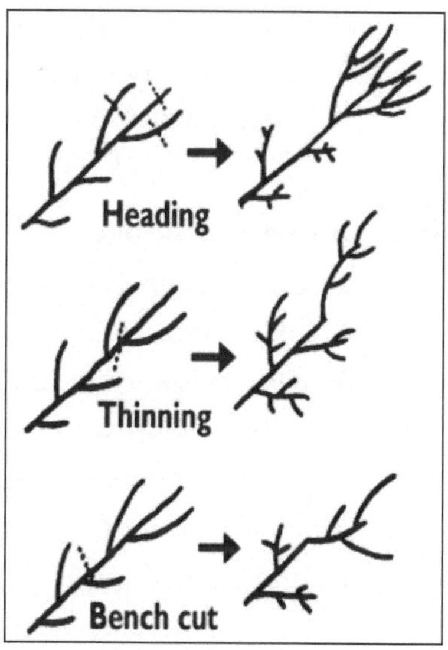

Systems of Pruning

Heading Back

When the terminal portion of branch/ shoot is removed and it encourages lateral growth from the remaining shoot.

- Heading back promotes the growth of lower buds as well as several terminal buds below the cut.
- When lateral branches are headed into one year old wood, the area near the cut is invigorated.
- The headed branch is much stronger and rigid, resulting in lateral secondary branching.

Thinning Out

When a shoot is entirely removed from the point of its origin and no re-growth is allowed to occur from the cut ends.

- Do not invigorate the tree.

Dehorning

Cutting away the main limbs or thick major branches.

Bulk Pruning

Heavy pruning all over the tree. For good fruit production only judicious heading back or thinning out should be done.

Season of Pruning

- Dormant Pruning
- Summer Pruning

Dormant Pruning

- Most often done during the winter- commonly referred to as dormant pruning.
- Dormant pruning is an invigorating process.
- Heavy dormant pruning also promotes excessive vegetative vigor,
- Timing of dormant pruning is critical. Pruning should begin as late in the winter as possible to avoid winter injury.

Why Dormant Pruning?

- Visible Branching Patterns (branches can be seen easily at this time).
- Other orchard operations are less pressing
- Less danger of pulling bark away from around pruning wounds
- Pruning goes well with cool weather.
- Increased growth response
- Has less dwarfing effect on the trees than summer pruning.
- Improved Equipment Access
- Fungus diseases are dormant too, so there's less danger of transmission

- ☆ Tree is just about to start the active growth that promotes wound-healing

Pruning is a Dwarfing Process

Pruning appears to boost growth because it encourages vegetative growth close to the pruning cut. However, a tree that has been pruned periodically always weighs less than a tree that hasn't been pruned.

Pruning Reduces Yield

Pruning eliminates fruit-bearing wood that has blossom buds on it. Although fruit quality is enhanced by pruning, output from pruned trees is almost always lower than output from unpruned trees. By maximizing the amount of leaf area per fruit, pruning enhances fruit size. Pruning enhances the light distribution all through the tree, which is crucial for the growth of fruit sugar levels and red colour.

Pruning Delays Fruiting

In young trees, pruning promotes vegetative development as opposed to reproductive growth. A tree that has not been pruned will always blossom and bear fruit earlier in its life than a tree that has. Young trees are pruned in order to encourage the development of desired branches and to build a sturdy tree structure which will support heavy crops as the tree matures. A tree's physiology shifts from vegetative growth to reproductive growth as it ages. It's critical to delay fruiting until the trees have almost completely filled their space in order to achieve high annual yields from mature trees. One technique for delaying young trees' fruiting is pruning.

Summer Pruning

Summer pruning is the deliberate removal of leafy branches while they are still growing. The timing, degree, strength, location, and variety of summer pruning affect the responses. In the 1980s, a number of studies studied summer pruning, and numerous broad conclusions regarding the process can be drawn. Summer pruning lowers within-tree shade, usually enhances fruit red color development, and occasionally enhances flower bud development. The leaves that create photosynthates (sugars) for the growth of all tree sections are removed during the summer pruning. Fruit size and sugar content might occasionally decrease as a result of summer pruning. Summer pruning prevents late-season trunk enlargement and root growth because it reduces whole-tree photosynthesis.

Advantages of Summer Pruning

- ☆ Better illumination of leaves and fruits
- ☆ Improved assimilation in centre of trees

- Better fruit colour, fruit quality and storage.
- Slower total growth of trees
- Less wood production
- Less total pruning effort and improved work efficiency
- Possibility of reducing crown volume of tree, better utilization ot space, more tree per acre
- Less danger of winter injury

Training

A process, by which a plant is tied pruned, fastened or staked to give a proper shape. It means developing a desired shape of the tree with particular objectives by controlling habit of growth. Training is start from nursery stage of plant. Some fruit crops like grape wines, ber, fig, guava *etc.* require training. Training young fruit trees is essential for proper tree development. It is better to direct tree growth with pruning. The goal of tree training is to direct tree growth and minimize cutting.

Tree development can be steered toward a particular shape and form through the practice of training. For appropriate tree development, young fruit trees must be trained. Tree training aims to reduce cutting while directing tree development. The main goal of training is to ensure that a young tree develops a sturdy framework of scaffold branches, admits more sunlight and air to the centre of the tree, and exposes as much of its leaf surface to the sun as possible. Training also aims to control the tree's growth so that different cultural operations, like spraying and harvesting, are carried out at the lowest possible cost, protects the tree from sunburn and wind damage, and ensure a balanced distribution of fruit-bearing parts on the main branch.

Aim to adapt training and pruning schemes to the cultivar to optimize tree manipulation and eventually to decrease field work (Lauri *et al.*, 2008).

Tree manipulation not only acts upon the within-tree light climate with known consequences on leaf functioning and fruit quality, but it also modulates infestation by pests and infection by pathogens (Simon *et al.*, 2006, 2007).

Objectives of Training

- To admit more light and air to the centre of the tree to expose maximum leaf surface to the sun
- To direct the growth of the tree so that various cultural operations such as spraying, ploughing, harvesting can be performed easily and at lower cost
- To protect the tree from sun burn and wind damage
- To secure a balanced distribution of fruit bearing parts of the tree

Principles of Training

- ☆ Formation of the mainframe work must be strong. The branches must be suitable spaced apart and the tree must be balanced on all the sides
- ☆ Never allow several branches to grow at one place or very near each other
- ☆ Careful training to main branches is very essential
- ☆ Another important point about training is that if two branches are growing at the same point try to train them to grow at a wider angle. Narrow angle is always weak

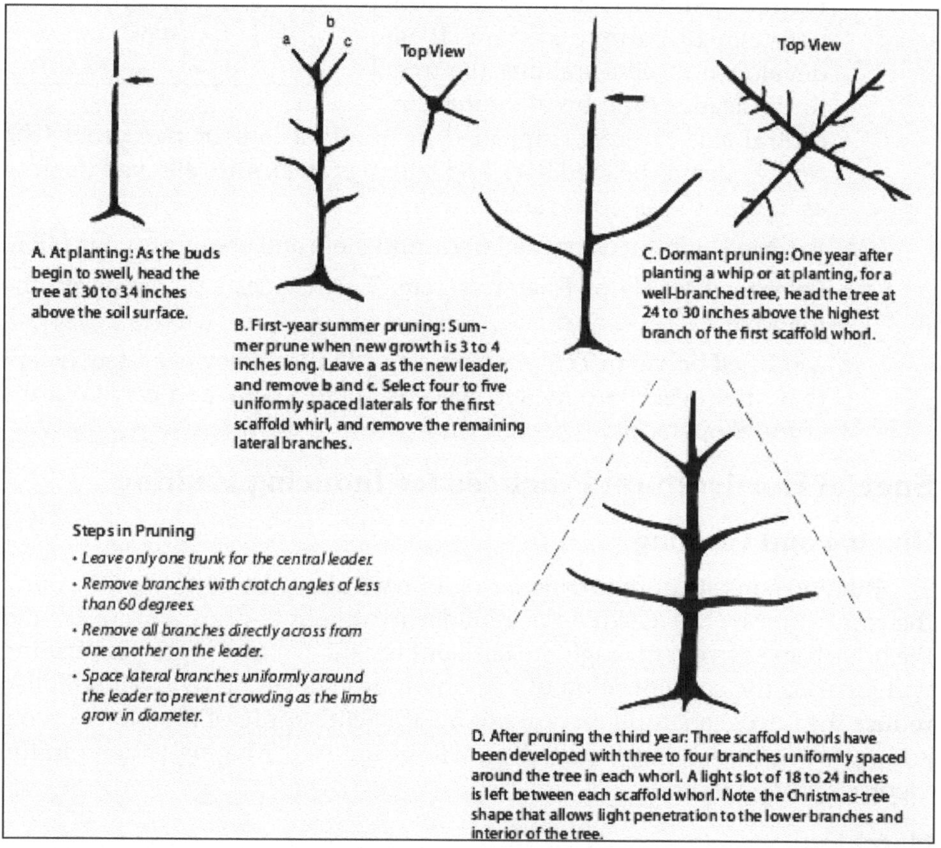

Figure 16.1: Pruning a Semi-Dwarf or Larger Central-Leader Tree when Planted as an Unbranched Whip.

Systems of Training

- ☆ **Central leader system:** In this system the central leader system branches are allowed to grow indefinitely so that it will grow more rapidly and vigorously than the side branches and tree became tall.

Such a tree bears fruit more near the top. The lower branches are less vigorous and less fruitful.

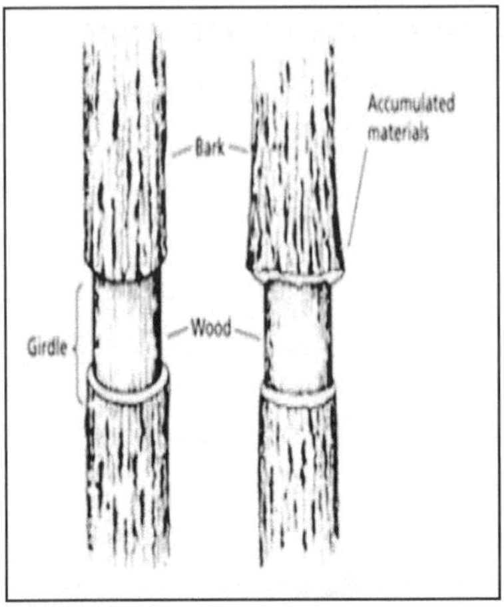

- ✫ **Open centre or vase system:** The main stem is allowed to grow only up to a certain height about 1.5 to 1.8 m and then it cut for development of lateral branches. It allows full sunshine to reach each branch.
- ✫ **Delayed open centre or modified leader system:** It is intermediate between the above system. It is developed by first training the tree to the leader type by allowing the central axis to grow unpruned for the first four or five years. The central stem is headed back and lateral branches are allowed to grow as in the open centre system
- ✫ **Bush system:** An unpruned tree multi-stem and dwarf growing habit.
- ✫ **Overhead trellis or Bower system:** When vines are trained on the mandap.
- ✫ **Modified bower or Telephone system:** Similar to bower system except that after every two meters space is kept to walk and carry out the cultural operation.

Special Horticultural Practices for Inducing Fruiting

Ringing and Girdling

Ringing consists of removing a ring of bark about 1 to 2 cm wide around the trunk of branches. Girdling is a milder treatment to draw a knife around the branch so as to cut through the bark but not the wood. Ringing or girdling will increase the concentration of carbohydrates above the twigs. It will also reduce the nitrogen supply because of insufficient supply of food to the roots and resulting in slow and limited growth of the tree. Ringing is done in the vigorous mango trees.

Notching

It is similar to ringing except that in notching only soil slip bark about 0.2 to 0.5 cm thick and 1.5 to 2.5 cm in length is removed just above or close to a dormant bud in slantwise so that the latex does not coagulate in the bud itself. The bud selected should be large and healthy which is produced on a perfect mature wood and has undergone dormancy. Generally, 3 to 4 buds in the middle portion of the selected shoot are best to operate on. Fig has responded to notching and it is practiced in fig cultivation. The season for notching the fig is August- September.

Smudging

Smudging is the practice of smoking the tree by burning brushwood on the ground and allowing smoke to pass through the center of the crown of the tree the smoking is discontinued as soon as the terminal buds begin to swell. It is not commonly followed in India. The practice of smoking the trees like mango is commonly employed in the Philippines to produce off-season crops. Smoking contains ethylene gas, which is responsible for the initiation of flowering.

Bending

Bending a branch downward, sometimes checks growth and causes accumulation of starch in the branch with greater flowering. This tends to increase carbohydrate concentration. In the case of bending the effect of apical dominance of the growing shoot is removed and auxins during translocation activate the dormant buds. The bending of a shoot towards the light. The auxin concentration is highest on the dark side of the stem and causes cells on that side to elongate, resulting in stem curvature.

Root Pruning

Root pruning results in less carbohydrate utilization of the top growth though there is a little more utilization of carbohydrates for root functions. There is an accumulation of carbohydrates due to the check of top growth, which results in fruit bud differentiation. The effect of root pruning is to check the vegetative growth and the plant became dwarf. Root pruning is generally included in bahar treatments given to fruit trees like Mosambi, santra, guava, pomegranate, lime *etc.*

Growth Correlations and Apical Dominance

The pace of height growth, the quantity and proportionality of branches and branch crotch angles all affect the shape and size of trees. Growth correlations take place within every shoot and throughout the entire plant in deciduous fruit trees, just like in other woody species. Shoot growth, branching and branch angle development are regulated by the correlative system, which is still not entirely understood.

Apical dominance is the most obvious phenomenon. Young leaves and terminal apices on developing shoots prevent the growth of axillary buds. One or two uppermost buds dominate and create long shoots in second-year shoots, while the remaining buds either become spurs or dormant. On either growing or dormant branches, pruning reduces apical dominance, frees buds from correlative inhibition, and modifies the structure and form of the tree.

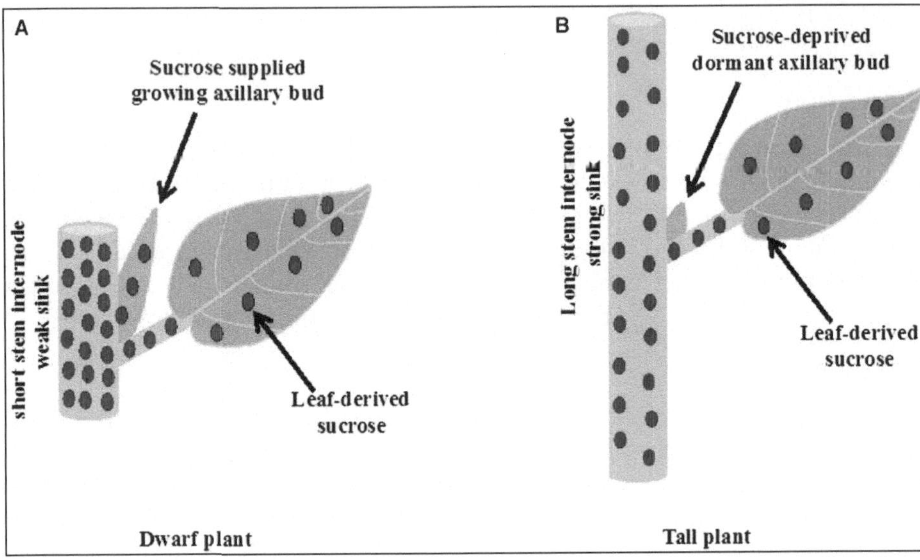

Pruning and Training in different Kinds of Fruit Trees

Guava

Light annual pruning is necessary after every harvest. There are three pruning times in guava that is Jan-Feb, May-June, and Sep-Oct. Guava bears on the current season's growth and flowers appear in the axils of leaves, therefore, it responds well to pruning. Pruning could prove to be the most effective method for eliminating rainy season crops and production of winter season guava.

Grapes

The yield of grapes is dependent to a larger extent on the method of pruning. Being a vine crop, training is essential for grapes. Branching habit 'Helicoid' systems of training commonly followed in grapes include head, Kniffin, overhead trellis, and bower. Grapevine is a past season bearer. In the mild tropical region, vines are pruned twice and the crop is harvested twice. As a result, five crops are harvested every two years. In North India pruning is generally completed by mid-January.

Pear

Like warm, sheltered growth areas. Propagated by grafting fruit on 2-year wood or older. Late winter or early spring. Winter, spur thinning or large branch removal. Spring, blossom or fruit thinning. Regular thinning to prevent overcrowding. Self-fertile.

Plum

Fruit at the base of year-old wood and along 2-year and older stems. Prone to disease and infections like silver leaf disease and cankers. Prune at bud break. Deadwood in late summer or Autumn. Use care to not prune back too much young growth. Propagated by grafting.

Sweet/Duke Cherry

Fruit on older wood and base of previous year's growth. Pruning in winter makes trees susceptible to silver leaf disease. Prune for training at bud burst in spring for new trees and in summer for established trees. Prune out diseased, dying or dead wood immediately. Remove shoots with narrow forks or V angles. Fruit thinning rarely needed.

CONCLUSION

Fruit trees are pruned during their latent or vegetative phases of growth by thinning or heading shoots and branches. Pruning affects the majority of physiological functions. Pruning and training can promote high yield, good fruit quality, and the balance between vegetative growth and fruiting, making it an essential orchard practice. Pruning and training are also necessary for healthy fruit development. Pruning trainers that are successful take note of how their trainee plants react to pruning and other plant management techniques. Understanding the physiology of plants and how pruning affects that physiology is necessary for efficient fruit production. To enhance the functionality of the tree structure and correct or maintain it. It is done to boost the plant's productivity and fruit quality, lengthen the tree's life, shape it into a manageable size, and to get the most out of the orchard. When a branch is removed, less potential leaf surface is created in addition to the removal of stored carbohydrates. It boosts growth near the cut, and improves fruit size, and nitrogen per growing point.

REFERENCES

Bhagawati, R., Bhagawati, K., Choudhary, K. V., Rajkhowa, J. D. and Sharma, R. (2015). Effect of Pruning Intensities on the Performance of Fruit Plants under MidHill Condition of Eastern Himalayas: Case Study on Guava. *International Letters of Natural Sciences*. Vol. 46. pp 46-51.

Jadhav, B. J, Mahurkar, V. K. and Kale, V. S. (2002). Effect of time and severity of pruning on growth and yield of guava (*Psidium guajava* L.) cv. Sardar. *Orissa Journal of Horticulture*. 2002. 30: 2, 83-86.

Lauri PÉ. 2008. Trends in Apple Training in France – An Architectural and Ecophysiological Perspective. *Acta Horticulturae* 772: 483-490.

Meland, M. (2009). Effects of different crop loads and thinning times on yield, fruit quality, and return bloom in *Malus x domestica* Borkh. 'Elstar'. *Journal of Horticultural Science and Biotechnology*. 84: Special Issue, 117- 121.

Simon, S., Lauri, P. É., Brun, L., Defrance, H., Sauphanor, B. (2006). Does fruit-tree architecture manipulation affect the development of pests and pathogens? - a case study in apple orchard. *Journal of Horticultural Science and Biotechnology* 81(4): 765-773.

Simon, S., Sauphanor, B., Lauri, P. É. (2007). Control of fruit tree pests through manipulation of tree architecture. Pest Technology 1(1): 33-37.

Training and Pruning Apples Trees, Virginia Cooperative Extension publication 422-021, http: //pubs. ext.vt.edu/422-021/

Chapter 17

Precision Farming in Horticulture ('Per Drop More Crop')

Amul Yadav[1]*, *Ankita Sharma*[2] *and Pragya Uikey*[2]

[1]*M.Sc. Scholar,* [2]*Ph.D. Research Scholar,*

Department of Horticulture, College of Agriculture, Jawaharlal Nehru Krishi Vishwa Vidyalaya, Jabalpur, Madhya Pradesh

**e-mail: amulyadav13@gmail.com*

INTRODUCTION

Precision farming or precision agriculture is about doing the right thing, in the right place, in the right way, at the right time Optimizing agricultural production inputs like water, seed, fertilizer, *etc.* to improve yield, quality, profit, decrease waste, and become eco-friendly in order to increase the accuracy of its applications, precision farming aims to match agricultural inputs and practices to crop and agro-climatic conditions. Agriculture, by its very nature, is a complicated and scientific field. Information and communication technology has digitally revolutionized agriculture as it has significantly expanded the broadband connectivity by deploying the 'Internet of Things' (IoT), big data and other innovative applications for the digital society. These interventions have come-up as E-agriculture solutions as modern farmers can willingly embrace the emerging technologies. To combat with the climate viagries that the farmers are experiencing, cloud computing helps to generate precise weather data by providing an insight to minimize production risk and ensure profit to build economic resilience (Sharma and Rehan, 2020).

What is Precision Farming in Horticulture?

Precision Farming in Horticulture sector provides an excellent opportunity in raising the income of the farmers throughout the year. Despite the potential and significance of growing a plethora of horticultural crops, the sector is yet to make a break-through in advancement of the productivity and market chain for a variety of reasons. This is attributable to the fact that most producers are not implementing the most advanced production technology, such as Precision Horticulture Practices regarding soil, water, and planting material choices. The non-judicious application of critical inputs and non-adoption of improved technologies lead to less production as well as productivity per unit area (Scharf, 2015 and Bisen *et al.*, 2020).

According to Professor Pierre C. Robert, who is regarded as the father of precision farming, the practice involves more than just the introduction of new technologies; rather, it represents an information revolution made possible by new technologies that results in a higher level, more precise arm management system. (Dutta *et al.*, 2011) [1].

Precision vegetable farming is a modern-day procedure utilized to improve crop productivity using the newest technologies, cloud computing, information technology (IoT), WSN, *i.e.*, artificial intelligence (AI) machine learning (ML). Thus far, the majority of research suggests that PA-based methods possess a good impact on productivity and sustainability. The farmer may do much more regular tillage, alter sowing rates, apply fertilizers depending on soil conditions, create even more crop protection programmes with enhanced accuracy, and comprehend the yield variance area as the true benefits of precision vegetable farming. However, the producer must be willing to modify his management styles in order for these gains to be really cost-effective in harvest production (Brar *et al.*, 2020). In order to improve crop performance and setting quality, precision farming involves the use of tools and concepts to manage temporal and geographical variability regarding all variables of gardening generation (Lee *et al.*, 2010). Utilizing location-specific high-tech interventions is necessary for optimal resource management in precision farming. For today's farming difficulties, such the need to combine efficiency with ecological issues, precision vegetable farming offers a novel solution utilizing a systems approach (Khan, 2013).

Precision Farming is the use of advances and standards to oversee spatial and worldly changeability related with all parts of green creation for improving harvest execution and condition quality (Nabi *et al.*, 2017). It is a farm input management system that is information and technology based that tries to employ technologies and concepts to discover, analyze, and manage geographical and temporal variability related to all elements of agricultural output. The application of precision farming techniques improved environmental quality, land resource protection, farm profitability, and sustainability. This production system has great potential to solve all the above associated problems of agricultural production systems (Bisen *et al.*, 2020).

- ☆ An integrated information and creation-based farming system that strives to minimize unexpected effects on wildlife and the environment while increasing long-term, site-specific and overall farm making efficiency, productivity, and profitability.
- ☆ Site-Specific Crop Management (SSCM) Precision Agriculture whereby decisions on resource application and agronomic practices are improved to better match soil and crop requirements as they Vary in the field.
- ☆ Site-Specific Crop Management (SSCM) is a form of precision agriculture that intends to better match crop and soil needs as they change in the field by improving resource management and agronomic methods.
- ☆ Farming by the foot, farming by satellite, site-specific organization its management, *etc.*

The following are the five components of precision farming:

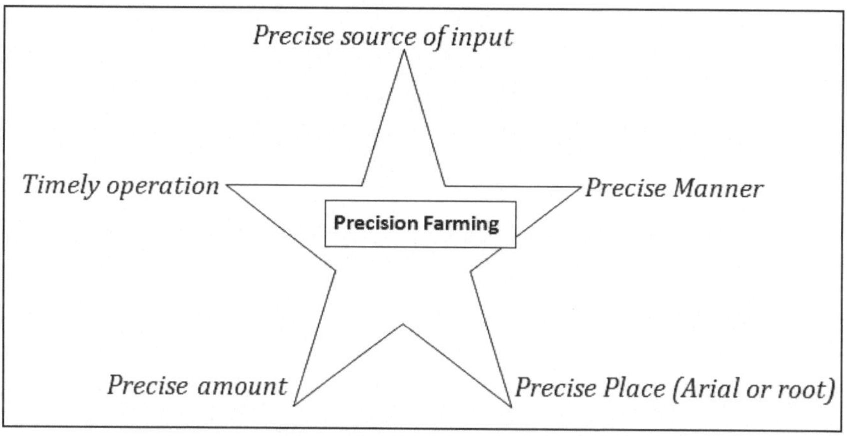

Figure 17.1: Components of Precision Farming (*Source*: Nikhil Thakur *et al.*, 2020).

Why Precision Farming?

1. To boost agricultural productivity in terms of profit.
2. To protect cultivable land from soil deterioration.
3. To reduce the usage of chemicals in agricultural production.
4. For efficient use of water resources.
5. To disseminate modern farm techniques that improve crop quality, quantity, and minimize cost of production of agricultural crops.

Development of Precision Farming

- ✰ The notion of precision farming initially emerged in the United States of America in the 1980s.
- ✰ Precision farming is an agricultural management approach that is based on modern information technology including GPS (Global Positioning System), remote sensing technologies, and geographic information systems (GIS)
- ✰ Precision farming in India is defined as the precise use of agricultural inputs for crop development in India is described as taking into consideration pertinent elements including soil, weather, and crop management techniques.
- ✰ It is actually information and technology-based farming system where inputs are managed and distributed on a site-specific basis for long term benefit (Himani B. Patel. 2016)

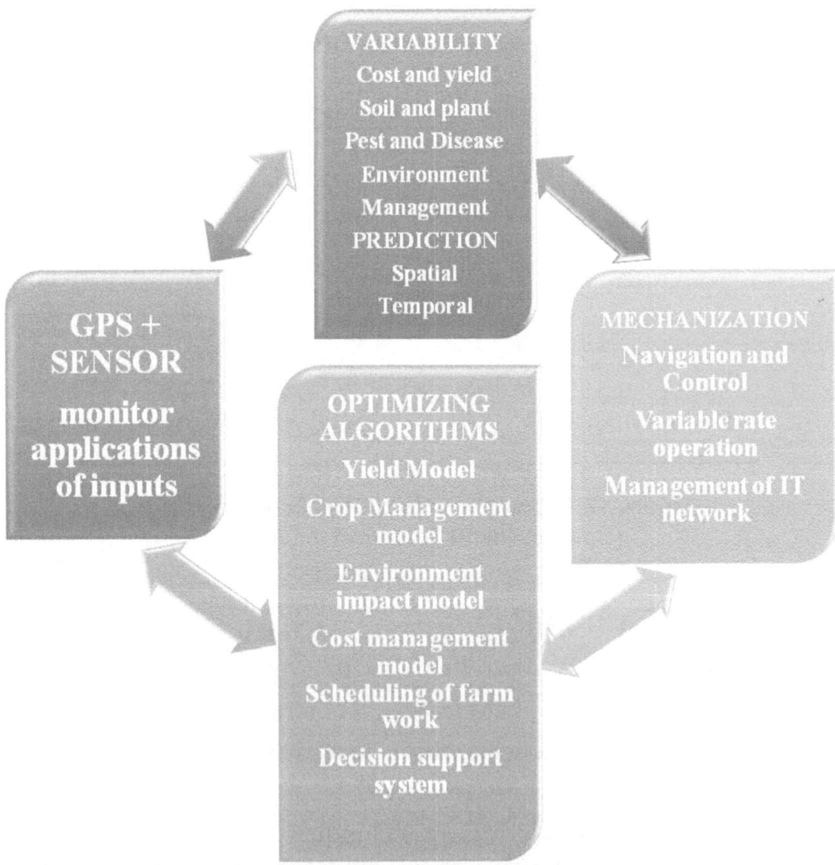

Figure 17.2:Development of Precision Farming (*Source*: Ankita Sharma *et al.,* 2020[6]).

Extent of Adoption of Precision Farming in India

Reception and achievement of PA in India will be determined by whether or not the selection techniques are properly designed. An arranged quantities of examinations and investigation are required before utilization of PA to Indian agri-business (Nikhil Thakur *et al.*, 2020).

There are 3 stages to enter PA age, to be specific present stage, middle of the road stage and future stage. Present stage includes uniform harvest and soil the executives, improvement of master labor and establishment for PA, promotion of PA idea by broad communications correspondence, course, workshop and so forth, though transitional stage will follow defined irregular testing inside zone, depiction of the board zone all through the nation approval of PC models with zone explicit information and future stage will include fine framework examining and detecting, utilization of zone explicit PC model to recreate the agronomic information conditions and exact detecting and the executives (Mondal and Basu, 2009).

Need for Precision Farming

One-third of the world's population is fed by hundreds of millions of smallholder farmers, an astounding performance given that much of this occurs in non-mechanized systems with scant to no access to information and practical solutions for improved farming while dealing with new production issues. Climate change has a negative impact on yields through thermal waves, droughts, and other extreme atmospheric conditions (Nelson *et al.*, 2009). Soils are getting depleted, experiencing degradation and erosion, arable land and labour for farming is lost due to urbanization (Satterthwaite *et al.*, 2010), and drying up of ground and surface waters. Exhausting Fossil fuel depletion has an impact on agricultural operations and elevates fertilizers costs. Global food security and rural livelihoods are seriously threatened by these challenges. The farmers having small holdings struggle to meet the demand and produce enough yields to support their families' basic needs and can barely keep up with their counterparts in the developed countries of the world who have access to high quality services (Sharma *et al.*, 2020).

Other contributing factors about the need for precision farming include the decline in overall productivity, the diminishing and degrading natural resources, stagnant farm incomes, a lack of eco-regional approach, shrinking and fragmented land holdings, global climatic variation, and limited non-farm employment opportunities.

Goal of Precision farming

1. **Improved sustainability and profitability**: By precisely balancing input amounts (seeding rate, variety, herbicide, and insecticide) with crop requirements, which can be influenced by weather, soil

properties (nutrient availability, texture, and drainage), and previous crop performance, the maximum profit can be achieved in each zone or site in a field. Precision farming works to maintain this profitability at the same time.

2. **Optimizing production efficiency:** In general, precision farming aims to enhance outcomes throughout a field. In the unavailability of a field with uniform yield potential, the determination of fluctuations in yield potential may enable opportunities to optimize the output amount at each site or within each "zone" (and, thus, a uniform yield objective).
3. **Optimizing the quality of product:** Precision farming seeks to enhance product quality by utilizing sensors that determine the crop's quality traits. Inputs are therefore applied in line with the data gathered from the sensors. When there are quality premiums, this can alter the amount of input required to maximize profitability and agronomic responsiveness in production systems.
4. **Use of chemicals and seeds efficiently:** PA involves efficient use of inputs *i.e.* chemicals, seeds *etc.* according to the yield potential of the soil. Zaman *et al.* (2005) have studied variable rate nitrogen application in Florida citrus based on ultrasonically-sensed tree size. Tree canopy sizes were measured real-time in a typical 17-ha Valencia grove with an automated ultrasonic sensor system equipped with Differential Global Positioning System (DGPS). Prescription maps for variable application of nitrogen fertilizer were generated from ultrasonically scanned tree sizes on a single tree basis using ArcView GIS and Midtech Fieldware. Leaf samples from trees with different canopy sizes, which had been fertilized at a conventional uniform rate of 270 kg N/ha/y, were analyzed for nitrogen concentration.
5. **Effective and efficient pest management:** Reducing agricultural production inputs will save money and the environment as a result of precision farming. Site-specific variable rate application places these chemicals (*e.g.*, herbicides, insecticides) where the issue emerges, as opposed to standard agricultural methods that apply them to the whole field.
6. **Energy, water and soil conservation:** Crop planning is the first step of a comprehensive precision agriculture strategy, which also includes soil conservation or minimally disruptive tillage practices. Additionally, water is distributed efficiently using techniques like drip irrigation, *etc.* PA also encourages energy saving because none of these situations require a lot of energy.
7. **Surface and ground water protection:** Through the effective utilization of inputs like chemicals and other materials, PA seeks to protect the environment. This stops their evaporation into surface water or discharge into ground water.

8. **Minimizing environmental impact:** Improvements in management choices that better match inputs to production demands naturally result in a reduction in the net loss of any applied input to the environment. This doesn't mean that there hasn't been or won't be any environmental harm from the production system; rather, it means that the risk of it happening is lower.
9. **Minimizing Risk:** Most farmers today consistently manage their risks, which may be seen from both an environmental and financial standpoint. Farmers typically use risk management in a production system by adding extra inputs, even when the unit cost of a particular input is seen as being "cheap." A farmer may use an additional spray, add additional fertilizer, buy more machinery, or hire more labour to guarantee that the food is produced, harvested, and sold in a timely way and to ensure a return (Kumar *et al.*, 2022).

Steps in Precision Farming Process

1. **Assessing variability:** Inputs must be used in precision farming in line with the current variability. Assessing in-field variability is therefore a very important initial step in precision farming. The instruments for assessing variability include remote sensing, the Global Positioning System, geographic information systems, and yield monitoring.
2. **Managing variability:** After determining the extent of geographical variability, these precision farming techniques try to control it by applying and manufacturing agricultural inputs only in the necessary quantities at precise times and locations, a method known as variable rate application.
3. **Evaluation of precision farming:** The following three evaluation-related concerns are crucial:
   a. Market return from the sale of the produce is the main consideration for economic viability.
   b. Maintaining the environment with a precision farming focus may enhance the conditions for the soil, water, and crops.
   c. Finally how far this technology can be transferred to other farmers (Himani B. Patel. 2016)

Elements of Precision Farming

1. Information
2. Technology
3. Decision support (management)

1. Information

- ☆ Crop characteristics like stage of crop, crop health, nutrient requirement *etc.*
- ☆ Detailed soil layer with physical and chemical properties, depth, texture, nutrient status, salinity and toxicity, soil temperature, productivity potential *etc.*
- ☆ Microclimate data (season and daily) about the canopy temperature, wind direction and speed, humidity *etc.*
- ☆ Surface and sub-surface drainage conditions.
- ☆ Irrigation facilities, water availability and other planning inputs of interest.

2. Technology

- ☆ PA is an integrated agricultural management system incorporating several technologies. The technological tools often include computer, GIS, GPS, spectral imagery, aircraft and vari-rate controllers *etc.* (Das *et al.*, 2017).

3. Decision Support (Management)

If there is no decision support system (DSS) to provide VRT suggestions, simply knowing about variability in the field will not alleviate any issues. The appropriate steps are for a DSS:

- ☆ Identify field-based environmental and biological conditions and processes that may be tracked and adjusted to improve crop output.
- ☆ Choose sensors and supplementary hardware to gather information on these processes and states.
- ☆ Acquire, transmit, and store the field data.
- ☆ Transform the data through processing and manipulation into knowledge and information that is helpful.
- ☆ Provide knowledge and information in a way that may be used to interpret it and make judgments.
- ☆ Decide on a course of action to alter the recognized condition or process in a way that benefits

Tools and Equipment in Precision Farming

Precision Farming is a combination of application of different technologies. All these combinations are mutually inter related and responsible for developments. The same are discussed below:

1. Global Positioning System (GPS)

GPS allows precise mapping of the farms and together with appropriate software informs the farmer about the status of his crop and which part of the farm requires what input such as water or fertilizer and/or pesticides *etc.* Having exact location information at any time permits harvest, soil and liquid dimensions being mapped. GPS receivers perhaps have with the spot or possibly mounted on implements enable owners to return to places to sample or perhaps deal with those areas. GPS component in precision vegetable farming is implementing the most significant factor. (Hilderth, 2003). Precision farming is an incorporated harvest management method that attempts to enhance inputs' kind and quantity with the actual crop requirements for small regions within a farm region. We have produced an interactive, portable telephone system to gain field info right away into a digital database containing yield, soil, road, water alongside contour maps (Zijp, 1994). Applications for GPS-based precision farming have been utilized for yield mapping, soil sampling, tractor steering, harvest scouting, and farm preparation. Farm owners can work in low visibility conditions, such as rain, sand, fog, and night, thanks to GPS. By combining improved soil amendment and fertilizer consumption, determining the cost-effective level of weed and insect control, and preserving the organic energy sources for future use, growers may reap extra rewards. Many growers now use GPS-derived products to streamline operations inside their farming businesses. By using GPS, farmers are able to return year after year to certain locations in the region to gather soil samples or even check on crop conditions.

2. Geographic Information System (GIS)

A sophisticated system made up of tools for the gathering, storage, and retrieval of data at will as well as for the analysis, transformation, and presentation of the geographical data for a particular purpose is known as a Geographic Information System (GIS), which is a crucial fundamental technology (Delgado *et al.*, 2019). Geography, environmental sciences, natural resources, forestry, agriculture, food, manufacturing, finance, and health services seem to be just a slew of the numerous domains whereby GIS is generating an influence. The use of GIS technologies for agriculture related applications at local, regional, national, and international sizes has grown significantly during the past few decades. In order to get a comprehensive understanding of a farm or region and to provide intervention or corrective actions for the crops and/or soils, these applications most frequently entail the use of GIS along with partner technologies like remote sensing, GPS, and data analytics (Soomro, 2015). Researcher utilized GIS features, such as the distribution of soil types, maps of soil texture, distribution of buried deep underground water levels, distribution of soil fertility, distribution of soil pollution, distribution of hydraulic conductivity of soil (Ks), slope (S), depth to water table (DTW), and distribution of electrical conductivity of groundwater (ECw), as well as information on the climate, topography, and satellite data, to identify an expansive range of interventions, dependencies, and impacts. Zolekar and Bhagat (2015) have used GIS-based MCDM model with IRS P6 LISS-IV images as input for the evaluation of agricultural practices in hilly regions. The rank of influential criteria was determined by correlation analysis and recommendations from scientific literature. The combined use of remote sensing and GIS turned out to be beneficial for land suitability evaluation. Abdel Rahman *et al.* (2016) combined the remote sensing and GIS technology to assess the soil fertility status. They have used the LISS III and IV images for land use classification and the RUSLE method for soil erosion estimation, collected the soil nutrient field data, and applied a geostatistical model to identify the spatial variation of soil erosion and nutrient availability.

3. Variable Rate Technology (VRT)

Variable rate application (VRA) in precision agriculture is an area of technology that focus on the automated application of inputs to a desired crop. The agricultural inputs application is based on information that is recorded by sensors, maps and GPS. These inputs contain things like fertilizers, chemicals, and seeds, and they all aid to augment the farmers crop production. There are several types of technology that are used in variable rate application for precision agriculture. They contain different modern agricultural tools like drones and satellite imaging, to IOT and artificial intelligence (AI). Applying fertilizers and chemicals is a common farming activity that can be entirely automated with the precise execution of variable rate application technology (VRT) (Parihar *et al.*, 2021)

4. Yield Maps

Yield maps are produced by processing data from adapted combine harvester that is equipped with a GPS, *i.e.* integrated with a yield recording system. Yield mapping involves the recording of the grain flow through the combine harvester, while recording the actual location in the field at the same time. Yield monitoring is utilized through precision agriculture because it aids in identifying the measures and describe the intra-field variability within a cropping system which is exactly what forms the basis of the concept of precision agriculture. Within a single field, the yield monitoring provides a wide range of statistics. This information is produced as a result of the intricate interactions among several elements that occur on a farm, including as farming practices, environmental factors, and climatic conditions. As a result, this information becomes a valuable resource when trying to comprehend the farm in order to use additional precision agricultural techniques.

5. Remote Sensors

These are generally categories of aerial or satellite sensors. They can indicate variations in the colours of the field that corresponds to changes in soil type, crop development, field boundaries, roads, water, *etc.* Arial and satellite imagery can be processed to provide vegetative indices, which reflect the health of the plant.

1. **Remote sensors for monitoring crop health:** Remote sensing data and images provide farmers with the ability to monitor the health and condition of crops. Reflected light that is invisible to the human eye can be detected via multispectral remote sensing. The chlorophyll in the plant's leaf reflects green light while absorbing the majority of the sun's blue and red-light waves. Stressed plants reflect different wavelengths of light than healthy plants do. Compared to stressed plants, healthy plants reflect more infrared radiation from the spongy mesophyll plant leaf tissue. Farmers will have more time to investigate the issue region and use a remedy if they can identify plant stress zones before they become noticeable.
2. **Remote sensors for monitoring water stress:** The use of remote sensors to directly measure soil moisture has had extremely limited success. Since synthetic aperture radar (SAR) sensors are sensitive to soil moisture, soil moisture has been measured using SAR sensors directly. SAR data must undergo substantial processing in order to eliminate surface-induced noise, such as terrain, vegetation, and soil surface roughness. A reduction in crop evapotranspiration is a sign of agricultural water stress or other issues, such plant disease or insect infestation. To measure field fluctuations, a crop water stress index (CWSI) model has been integrated with remote sensing photos.
3. **Remote sensors for weed management**: Using aerial remote sensing to track and identify dispersed weed populations has not yet proven to be very effective. Some challenges include the fact that weeds frequently spread across a crop with a similar spectral makeup and thus extremely large - scale high-resolution images will be needed for detection and identification. Remote sensors are directly mounted on the sprayer equipment when using machine vision technology systems to find and classify weeds. High spatial resolutions are facilitated by being adjacent to the crop. Machine vision systems provide the real-time capabilities required to operate sprayer equipment that may be employed in the field.
4. **For abiotic stress:** Abiotic stress such as deficiency of moisture can also be identified through remote sensing technology. Several techniques such as high spatial resolution multispectral and thermal airborne imagery were employed to monitor crown temperature and the Photochemical Reflectance Index (PRI) in peach orchards. Different irrigation regimes included sustained and regulated deficit irrigation strategies, were utilized to induce stress. The results depict that a difference in reflectance pattern between well irrigated and stressed plants can be observed (Saurez *et al.*, 2010)

5. **On Quality:** One of the main factors that the producer considers is quality. The quality of grape wine, which is closely associated to the vigour of the vine, was evaluated by Johnson *et al.,* in 2001. In a 3ha commercial vineyard of Chardonnay wine grapes, sub-block management zones were created using remotely sensed vegetation index images. Following ground-based studies, it was obvious that low and high vigour zones could be distinguished by their biomass (mainly shoot vigour), vine water status, and, most significantly, their characteristics in terms of fruit and wine. The creation of distinct wine lots from a block that was previously managed as a single management unit was made possible by harvesting by vigour zones. This feature alone has importance since it gave the winemaker more freedom in the final blending process.

Research on apple orchard area estimation using remote sensing and agro-metrology land-based observation was performed by Mustaq and Asima (2014) in Pulwama district of Kashmir valley. Landsat and AWIFS digital data were utilized to monitor and estimate area under apple orchards. It was found that majority of apple orchards (89.82 per cent) were concentrated between elevation range 1500-2000 mts, elevation above (2000m) occupied 10 per cent area while 0 per cent under lower elevations

6. Proximate Sensors

These sensors can be used to measure soil parameters such as N status and soil pH) and crop properties as the sensor attached tractor passes over the field.

7. Precision Irrigation Systems

Recent developments are being released for commercial use in sprinkler irrigation by controlling the irrigation machines motion with GPS based controllers. Wireless communication and sensor technologies are being developed to monitor soil and ambient conditions, along with operation parameters of the irrigation machines (*i.e.* flow and pressure) to achieve higher water use efficiency.

8. Precision Farming on Arable Land

The use of PA techniques on arable land is the most widely used and most advanced amongst farmers. CTF (contolled Traffic Farming) is a whole farm approach that aims at avoiding unnecessary crop damage and soil compaction by heavy machinery, reducing costs imposed by standard methods. Controlled traffic methods involve confining all field vehicles to the minimal area of permanent traffic lanes with the aid of decision support systems. Another important application of precision agriculture in arable land is to optimize the use of fertilizers especially, Nitrogen, Phosphorus and Potassium.

In a study conducted across 20 districts in the four states of Gujarat, Madhya Pradesh, Uttar Pradesh, and Bihar, Chaudhari *et al.* (2019) developed a research technique for the area evaluation of three key fruit crops, including the banana, mango, and citrus. A suitable bio-window for analyzing various crop types was chosen, and mapping of crops was carried out using object-based classification of high-resolution remote sensing data (Resource sat LISS III and/or LISS IV, Cartosat - 1 PAN) and pixel-based hybrid classification, which combines unsupervised ISODATA clustering with supervised MXL classification. Their accuracy was then evaluated, and they were compared to departmental reported statistics.

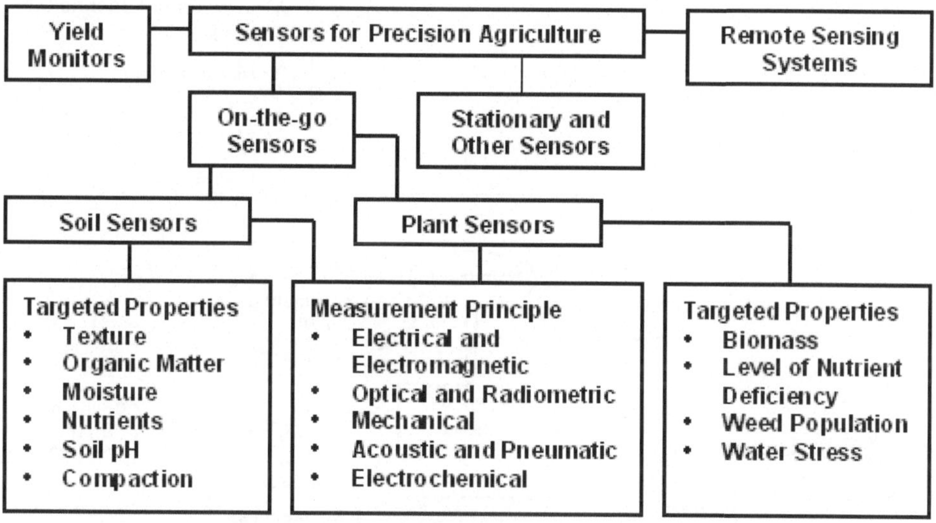

Figure 17.3: On-the-go Soil and Plant Sensors as Part of the Family of Sensors Used in Precision Agriculture (*Source*: Achim Dobermann *et al.*, 2004).

Classification of Precision Farming Technologies

Data Collection Technologies	*Data Process & Decision-Making Technologies*	*Application Technologies*
☆ Soil sampling and mapping	☆ GPS-GIS based farm management	☆ Variable-rate technology
☆ Yield monitoring and mapping	☆ Agricultural mapping software	☆ Yield monitors
☆ Remote sensing	☆ Geoinformatics	☆ Agricultural robots
☆ Global satellite positioning	☆ Geostatistics	☆ Wireless data logger and Sensor catalogue
☆ Geospatial Technology	☆ Crop modelling	☆ Global Navigation Satellite system (GNSS) based guidance

Data Collection Technologies	*Data Process & Decision-Making Technologies*	*Application Technologies*
☆ Field/crop scouting	☆ Artificial Intelligence primarily based Controlling systems	☆ Hyperspectral sensor-based applications ☆ Automated control system through Greenhouse and Polyhouse cultivation ☆ Precision Laser Land leveller ☆ Mulching ☆ Low Tunnel Technology/ Walking Tunnels ☆ Micro irrigation ☆ Site-specific Nutrient Management (SSNM) ☆ Drone Technology

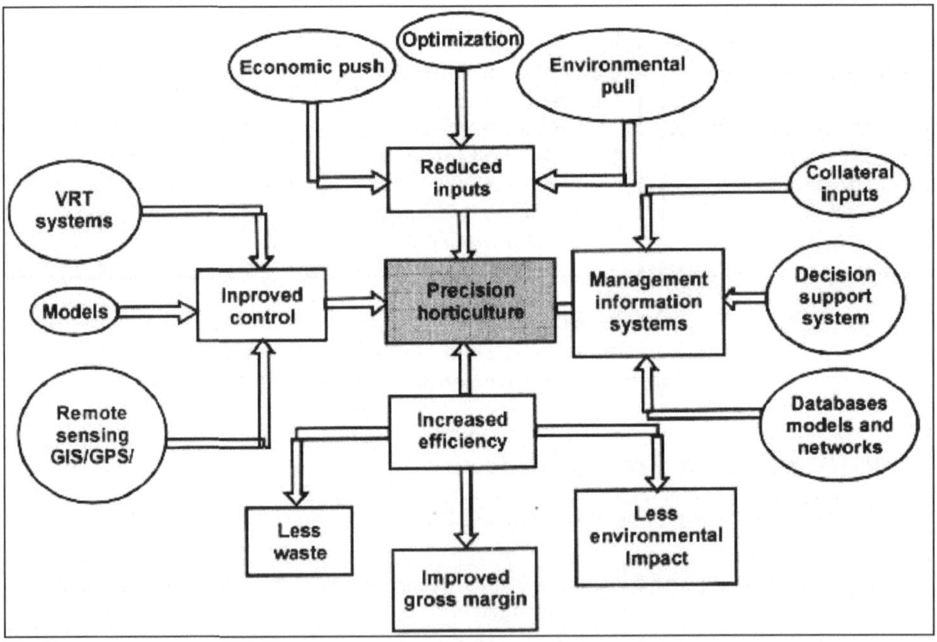

Figure 17.4: Technologies for Precision Farming
(*Source*: Hind Agricultural Research and Training Institute).

Insect Pest and Disease Management in Precision Farming

☆ Diseases and insect pests pose serious problems for crop security. Improved sensors for precision farming are constantly being developed as a result. Such contemporary technology consists of pest detection sensors that identify the presence of disease and insect pests on crops. In essence, the sensors deliver field data in real-time.

- ☆ **Net houses:** To protect plants from insects, finer mesh nets with smaller holes can be utilised. To manage leaf miners, striped flea beetles, aphids, and jassids, use nylon netting with a 30-mesh size (30 holes per inch or 0.75 mm holes), consistent weaving, and long-lasting quality.

Sensors for Accurate Insect Pest Detection

For the identification of insect pests on crops, farmers might employ a variety of sensors. They range from the most basic to the most sophisticated job principles. The following are some of the popular sensor types:

1. **Low-power Image Sensors**: Based on a low image sensor, the low-power image sensor is a wireless autonomous monitoring system. The wireless sensor is installed in a single trap, and it periodically captures images of the contents of the trap and sends them wirelessly to a control centre. The amount of pests discovered at each trap is then calculated using sent photos. A farmer may schedule when to begin applying crop protection and in which field areas based on the quantity of insects in the region.
2. **Acoustic Sensors**: An acoustic sensor is an insect pest detection sensor which works by monitoring the noise level of the insect pests. How does it work? Wireless sensor nodes connected to a base station are placed in the field. When the noise level of the pest crosses the threshold, a sensor transmits that information to the control room computer, which then accurately indicates the infestation area.
3. The presence of insect pests can also be tracked using sensors that measure the Leaf Area Index (LAI). Leaf destruction is caused by insect pest feeding. Plants lose chlorophyll as a result of this. This results in a decrease in total leaf area and, as a result, a decrease in the plant's ability to photosynthesize. The sensor can detect an insect attack early on by measuring the leaf area index and alerting farmers to take appropriate action. This sensor calculates the leaf area index in real-time using radiation measurements and other parameters. This type of sensor is also used to detect crop diseases.

Eike *et al.* (2009) measured the visible and near infrared reflectance of 1153 leaves and 392 canopies in 11 peach orchards in California to assess the viability of spider mite damage detection in orchards. In order to create normalized difference indices, important wavelength pairs discovered by partial least squares regression were merged. The relationship between these indicators and mite damage was assessed. Significant correlations were found between mite damage and two spectral areas for canopies (at blue and red wavelengths) and eight spectral regions for leaves. Index values had a linear relationship (R2=0.47) with mite damage, making it possible to locate mite hotspots in orchards. To use this approach for early mite identification, nevertheless, greater aerial imaging standardization and taking perturbing environmental conditions into account would be required.

Sensors for Early Crop Disease Detection

If crop illnesses are not promptly and effectively treated, the production can be drastically reduced, threatening the safety of the world's food supply. For this reason, every farmer's most crucial responsibility is to protect themselves against sickness. Farmers adopt contemporary farm practises to safeguard their crop since early disease identification may be effectively controlled. Both direct and indirect illness detection techniques are part of these measures.

1. **Thermography Disease Detection Method:** Thermography sensors track the variations in plant leaves' and the canopy's' surface temperatures. Infrared radiation emitted by the plant's surface is captured by the sensor.
2. The decreased transpiration will cause the plant surface temperature to rise if there is a pathogen infection. The sensor can determine whether a disease is present based on a temperature change. The illness's alterations can be detected by thermography sensors before the condition even manifests.

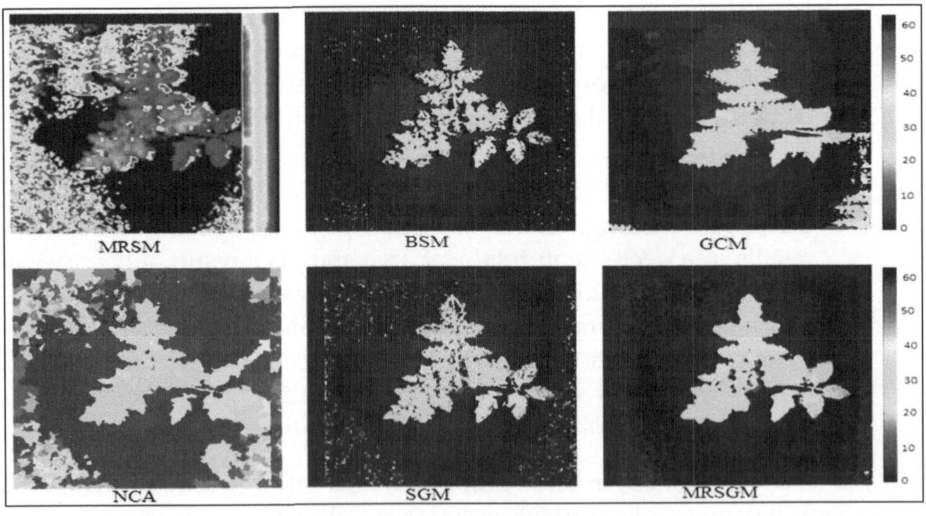

Figure 19.5: Disease Detection by Thermography in Tomato (*Source*: Raza S-e-A *et al.,* 2015).

Precision disease control is limited due to its high sensitivity to the change of environmental conditions during measurements. Another problem is that the thermography method can't be used to identify the type of infection.

3. **Fluorescence Disease Detection Method:** The fluorescence technique is used by sensors to detect the fluorescence of chlorophyll on the leaves, incoming light, and changes in the fluorescence characteristics. Through monitoring variations in chlorophyll and photosynthetic activity, it can identify the presence of pathogens.

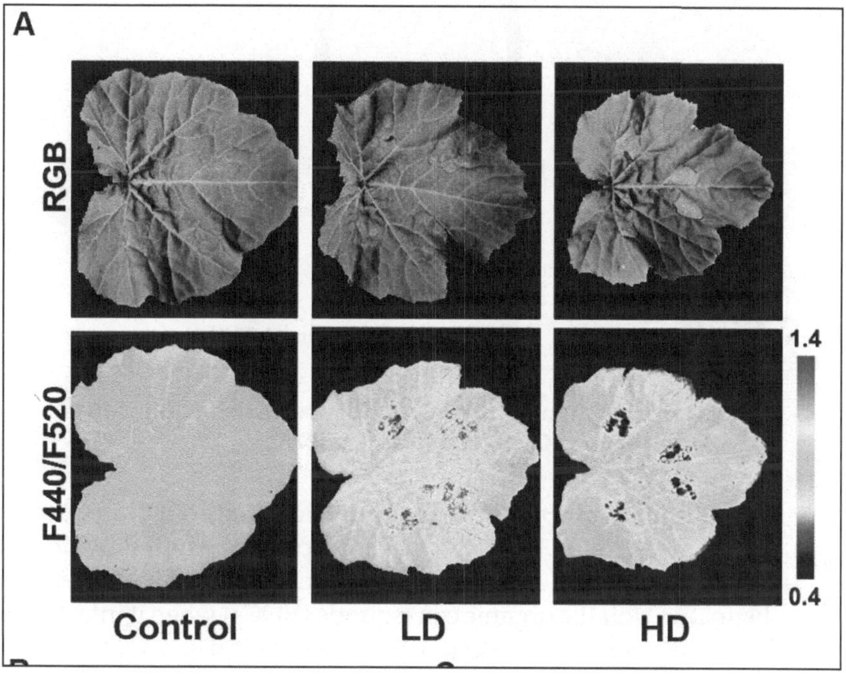

Figure 19.6: Disease Detection by Fluorescence Method in Zucchini Leaves (*Source*- Pérez-Bueno ML 2016).

Although fluorescence measurement provides sensitive detection of abnormalities in photosynthesis, the practical application of this technique in a field setting is limited.

4. **Hyperspectral Disease Detection Method:** When measuring plant health, sensors using the hyperspectral technique employ a broad spectrum, spanning 350 to 2500 nm. When an infection occurs, biophysical and biochemical characteristics change, and these variations in reflectance are what are measured. In order to provide more thorough and precise information regarding plant health, hyperspectral cameras capture data in three dimensions using the X, Y, and Z axes for spatial and spectral information, respectively. In most

cases, sensors are mounted to an unmanned aerial vehicle to monitor a vast field area (UAV).

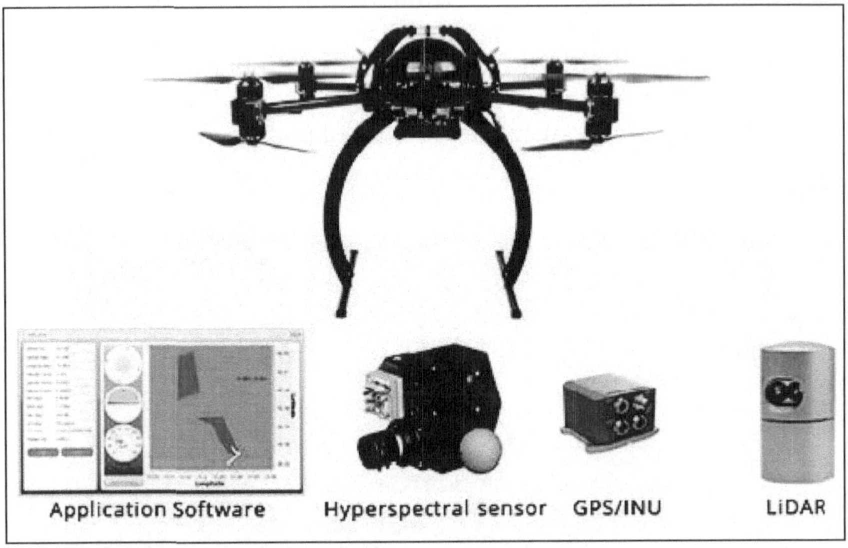

Figure 19.7: Equipment for Crop Disease Ddetection with Gyperspectral Sensor.

5. **Gas Chromatography Disease Detection Method:** This non-optical sensor measures volatile chemical compounds emitted by sick plants to help identify agricultural diseases. Each species of pathogen releases distinctive volatile organic compounds (VOCs) when it infects a plant. The same phenomenon occurs when mechanical harm stresses a plant. Gas chromatography-based sensors can effectively determine the kind and nature of infection in this regard.
6. This method's sole drawback is the requirement for sampling previously obtained volatile organic molecules for a longer period of time prior to data processing, which significantly restricts its use in the field.

Drawbacks of Precision Farming

- ☆ High cost of application.
- ☆ Lack of technical expertise/labour, knowledge and technology.
- ☆ Not feasible/costly for small land holdings.
- ☆ Heterogeneity of cropping systems.

Future Perspective

- ✰ Formation of farmers co-operatives and union is necessary since many of the precision agriculture tools are of high cost (GIS, GPS, RS *etc.*).
- ✰ Creating awareness amongst farmer about harmful effects of applying imbalanced doses of farm inputs like irrigation, fertilizers, insecticides and pesticides.
- ✰ Educating farmers about technical aspects of precision farming.

CONCLUSION

In many developing nations, including India, precision farming is still only a theory. To encourage its implementation, technical and strategic assistance from both the public and commercial sectors are crucial. Exploration, analysis, and execution are the first three steps of a successful adoption, though. Environmental and economic concerns that now surround production agriculture can be addressed by precision farming. Cost-effectiveness and the most effective methods to employ the information required to guide new technologies are still up for debate. Incorporating a methodical approach, precision farming offers a fresh perspective to modern agricultural problems like the need to strike a balance between production and environmental considerations. Advanced information technology serves as its foundation. It entails documenting and simulating soil and plant species variation as well as combining agricultural methods to meet site-specific requirements. It aims at increased economic returns, as well as reduced cost at field level.

REFERENCES

AbdelRahman MA, Natarajan A, Srinivasamurthy CA, Hegde R. (2016). Estimating soil fertility status in physically degraded land using GIS and remote sensing techniques in Chamarajanagar district, Karnataka, India. The Egyptian Journal of Remote Sensing and Space Science. 19(1): 95-108

Achim Dobermann, Simon Blackmore, Simon E. Cook, and Viacheslav I. Adamchuk. (2004). "New directions for a diverse planet". Proceedings of the 4th International Crop Science Congress, 26 Sep – 1 Oct 2004, Brisbane, Australia. Published on CDROM.

Arora S. (2015). Precision agriculture and sustainable development. Kurukshetra. 54(2): 18-22.

Brar NS, Kumar T, Kaushik P (2020) Integration of Technologies Under Climate Change for Profitability in Vegetable Cultivation: An Outlook.

Chaudhari K, Nishant N, Upadhyay G, More R, Singh N, Vyas SP. (2019) Crop inventory of orchard crops in India using remotely sensed data. The International Archives of the Photogrammetry, Remote Sensing and Spatial Information Sciences. XLII-3/W6.

Das U, Pathak P, Meena, Mallikarjun N. (2017). Precision Farming a Promising Technology in Horticulture. International Journal of Pure and Applied Bioscience. 6: 1596-1606.

Delgado JA, Vandenberg B, Neer D, D'Adamo R. (2019). Emerging nutrient management databases and networks of networks will have broad applicability in future machine learning and artificial intelligence applications in soil and water conservation. Journal of Soil and Water Conservation.74(6): 113A-118A

Dutta, R., Stein, A. and Bhagat, R.M. (2011). Integrating satellite images and spectroscopy to measuring green and black tea quality. Food Chem., 127 (2): 866-874.

Eike L, Adam H, Minghua Z, Walter JB, Cecil D. (2009). Remote sensing of spider mite damage in California peach orchards. Int. J App. Earth Observation and Geoinformation.11(3): 244-255.

Hildreth JC (2003) The Use of Short-Interval GPS Data for Construction Operations Analysis. PhD Thesis, Virginia Tech

Himani B. Patel. (2016). Precision farming – a new technique for horticultural crops. *Innovative Farming,* **1**(2): 35-37.

Khan AM (2013) Precision farming technique for water resource management. PhD Thesis, Aligarh Muslim University

Kumar D, Gangwar V, Kumar A. (2022), Utilization of Precision Farming in Vegetables, Agriculture and Food: E-Newsletter, Volume 04 - Issue 08 - August 2022: 330-332.

Lee W-S, Alchanatis V, Yang C, Hirafuji M, Moshou D, Li C (2010) Sensing technologies for precision specialty crop production. Computers and electronics in agriculture 74: 2-33

Mondal P, Basu M, Bhadoria PBS. (2011) Critical Review of Precision Agriculture Technologies and Its Scope of Adoption in India. American Journal of Experimental Agriculture. 1: 49-68.

Mushtaq G, Asima N. (2014). Estimation of apple orchard using remote sensing and agro-meteorology land-based observation in Pulwama district of Kashmir valley. Int. J Remote sensing and Geo. Sci; 3(6): 2319-3484.

Nabi A, Afroza B, Mushtaq F, Malik A. (2017) Precision Farming in Vegetables. Journal of Pharmacognosy and Phytochemistry. 6: 370-375.

Nelson, G.C., Rosegrant, M.W., Roertson, R., Sulser, T., Zhu, T., Ringler, C., Msangi, S., Palazzo, A., Batka, M., Magalhaes, M. (2009). Climate Change: Impact on Agriculture and Costs of Adaptation; International Food Policy Research Institute (IFPRI): Washington, DC, USA.

Parihar DS, Awasthi V, Mahadik A. (2021). Variable Rate Application (VRA) for Precision Farming. Agriculure and Environment, E-Newsletter.

Pérez-Bueno ML, Pineda M, Cabeza FM and Barón M (2016) Multicolor Fluorescence Imaging as a Candidate for Disease Detection in Plant Phenotyping. *Front. Plant Sci.* 7: 1790. doi: 10.3389/fpls.2016.01790

Raza S-e-A, Prince G, Clarkson JP, Rajpoot NM (2015) Automatic Detection of Diseased Tomato Plants Using Thermal and Stereo Visible Light Images. PLoS ONE 10(4): e0123262. https: //doi.org/10.1371/journal.pone.0123262

Satterthwaite, D., McGranahan, G., Tacoli, C. (2010). Urbanization and its implications for food and farming. Philos. Trans. R. Soc. B Biol. Sci. 365, 2809–2820.

Sharma A and Rehan (2020). Precision horticulture- a site specific and modern approach to optimize production, In Secondary Agriculture: Perspectives and Approaches, Learning Media Publication, Chapter-3: 30-48.

Sharma A and Rehan. (2020). Big Data-An Evolution to Revolutionize A Climate Smart Agriculture Industry. Agriculture Observer. (16).

Soomro TR. (2015). GIS enabling smart agriculture. Smart Agriculture: An Approach towards Better Agriculture Management. 8: 1-6

Suarez L, Zarco TPJ, Gonzalez V, Berni JA, Sagardoy R, Morales B. (2010). Detecting water stress effects on fruit quality in orchards with time-series PRI airborne imagery. Remote Sensing Env. 114(4): 286-298.

Thakur N, Dogra BS, Meenakshi and Kumar R. (2020). Precision farming-Concepts and Practices, Journal of Pharmacognosy and Phytochemistry 2020; Sp 9(5): 403-407

Zaman QU, Schumann AW, Miller WM. (2005). Variable rate nitrogen application in florida citrus based on ultrasonically-sensed tree size. Am. Soc. Agril. Engineers. 21(3): 331-335.

Zijp W (1994) Improving the transfer and use of agricultural information: A guide to information technology. The World Bank.

Zolekar RB, Bhagat VS. (2015) Multi-criteria land suitability analysis for agriculture in hilly zone: Remote sensing and GIS approach. Computers and Electronics in Agriculture.118: 300-321.

Chapter 18

Processing and Value Addition of Minor Horticultural Crops

Muzeev Ahmad[1]*, *Khursheed Alam*[2], *Rizwan Ali*[3], *Vikas Kumar*[4] *and Astha Vishwaraj*[5]

[1]*Assistant Professor,* [3]*Ph.D. Research Scholar,*

Department of Agriculture, Integral Institute of Agricultural Sciences and Technology, Lucknow – 226 026, Uttar Pradesh

[2]*Ph.D. Research Scholar,* [5]*M.Sc., Student,*

Department of Vegetable Science, Sardar Vallabhbhai Patel University of Agriculture and Technology, Meerut – 250 110, Uttar Pradesh

[4]*Ph.D. Research Scholar, Department of Vegetable Science, Banda University of Agriculture and Technology, Banda – 210 001, Uttar Pradesh*

**e-mail: muzeev1@gmail.com*

INTRODUCTION

Despite the huge production of horticultural crops, India processes only about 2.2 per cent of the total production. A considerable number of fruits and vegetables produced are lost due to improper postharvest management and a lack of appropriate processing technologies. Out of the total production of about 270 million tonnes of fruits and vegetables in the country, only 1 million tonnes of value-added processed products are produced. About 20 per cent of the products worth crores are wasted every year. Value addition is the process of changing or transforming a product from its original state to a more valuable state (Sharma *et al.*, 2014). Various value-adding technologies such as processing and preservation techniques, dehydration and drying technology, freezing technology, packing, labelling *etc.* can be applied to agricultural produce to increase its value. The total value of any agricultural produce is increased by performing certain postharvest processing operations rather than selling it as such after harvest (APO, 2009. About 60 percent of the consumers live in the rural sector, the processed food is transported back to where it was produced at much higher prices and after incurring substantial losses. Had the produce been processed in the production catchment, the consumers there would have accessed the processed food at lower prices, postharvest losses would have been avoided and more employees would have been created in the rural areas. Rural folk can play an important role in postharvest management right from harvesting and can salvage a large part of the loss. It will also help in higher availability of this dietary material, which comes under protective foods at lesser cost and improve the nutritional status of the vast population of our country.

Status of Processing Industry

In India, less than 3 percent of fruits and vegetables produced are commercially processed as against 65 percent in the US, 70 percent in Brazil, 78 percent in the Philippines, 80 percent in South Africa and 83 percent in Malaysia. The value addition is only 7 percent in India as against 23 percent in China and 88 percent in the UK Table 18.1.

Table 18.1: Extent of Postharvest Losses (Estimates) in Horticultural Crops

Sl.No.	*Crops*	*Quantity (Mt.)*	*Loss (per cent)*
1	Fruits	55.00	20
2	Vegetables	95.00	25
3	Floriculture	-	40
4	Tuber Crops	32.19	20
5	Tuber Crops	3.02	12

Value-Added Products of Fruits

Minor fruits are treasure houses of several health-promoting substances besides possessing acceptable sensory profiles. Value addition to such fruits would not only promote the economic status of people but also promote the health and nutrition status of people consuming such fruits. Value addition through processing and preservation must be considered as an important alternative for reducing the postharvest losses of this nutritive fruit and for serving in the off-season. There is a great scope for processing and value addition to the minor fruits. There are numerous minor fruits that need to be explored for value addition for commercial processing Table 18.2.

The main processed products consumed by people are jam, ready-to-serve fruit drinks, chutneys, candies, pickles, squashes, concentrate *etc.* (Roy, 2000; Khurdiya, 2001; Singh *et al.*, 2008). Various processed products like canned jackfruit bulbs in syrup, squash, raw jack pickle, roasted jack seeds, jack seed flour and candied jackfruit, have been prepared from Jack fruit (Berry and Kalra, 1987; Chadha and Pareek, 1988; Das and Prakash, 2009). Various processed products such as nectar, squash, slab, toffee, powder, *etc.* can be made with bael pulp. Ber can be processed to prepare murabba, candy, dehydrated ber, pulp, jam, and ready-to-serve beverages (Khurdiya, 1980; Pareek, 2001). Jamun fruits can be processed into excellent-quality fermented and non-fermented beverages. Besides that, good quality jelly, jam, and leather can be prepared. A good quality jelly can also be prepared from its fruits. The seeds can be processed into powder, which is very useful to cure diabetes (Khurdiya, 2001).

Table 18.2: Value Addition in Jackfruit

Stage of Fruit	*Value Added Products*
Unripe fruit	Pickle, Chips, Papad, Brined Jackfruit, RTC Jackfruit, Dehydrated Jackfruit, Culinary preparations, Cutlets, *Biryani*
Half ripe fruit	Candy, Preserve
Fully ripe fruit	Jam, Leather, Rind Jelly, Squash, Nectar, Canned Bulbs, RTE Bulbs, RTS Drinks, Chutney, Toffee, Wine, Halwa, Kheer, Gulab Jamun, Icecream, Custard, Cake
Seeds	Seed powder, Starch flour, culinary preparation, *Pakoda, Kheer*

Source: Bhuyan *et al.*, 2013.

Table 18.3: Value Addition in Jackfruit

Plant Part	*Value Added Products*
Edible value-added products	
Inflorescence sap	Toddy, Jaggery, Sugar, Honey, Wine
Fruit	Toffee, Spread, Jam, Pickle, Sweets (Burfi), Beverages (RTS, Squash, Nectar)
Kernel	Canned products
Non-edible value-added products	
Leaf	Mats, Baskets, Fans, Hats, Umbrellas, Buckets, Writing Material, Fence, and Fibre extracted is used to make brushes and handicrafts
Fruits	Fiber extracted is used to make toys and fancy items
Stem	Sheds construction, as timber

Out of this, a significant portion goes to waste due mainly to its perishable nature and seasonal glut. The postharvest losses in jackfruit are around 30-35 per cent during the peak season (Lakshminarayan, 2017). Jack fruit has great potential for value addition and more than 100 items can be prepared from jack fruit right from the immature stage to the well-ripened stage (Table 18.5). The importance of the fruit, seed and rind is known very little to the growers and consumers. Hence, value addition is important to utilize the surplus fruits available during the season as well as to improve the livelihood of the farmers by producing value-added products.

Value Addition in Palmyra Palm Every part of the palmyra palm can be used for value addition and can be processed into a range of edible as well as non-edible products.

Karonda (*Carissa carandas* L.)

Karonda is an indigenous fruit of India and belongs to the family Apocynaceae. It is a widely used medicinal plant by tribals throughout India and popular in various indigenous systems of medicines like Ayurveda, Unani and Homoeopathy and is a good appetizer. Unripe fruits are very sour at maturity but sourish-sweet when ripe. The fruit is a globose berry, oblong, broad-ovoid, or round, 125 mm to 250 mm long. It has thin tough, green to purplish-red smooth, glossy skin turning dark-purple or nearly black when ripe. Karonda fruits are rich sources of iron and an excellent source of vitamin A, C and B complex, fiber, carbohydrates, and minerals such as calcium, phosphorous, potassium, sodium, and sulfur. Karonda fruit usually contains 83.67 percent moisture, 2.3 percent protein, 1.77 percent fat and 4.7 percent carbohydrate. It also contains a good amount of pectin (Peter, 2007). Fresh karonda fruits are generally not consumed because of their highly acidic and astringent properties; therefore, karonda is not a popular table fruit. But it has got great potential in processed forms. The fruits may be eaten as a dessert when ripe or used in the preparation of fruit products such as candies, jelly, squash, and chutney. Moreover, the storage life of a karonda is very short because of its soft flesh and high moisture content. The unripe fruit harvested at maturity can be stored for 5 to 7 days at room temperature, but at ripe stage, it can be stored only up to two days. Although recently value-added commercial preparations are made for domestic use and for export by food processing companies, the plant has largely remained an underutilized species (Manivasagan *et al.*, 2006). Karonda has a great potential for value addition and several value-added products can be prepared from unripe as well as ripe fruits (Table 18.4). Some of the value-added processed products that can be prepared from karonda are discussed below with their method of preparations

Table 18.4: Value Addition in Karonda

Stage of Fruit	*Value Added Products*
Unripe fruit	Pickles, Candy, Chutney
Fully ripe fruit	Jam, RTS, Nectar, Squash, Canned fruits, Syrup, Fruit powder

PIAR/Chironji (*Buchanania lanzan* Spreng)

Buchanania lanzan Spreng (common name - Char, Chironji), belongs to the family Anacardiaceae. The Chironji tree is commercially very useful. It bears edible fruits that are sweet, nutritious, laxative, and useful for fever and excessive thirst (Anupama, 2015). The seed which is present in the fruit is deshelled to get the kernel. The kernel is called Chironji nuts. They are small, brownish and have a nutty taste. Fruits have maximum values of oil (61.6 per cent to 62.0 per cent), protein (50.2 to 48.0 per cent) and sugar (3.90 to 3.82 per cent) [Sharma, 2012]. The fruit fetches a much lower price for the collectors than the kernel and hence, collectors usually grind the seed to obtain the kernel, 5 kg of fruit can yield almost 1 kg of *chironji* kernels (Singh *et al.*, 2016). Seeds/kernels of *Buchanania lanzan* are nutritional, palatable and used as a substitute of almonds in confectionery. Chironji nuts contain 3.0 per cent moisture and are rich in lipid/fat (59.0 per cent), protein (19.0–21.6 per cent), starch/carbohydrate (12.1 per cent), fibre (3.8 per cent), minerals like calcium, phosphorus, iron and vitamins like thiamine, ascorbic acid/vitamin C, riboflavin, niacin (Anupama, 2015). Chironji nuts are used as a brain tonic and they are also good for the skin due to their richness in fat, vitamins and minerals (Table 18.5). These nuts are antioxidants and improve body immunity to fight infections. The oil extracted from kernels is known as 'char' and is used for curing skin diseases and removing spots/blemishes from the face. This oil contains palmitic and oleic acids. It is applied to glandular swellings of the neck (Siddiqui *et al.*, 2014).

Table 18.5: Value Addition in Chironji

Plant Part	*Value Added Products*
Fruit	RTS, Nectar, Squash, Fruit powder, Wine
Kernels	As dry fruits in sweets like halwa, kheer, laddu, paak *etc.*, Chironji oil

Indian Hog Plum/Amra (*Spondias mangifera* Willd.)

Spondias mangifera Willd. (syn., *S. cytherea* and *S. pinnata*), Malay apple or golden apple or Indian hog plum or Amra, belongs to the family Mangiferaceae, and is an equatorial or tropical tree, with edible fruit containing a fibrous pit. In India, the tree is naturally found to occur in the foothills of the Himalayas, Uttar Pradesh, Bihar, Jharkhand, Chattisgarh, Uttarakhand and many forest areas. The fruits are 2.5 to 3.5 in (6.25–9 cm) long and grow in bunches of 12 or more. The fruits are sour and tasty. Over several weeks, the fruit fall to the ground while still green and hard, turning golden-yellow as they ripen. Its fruit may be eaten raw; the flesh is crunchy and a little sour. Leaves are used as an ingredient for making herbal hair lotion and fruits are used against dysentery and dyspepsia (Table 186). All parts of the plant have got many traditional and ethnobotanical uses (Singh *et al.*, 2014)

Table 18.6: Value Addition in Amra

Plant Part	*Value Added Products*
Unripe fruit	Chutney, Pickles, *Panna,* Stews, Dried fruit Powder
Ripe fruit	Pickles, Jams, Jelly, Chutney, RTS, Nectar, Squash
Flower	Salad, Cooked as a vegetable
Flower bud	Pickle
Leaves	Flavouring in stews

Value-Added Products of Vegetables

The production of frozen peas, garlic and ginger paste, tomato puree, mango pulp *etc.* has been taken up in a big way only recently in India. Some of the popular processed or value-added products in the country, dried onions and garlic powder, ginger and garlic paste, jams of mixed fruits, pineapple, and strawberry; juice and concentrates of apple, pineapple, orange and litchi, squashes of litchi, canned pineapple, canned beans, frozen beans, cauliflower, and okra; pickles of mango, lime, chilies and mixed fruits and vegetables; tomato ketchup and puree; mango fruit drink and nectar, chili sauce, mango chutney *etc.* In addition to offering a higher return, value-added products can open new markets, create recognition for a farm and expand the market season. The processing centers in the region can take up diversification of the processing line to develop newer products listed as under (Table 18.7).

Table 18.7: Processed Products from Vegetables

Sl.No.	*Vegetables*	*Existing Products*	*Newer Products*
1	Bitter gourd	Dehydrated	Juice
2	Carrot	Juices, dehydrated	Colour extraction, osmotically dried shreds
3	Cauliflower	Dehydrated	Minimally processed
4	Chili	Chutneys, pickles, dehydrated	Puree, essential oil
5	Cucumber	Slices, whole in a brine solution	Pickled
6	Mushroom	Canned, dehydrated, ketchup, pickle	Phytochemicals
7	Okra	Minimally processed, dehydrated	Coated and battered
8	Onion, garlic	Dehydrated, paste	Ready-to-use gravy base
9	Shelled beans and peas	In brine, frozen, canned	Protein isolates
10	Tomato	Juice, puree/concentrate	Ready-to-use gravy base, lycopene extraction
11	Potato	Chips, starch	Alcohol

Value-Added Products of Flower

The value-added products from flowers include dry flowers and potpourri, essential oils, flavours and fragrances, pharmaceutical and nutraceutical products, pigments, and natural dyes, gulkand, rose water and insecticidal and nematocidal compounds. Spices are processed for essential oils, oleoresins, natural colours and spice extracts. Mushrooms are processed into fresh, dried, and pickled forms (Table 18.8a).

Table 18.8a: Value-Added Products of Flower

Crop	*Value Added Products*
Chrysanthemum	Knockdown effect on Lepidoptera insects, bouquets
Marigold	Essential oil and oleoresins for perfumery, aggarbattis, repellent for flies and mosquitoes, natural colorants to be used as dyes, poultry feed, pharmaceuticals, herbal products, bouquets, loose flowers
Rose	Food flavor, lotions, eye washes, tobacco flavouring, gulkand, herbal products, dried flower arrangements, bouquets
Tuberose	High-grade perfumes, lotions
Crop	Value added products
Fenugreek	Ayurvedic medicines, health products
Garlic	Oleoresins, prebiotics, medicinal compounds, beauty products
Ginger	Oleoresin, medicines
Turmeric	Curcumin (natural colour), ayurvedic medicine

Flower Ornaments

Various flower ornaments have been used by people from time immemorial. The commonly used floral ornaments are garland, Veni/gajra, wreaths *etc.* Garlands are being used in every type of function. It is so common that it has become a part of our lifestyle right from birth to death (Table 18.8b).

Table 18.8b: Value-Added Products of Flower

Flowers	*Cosmetics*
Rose	Soaps and body creams
Calendula	Moisturizer
Verbena	Acne creams, but also make deodorants and hand wipes
Pansy	Face washes, cleansers, and face mists
Hibiscus	Face cream/wash

Cosmetics from Flowers

Generally, botanical products from flowers and other plants are a rich source of vitamins, antioxidants, essential oils and oils, hydrocolloids, proteins, terpenoids and other bioactive compounds (Dubey *et al.*, 2004). Flower plant extracts from plants or wastes from plants processed industrially can be used for cosmetics such as. The use of plant extracts in the cosmetic formulation is increasing, mostly because of the poor image that animal-derived extracts have acquired during the past few years. A cosmetic formulation, including active principles of strictly natural origin, is designed to protect the skin against exogenous or endogenous harmful agents, as well as to balance again the dermal homeostasis lipids altered by dermatosis and aging.

CONCLUSION

The minor fruit species being rich in nutritional, medicinal, and processing qualities can play a very significant role in the livelihood security of rural communities through enhanced household income, employment generation, and environmental protection. Many of these minor fruits have great potential for commercial cultivation. Some small-scale industries are also using some minor fruits for processing pickles, chutney, jam, and jelly *etc.*

There is vast opportunity in the value-added floriculture trade. The consumption pattern is diversifying towards value-added products such as dry flowers, essences, perfumes, beverages, and other by-products from flowers for more profit. Moreover, due to the Covid-19 pandemic, the floriculture sector is facing hardship and trading of fresh flowers has become a big challenge. Under these circumstances, it is important to diversify into value addition where ever possible for alternative income through various feasible products like gulal, agarbattis/dhoopbattis making, Poultry feed, gulkand and rose water, pigment and dyes extraction, essential oil extraction, dry flower for those flowers which are eligible for such conversion and Vermicompost for recycling the spoiled flowers. These nuts are antioxidants and improve body immunity to fight infections. The oil extracted from kernels is known as 'char' and is used for curing skin diseases and removing spots/blemishes from the face. This oil contains palmitic and oleic acids. It is applied to glandular swellings of the neck (Siddiqui *et al.*, 2014).

REFERENCES

Anupama, P. (2015). Chironji Health Benefits and Medicinal Uses. Retrieved from accessed.

Berry, S. K. and Kalra, C. L. (1987). Cultivation and processing of jackfruit. Indian Food Pack., 42(5): 62-67.

Bhattacharyya, B. K. and Bhattacharjee, D. (2007). Bactriocin: A biological food preservative. *J. Food Sci. Technol.*, 44(5): 459-464.

Bhuyan, M. A. J., Saha, M. G. and Rahman, M. A. (2013). Value Addition to Jackfruit Fruits for Livelihood: Production Technology and Management Practices (*Artocarpus heterophyllus* Lam.) through Integrated Processing and

Preservation. In: Workshop on Valorisation of Traditional Processing of Indigenous and Underutilized Fruits. Institute of Technology of Cambodia, Phnom Penh, Cambodia, January 14-16.

Chadha, K. L. and Pareek, O. P. (1988). Genetic resources of fruits crops: achievements and gaps. *Indian J. Plant Genet. Resour.*, 1: 43-48.

Das, S. C. and Prakash, J. (2009). Minor Fruits: A Livelihood Opportunity for the Tribal Peoples of Tripura. *ISHS Acta Horticulturae*, 890: II International Symposium on Pomegranate and Minor - including Mediterranean - Fruits: ISPMMF2009.

Dubey, N.K., Kumar, R. and Tripathi, P. (2004). Global promotion of herbal medicine: India's opportunity. Current Science, 86, 37–41.

Khurdiya, D. S. (2001). Postharvest management of underutilized fruits for processed products. Winter school on the exploitation of underutilized fruits, pp. 291–298.

Lakshminarayan, N.G. (2017). Jackfruit-Based Value-Added Products. Retrieved from lakshminarayan, as accessed. https: //www.linkedin.com/pulse/ jackfruit-based-value-added-products-n-

Majumder, B. C. (2004). Minor fruit crops of India- tropical and sub-tropical. Daya Publishing House, New Delhi.

Manivasagan, S., Rana, G. S., Kumar, S. and Joon, M. S. (2006). Qualitative changes in karonda (*Carissa carandas* Linn.) candy during storage at room temperature. *Haryana J. Hort. Sci.*, 35(1 and 2): 19-21.

Peter, K. V. (2007). Under-utilized and under-exploited horticultural crops. New India Publishing House, New Delhi

Sharma, A. (2012). Scientific Harvesting for Quality Seed Collection of *Buchanania lanzan* Spreng for its Conservation and Sustainable Management- Case Study of Chhindwara, Madhya Pradesh, India. *International Journal of Bio-Science and Bio-Technology*, 4(1): 65-74.

Sharma, J. P., Upadhyay, S., Chaturvedi, V. K. and Bharadwaj, T. (2014). Enhancing Farm Profitability through Food Processing and Value Addition. In: Shukla, J. P. (Ed.). Technologies for Sustainable Rural Development: *Having Potential of Socioeconomic Upliftment* (*TSRD-2014*), published by Allied Publisher Pvt. Ltd., India. p. 88.

Siddiqui, M. Z., Chowdhury, A. R., Prasad, N. and Thomas, M. (2014). *Buchanania lanzan*: a species of enormous potential. *World J. Pharm. Sci.*, 2(4): 374-379.

Singh, D., Wangshu, L. and Prahalad, V. C. (2008). Processing and marketing feasibility of underutilized fruit species of Rajasthan. *IAMO Forum, pp.* 1-12.

Singh, M. K., Das, B. K. and Patidar, P. (2016). The effect of methanolic extract of *Buchanania lanzan* Spreng seeds on hematological indices. *Indian J. Pharmacol.*, 48: 214-215.

Chapter 19

Production of Quality Seed in Vegetable Crops

Pallavi Soni

Ph.D. Research Scholar, Department of Horticulture, Vegetable Science, College of Agriculture, IGKV, Raipur, Chhattisgarh

**e-mail: pallavisoniaka@gmail.com*

INTRODUCTION

Vegetables are grown on 1.2 per cent of the nation's total agricultural land. If properly cultivated, the majority of vegetables have the ability to generate a yield that may range from 5 to 10 times greater than to any cereal crop. Vegetables are now being produced in India is quite low. To enhanced seed and cultural production, improved techniques for plant protection are necessary, among these high-quality seeds are the most crucial element (Thooyavathy *et al.*, 2013). Vegetables is also known as protective food because vegetables are a rich source of fiber, potassium, vitamin A and C, folate, as well as other essential minerals and vitamins. A minimum amount of 400g per day or five portions of vegetables is beneficial for health (FAO, 2021). Vegetables play a prominent role in the nutritional security of the Indian population and economy of the majority of small and marginal farmers. For success in vegetable growing enterprise, high quality seed is not only highly desirable but also a statutory requirement. Indian vegetable seed industry is going enormously annually. To meet the requirement of the farmers for seed of good quality and genetic purity, it should be produced by using the standard agro techniques under strict seed certification standards (Singh and Asati, 2013). During flowering and fruit setting, vegetables require a particular temperature and other environmental conditions. While some vegetables are grown in one region of the country but their seed production is limited to another part of the country. The elimination of such microclimatic conditions demands a protected environment. Still the locally accessible genotypes or open pollinated cultivars continue to dominate vegetable production in our country. This is primarily a result of farmers' ignorance and substandard extension efforts. Increasing sales of vegetable seeds have a favorable impact on the Indian economy in terms of generating income, jobs, and foreign currencies on the global market.

Before being distributed to growers and consumers, seedlings and seeds of crops are continuously multiplied in accordance with set criteria and specifications to assure genetic identity, maintain varietal purity, and meet specific quality standards. Multiplication of seeds is usually a part of a legally sanctioned system for quality assurance of seed production in several countries. Since India declared its independence, the government has liberalized and promoted the seed trade. In India, a number of private seed companies with international bases are actively engaged in the production of vegetable seeds, dragging the public sector far behind (Koundinya and Kumar, 2014). According to estimates, 20–25 per cent of productivity of vegetables is influenced by the quality of the seeds (Anonymous, 2013). Mankind long ago acknowledged the value of high-quality seed. The Rigveda of ancient India mentions the importance of viable seed for the prosperity of the human race. It is referred to as "Subeejam Sukshetre Jayate Sampadyate" (Poonia, 2013) in the primal manusmriti, which literally translates as "A good seed in a good field will triumph and prosper." The National Seeds Corporation (NSC) was founded in 1963 (Gadwal, 2003), with the objective of encouraging the healthy growth of India's seed sector.

The primary responsibilities of NSC are to set up a sufficient system of quality control inspection for the storage, processing, and marketing of seeds. It also focuses on the function of producing foundation seed for varieties and multiplying seed from previously released varieties. Significant progress has been made in fruit and vegetable development over the past decade, and 15.97 million hectares of both crops have been planted, accounting for about 67.15 and 90 per cent, respectively, of all horticulture area and production (Sarolia *et al.*, 2018). There are few limitations on the system of producing quality seeds, such as high seed production costs, technological issues, and stringent regulations that have shattered the vegetable seed sector in India (Sharma, 2011). The main obstacles to the growth of a sustainable horticultural business are the inadequate supply of high-quality seed and planting materials.

Scope of Vegetable Seed Production in India

Since age's Indian farmers were mostly dependent on traditional varieties, therefore seed requirements were not met through farm saved seed. The impact of investments in horticultural research can be realized only then when farmers have better access to high quality seed and planting materials of the improved varieties at affordable prices and at their production sites. Higher productivity and easy access to the produce can be ensured by using high-quality seeds and planting materials for horticulture commodities in the recommended zones. A significant number of crop genotypes suitable for hot, arid agro climates have been developed. However, it is still difficult to find their seed and planting materials. The effort on producing high-quality seed and planting material of various plants has been stepped up by ICAR through its network on planting material production.

Quality Seed and its Attributes

Quality seed is defined as being genetically pure, varietally pure, having a high rate of germination percentage, being free from disease and pests, and having recommended moisture content. Strong germination, quick growth, and robust development are all ensured by high-quality seed for good plant stand in field. Poor quality seed causes yield declines due to excess population which reduce profitability. Quality seed is a vital input of crop production because it is the cheapest input and the basis for agricultural advancement. The quality of the seed used for sowing has a significant impact on crop output, and also on the other inputs interact to crop production. The cost of seed is substantially lower than that of other inputs despite the fact that it is very important for growing crops. It is predicted that using high-quality seeds with improved varieties will increase yield by 20–25 per cent (Hasanuzzaman, 2015). In order to increase the variety's potential yield, high-quality seeds are essential.

Classes of Seed

1. **Nuclear seed**: This is produced by the initial breeder, institute, or State Agriculture University (SAU) from fundamental nucleus seed stock, nuclear seed is one hundred per cent genetically pure and physically pure. The producing breeder issues pedigree certificates (Selvakumar, 2014). It does not require any tag.
2. **Breeder seed**: According to the directive of the Department of Agriculture and Cooperation (DOAC), Ministry of Agriculture, Government of India, the offspring of the nucleus seed multiplied in a large area under the supervision of plant breeders, institutes, and SAUs and were observed by a committee made up of representatives from the state seed certification agency, national and state seed corporations, the ICAR nominee, and the concerned breeder. This seed is also one hundred per cent genetically and physically for making foundation seed. The producing breeder issues a certificate for this kind of seed that is golden yellow in colour. It needs a golden yellow tag that is 6 cm wide and 12 cm long.
3. **Foundation seed**: the progeny of breeder seed generated by reputable public and private sector seed producing organizations under the supervision of seed certification authorities in a manner that ensures

its quality is upheld in accordance with established field and seed standards. For foundation seed, white colour tag has been issued by seed certification agency.

4. **Certified seed:** The offspring of foundation seed is created by licensed seed producers under the watchful eye of organizations that certify seeds in order to maintain minimum requirements for seed quality. For this category of seed, the seed certification body issues a blue-coloured certificate. It needs an azure blue tag that is 15 cm long and 7.5 cm wide.
5. **Truthfully Label (TL) Seed:** It is the product of either registered or certified seed, under the supervision of the farmer. It needs an opal green tag that is 10 cm wide and 15 cm long.

   * *The foundation and certified seeds could be replicated at stages I and II, but the number of generations after breeder seed cannot exceed three stages.

Table 19.1: Difference between Certified Seed and Truthful Labelled Seed

Certified Seed	*Truthful Labelled Seed*
It is optional to be certified. Quality assured by certification authority.	Sincere labelling is mandatory for varieties that have been notified.
Only limited to notified kinds	Both the notified and the released varietiestested for germination and physical purity
Both the minimum field and seed standards must be fulfilled.	Quality is assured by the supplier or manufacturer
Seed inspectors and seed certification officers may collect samples for examination.	Only seed inspectors can collect samples to evaluate the quality of the seeds.

Principle and Methods of Quality Seed Production of Vegetable Crops

In order to maximize the benefits from the introduction of new superior crop plant types, special attention must be paid to the maintenance of genetic purity and other attributes of seeds during seed production. The quality seed production is pre-dominantly influenced not only by genetic factors but also environmental conditions in which its multiplication is undertaken. Hence, emphasis should be concentrated on priority factors affecting seed quality and yield component.

For maintaining genetic purity of crop-genotype, its basic seed production cycle should be lesser and more emphasis should be given on storage facility of breeder/foundation seeds in vegetable crops. Based on storability there are two forms *i.e.* orthodox and recalcitrant seed and most of vegetable crops belonging to orthodox in nature (5-8 per cent moisture) and can tolerate desiccation and stored for longer period without deteriorating viability (Kumawat *et al.*, 2020).

Agronomic Principles of Seed Production

- ☆ Selection of Suitable Agro climatic Zone
- ☆ Selection of Land
- ☆ Isolation Distance of the Seed Crop
- ☆ Preparation of the land
- ☆ Selection of the variety
- ☆ Procurement of seed
- ☆ Seed: Seed Treatments, time of sowing/planting, seed rate, method of sowing, depth of sowing, rouging and pollination.
- ☆ Intercultural operation: - Disease-Pest Control, Nutrition of the Crop, Irrigation, Harvesting, seed extraction, drying, grading
- ☆ Inspection of seed crop
- ☆ Field standard
- ☆ Seed standard: Sampling, Purity Test, Germination Test, Seed Viability Test and Moisture Content Test.
- ☆ Packing
- ☆ Labelling storage

Table 19.2: Crop Monitoring and Rouging Stages for Quality Vegetable Seed Production

Crop	Field and crop monitoring stages (3–4 inspections)
Tomato	First before flowering, second during fruit setting and un-ripe fruit stage and third at mature/ripen fruit stage
Brinjal	First before flowering, second during fruit setting and tender fruit harvest stage and third at mature/ripen fruit stage
Chilli	First before flowering, second at fruit setting and immature fruit harvest stage and third at ripen fruit stage
Okra	First before flowering, second at fruit setting and tender fruit harvest stage and third at mature/seed fruit stage
Pea, beans and cluster bean	First before flowering stage, second during pod setting and tender/edible pod harvest stage, and third at mature/seed pod stage
Cucurbits	First at 2-4 true leaf stage and start of flowering, second at fruit setting and tender/marketable fruit harvest stage, and third at fruit maturity/ripen seed fruit harvest stage

Source: Sarolia, D. K., Samadia, D. K., Choudhary, B. R. and Singh, D. (2018). Production of quality seed and planting materials.

Table 19.3: Field Standard

Crops	*Isolation Distance*		*Off Types*		*Seed borne Diseases Affected Seeds*	
	Foundation Seed	*Certified Seed*	*Foundation Seed*	*Certified Seed*	*Foundation Seed*	*Certified Seed*
Tomato	50m	25m	0.10 per cent	0.20 per cent	0.10 per cent	0.50 per cent
Brinjal	300m	150m	0.10 per cent	0.20 per cent	0.10 per cent	0.20 per cent
Chilli/ Capsicum	500m	250m	0.10 per cent	0.20 per cent	0.10 per cent	0.50 per cent
Okra	500m	250m	0.10 per cent	0.20 per cent	-	-
Bitter gourd	1000m	500m	0.10 per cent	0.20 per cent	-	-
Cucumber	1000m	500m	0.10 per cent	0.20 per cent	-	-
Carrot	1000m	800m	0.10 per cent	0.20 per cent	0.10 per cent	0.20 per cent
Raddish	1000m	500m	0.10 per cent	0.20 per cent	0.10 per cent	0.50 per cent

Source: Thooyavathy, R. A., Perumal, K., Suresh, V. and Vijayalakshmi, K. (2013). Seed Production Techniques for Vegetables.

Table 19.4: Seed Standard

Crops	*Physical Purity*		*Germination Per cent*	*Moisture Content*
	Foundation Seed	*Certified Seed*		
Tomato	98 per cent	98 per cent	70 per cent	8 per cent
Brinjal	98 per cent	98 per cent	70 per cent	8 per cent
Chilli/Capsicum	98 per cent	98 per cent	60 per cent	8 per cent
Okra	99 per cent	99 per cent	65 per cent	10 per cent
Bitter gourd	98 per cent	98 per cent	60 per cent	7 per cent
Cucumber	98 per cent	98 per cent	60 per cent	7 per cent
Carrot	95 per cent	95 per cent	60 per cent	8 per cent
Raddish	98 per cent	98 per cent	70 per cent	6 per cent

Source: Thooyavathy, R. A., Perumal, K., Suresh, V. and Vijayalakshmi, K. (2013). Seed Production Techniques for Vegetables.

Hybrid Seed Production Techniques

Because of the greater adaptability and stability for excellent performance at the production sites, public sector hybrids are now understood to be on par with or better than hybrids marketed by a variety of private seed companies in India. In India, large-scale F_1 seed production may be carried out using manual emasculation and pollination procedures, and it is economically possible in plants like tomato, brinjal, bottle gourd, bitter gourd, and sponge gourd, ridge gourd, pumpkin, summer squash, watermelon, muskmelon, and cucumber are examples of plants in which a single manually pollinated crossover fruit produces a significant number of F_1 seeds. Nevertheless, relatively little seed is required to plant one hectare of these vegetables, and cutting-edge methods like plug-seedling cultivation in nurseries or cucurbit single-seed sowing under drip might reduce the cost of F_1 seed.

Production of Quality Seed in Vegetable Crops under Protected Environment

Vegetable crops face enormous obstacles during the production of commercially quality seeds in an open environment.

- ✰ Avoidance of recommended isolation distance/**Components/ characters/attributes of quality seeds**
- ✰ Inappropriate quality in any one of the characteristics of seed may result in low yield and poor quality of the output.

Northern Plains of India

Although the soil in the northern plains is very fertile, year-round outdoor vegetable cultivation is not practicable due to temperature swings between 0 and 48°C for the production of high-quality seeds in such locations, protected structures have proven to be a preferable choice.

Upper Regions of Himalaya

From November to March, the temperature in these regions ranges from -5 to -30 °C. Here, the constant snowfall makes it impossible to grow vegetables in an open field. Because of the protective architecture, it is feasible to grow vegetables all year round because the temperature variations are reduced.

Raising off Season Nurseries in Cucurbits

The cost of hybrid seed is quite expensive. It is essential that every seed must germinate. Controlled conditions are necessary for 100 per cent seed germination.

A moderate climate is required for summer squash to flower, set fruit and develop fruit, and produce seeds. Therefore, throughout the summer, its seed production is only limited to mountainous regions of north India. A low- and medium-cost greenhouse can now produce seeds of the summer squash varieties "Australian Green" and "Pusa Alankar" in the north Indian plains. Similar to this, the extremely lucrative crops tomato, capsicum, and cucumber also produce their seeds in a protected environment. When growing cross-pollinated vegetables like onions, cauliflower, and cabbage in a greenhouse without providing isolation space, you can maintain the purity of several kinds or lines.

Factors Affecting Quality Seed Production

Climate Factors and Physiographic Factors

Whether directly or indirectly, effect the plant performance such as photosynthesis and respiration play a role in the yield potential of vegetable seeds; therefore, the climate parameters that influence these processes work must be at their best. Every crop has certain climatic needs. A crop must be cultivated in an environment that complies with these specifications in order to achieve the maximum potential yield per unit of land. With few modifications, a crop that is ideally suited to a certain climate can be grown (AVRDC, 1990). Photoperiod Temperature, Rainfall, Wind, Altitude, Topography, and Steepness of Slope are the main environmental elements that affect the formation of vegetable seeds.

Agronomic Factors

While selecting a field for vegetable cultivation, the three main factors site topography, soil type, and water availability should be taken into consideration (quantity and quality) (Masabni *et al.*, 2009).

1. Method of Seed Production

Timing of planting seeds is crucial. There are numerous planting techniques, such as the "seed-to-seed" technique, which actually involves planting biennial seed and harvesting the seed crop the next season. Onions and carrots are usually grown in similar ways. The second approach is known as "root-to-seed," and it entails sowing biennial seeds and transferring the seed crop the next season (Sexton, 2001). For the purpose of producing the seed crop, an appropriate seed class and an approved source are required (Rashid and Singh, 2000).

2. Disease and Pest Management

The finest management practices are those that incorporate a variety of techniques and sustain the health of vegetable seed production. It is crucial to take into account the crop's economics, the cost of the management strategy, the region's history as a production site, the weather and climatic conditions, and any possibility for disease or pest development. Integrated management techniques give producers a way to lower the risks of suffering financial losses (Colquhoun *et al.*, 2015).

3. Harvesting, Processing, Extraction and Drying of Seeds

Seed quality is often determined at the harvesting stage. Thus, it is generally agreed that seeds should be extracted only after they have fully developed in the fruit or when their moisture content has fallen below 25 per cent. All process operations such as winnowing, cleaning, separation, drying, and packaging are carried out precisely and on schedule after harvest (Kuo *et al.*, 2006).

4. Storage Period of Vegetables Seed

The finest environment must be maintained for the seed after it has been extracted and dried in order to retain the highest levels of seed quality characteristics. Temperature and relative humidity together have the practical impact of reducing the potential viability or lifespan of seeds throughout the span of their storage life. In order to achieve ideal storage conditions, controlling RH and its subsequent impact on seed moisture content is more important than controlling storage temperature (McDonald and , 2001).

The most and least sensitive seeds to storage conditions are tomato and onion seeds.

Quality Control

According to the Seed Act of 1966, which regulates seed quality in India, seeds of registered or recognized varieties sold to farmers must adhere to minimum germination and physical purity criteria. It needs to be packaged in an appropriate container with the appropriate label. The label and packaging must have the following information: crop, variety, germination per cent, physical purity, date of testing, packaging lot and date, unit weight and price, name of seed production agency or institution, and other necessary details. The germination stated on the label is valid for nine months from the test date and must then be revalidated in every six months (Sarolia *et al.*, 2018).

Seed Certification Principles and Procedure

Seed Certification

A system of quality control for the production and multiplication of seeds that has been authorized by law is seed certification. It designed to maintain and make available continuous supply of high-quality seed and propagating materials of notified kinds and varieties of crops. It ensures the physical identity and genetic purity. Official sources supervise and examine seeds destined for domestic and foreign markets to ensure consistently excellent quality for customers. Field inspection, seed quality tests, and pre and post-quality checks are all included. While labelling is mandatory, certification is an optional process (Hasanuzzaman, 2015).

Phases of Seed Certification

- ✰ Receipt and scrutiny of application
- ✰ Verification of seed source
- ✰ Field inspection to verify field standard
- ✰ Postharvest seed inspection
- ✰ Seed testing to verify seed standard
- ✰ Issue of certificate, tagging, labelling and sealing

Seed Production Agencies

Government Players in Indian Seed Industry

Government seed producing and R and D institutes include:

1. National Seeds Corporation
2. State Seed Corporations
3. ICAR institutes
4. State Agriculture Universities

Vegetables like tomato, cabbage, brinjal, chilli, okra, and cucurbits are the main focus of private seed enterprises. As a result, the public sector is left to handle the production of actual seed or planting materials for crops like potatoes, sweet potatoes, and other tuber and bulb crops.

Private Players in Vegetable Seed Industry in India

Several small and medium-sized businesses have now started to operate under this initiative, which covers seed production for the majority of solanaceous and cucurbitaceous crops for the domestic market and exports. These are a few of these companies.

- Namdhari Seeds
- Oriental Biotech
- Tropica, Exim
- Krishidhan Seeds Mahyco
- Indo American Hybrid Seeds
- Golden Seeds
- Unicorn Biotech

Firms in the US, Europe, and Japan are the key markets for custom production for export. Nearly 70 per cent of all seed shipments are for vegetables (Walia and Lalotra, 2021).

Challenges and Opportunities of Vegetable Quality Seed Production

Challenges

- Inadequate infrastructure and expertise for seed marketing
- High Cost and Vague Market Demand
- Exposure to Diseases and Pests Due to the Quarantine Issue
- Absence of Clearly Defined Seed Strategy
- Poor Extension Services
- Private businesses frequently focus on lucrative crops

- ✰ Insufficient Understanding and Lack of Awareness Regarding Seed Production

Opportunities

- ✰ Varied Agro-Climatic Conditions
- ✰ Promotion of seed production units or agencies for the growth and security of plant varieties
- ✰ Development of infrastructure, processing, and marketing of seeds, the export of seeds and planting materials,
- ✰ Empowering rural youth and women to employ biotechnology methods for seed production, certification, marketing, and germplasm conservation.
- ✰ Government initiatives in rural infrastructure and general agricultural policies;
- ✰ High seed demand and limited seed supply
- ✰ Governments and assisting agencies are currently revaluating their strategies.
- ✰ The availability of trained human resources
- ✰ Existence of Prominent Regional and National Seed Production Projects
- ✰ The presence of several cooperatives and business agents

Future Prospects of Vegetable Seed Industry

Vegetable seed production has a huge potential to generate income. Vegetable quality seed development demands a significant amount of manual labour for pollination and emasculation. Vegetable seed production hubs must be diversified to non-traditional, highly productive areas. Funds from different government programmes, including MNREGA and the Tribal Sub plan, should indeed be made available to rural communities so they can undergo training in seed production. Entrepreneurship growth at the village level has to be supported. One of the key methods put forth by the government to double farmers' income by 2022 is the promotion of ancillary activities and the diversification of farm enterprises. Vegetable seed production can play a significant part in this strategy. India can lead the globe in the export of vegetable seeds and usher the rural prosperity because to its diversified climatic and soil conditions and massive human resource base.

CONCLUSION

A major economic sector in India is the production of high-quality seeds for the vegetable seed industry. India is fortunate to have a wide range of agro-ecologies ideal for producing many sorts of vegetables, including tropical, subtropical, and temperate veggies. In countries like India, where the majority of people are employed in agriculture, vegetable seed companies will always have a high chance of success and will remain crucial to the economy of such countries. Giving farmers access to high-quality seeds at reasonable prices and in a timely manner is more crucial nowadays. To ensure the supply of high-quality seeds and to protect farmers against fake seeds, stringent adherence to seed legislation is required. The government should impose fewer restrictions on the import and export of high-quality seeds. Political reasons must not be used in the formulation or implementation of policies. To compete with private seed businesses and offer farmers high-quality seeds at lower prices, the public sector in R and D needs to be strengthened. Consequently, collaboration between the governmental and private sectors could aid in India's development of high-quality vegetable seeds. According to several pre-established agreements, the public and private sectors should exchange germplasm and other inputs.

REFERENCES

Anonymous, (2013). Indian Agriculture: Performance and Challenges, State of Indian Agriculture 2012-13. Directorate of Economics and Statistics, Department of Agriculture and Cooperation, Ministry of Agriculture, Government of India, New Delhi, 1-22.

AVRDC, (1990). Vegetable Production Training Manual. Asian Vegetable Research and Development Center, Shanhua. 447.

Colquhoun J. B, Gevens A. J., Groves R. L., Helder, D. J., Jensen B. M., Nice G. R. W. and Rumark M. D. (2015). Commercial Vegetable Production in Wisconsin. 2.

Gadwal, V. R. (2003). The Indian seed industry: Its history, current status and future. Current Science, 84(3), 399-406.

Hasanuzzaman, M. (2015). Seed Quality. https: //hasanuzzaman.weebly.com/uploads/9/3/4/0/934025/seed_quality.pdf

Koundinya, A. V. V. and Kumar, P. P. (2014). Indian vegetable seeds industry: status and challenges. *International Journal of Plant, Animal and Environmental Sciences.* 4(4): 62-69.

Kumawat, S., Shekhawat, K., Kumar, A. and Yadav, B.D. (2020). Importance-of-quality-seed-production-in-agriculture.https: //www.krishisewa.com/seed-production/1123-importance-of-quality-seed-production-in-agriculture.html

Kuo, C. G., Malvyanona, R. F. and Kalb, T. J. (2006). Increasing Market Oriented Vegetable Production in Central Asia and the Caucasus through Collaborative Research and Development. AVRDC. The World Vegetable Centre, Shanhua, Taiwan. 250.

Masabni, J., Daniello, F. and Conter, S. (2009). The Texas Vegetable Growers Hand Book, 4th ed. Extension Vegetable Specialist, Texas A and M University.

McDonald, M. B. and Copeland, L. O. (2001). Principles of Seed Science and Technology 4thed. Kluwer Acadamic Publication. Norwell, MA.USA. 488.

Poonia, T. C. (2013). History of Seed Production and its Key Issues. International Journal Food, Agri. and Vet. Sci, 3(1), 148-154.

Rashid M. A. and Singh D. P. (2000). A Manual on Vegetable Seed Production in Bangladesh. AVRDC-USAID- Bangladesh Project-Horticulture Research Centre Bangladesh Agricultural Research Institute. Karshaf Printers (Pvt.) Ltd.

Sarolia D. K., Samadia D.K., Choudhary B.R. and Singh D. (2018). Production of quality seed and planting materials. CIAH/Tech./Pub. No. 75. ICAR-Central Institute for Arid Horticulture, Bikaner, Rajasthan, India.

Selvakumar, R. (2014). Glaustas Olericulture. New Vishal Publication, New Delhi 1st ed. 719-766.

Sexton, P. (2001). Vegetable Seed Production 2nd Edition. CABI Publishing. New York.

Sharma, J. P. (2011). Indian vegetable seed Industry: Potential and problems. In Quality seed production of vegetable crops technological Interventions, Kalyani Publishers, 60-78.

Singh, P. and Asati, B. S. (2013). Seed production technology of vegetable. Daya publishing house. Astral international Pvt. Lmt. Delhi.

Thooyavathy R. A., Perumal, K., Suresh, V. and Vijayalakshmi, K. (2013). Seed Production Techniques for Vegetables. Centre for Indian Knowledge Systems, Chennai Seed Node of the Revitalising Rainfed Agriculture Network. 1-54.

Walia, A. and Lalotra, A. (2021). Aspects of the Indian Vegetable Seed Industry. Just agriculture 1 (11): (e-ISSN: 2582-8223).

Chapter 20

Role of Rootstocks in Fruit Production

Anuj Pal[1], Kaushlendra Pratap Singh[1], Vibhu Pandey[2], Devendra Pal[1] and Rishabh Shukla[1]*

[1]*Ph.D. Research Scholar, Department of Horticulture,*

[2]*Ph.D. Research Scholar, Department of Fruit Science,*

Sardar Vallabhbhai Patel University of Agriculture and Technology, Meerut – 250 110, Uttar Pradesh

**e-mail: siranuj1749@gmail.com*

INTRODUCTION

Horticulture is being considered as a viable industry for increasing agricultural production, improving household nutritional security, and generating income through diversification, employment, value addition, and export. Despite the huge accomplishment gained in the horticulture sector, there are still numerous obstacles. Apart from new issues, low productivity per unit area remains a concern in most horticulture crops, with climate change affecting productivity even more. These challenges in the domain of biotic and abiotic stresses and their influence on various horticulture crops must be addressed. Rootstocks have a substantial impact on fruit crop development by altering canopy architecture, nutritional intake, flowering, yield, and fruit quality in various fruit crops **(Rom, 1987).** It can also deal with biotic and abiotic stresses such as soil pathogens, heat, salinity, and nutritional stress **(Reddy, 2003).** Fruit crops are regularly cultivated under unfavorable soil and environmental circumstances such as heat stress, drought, flooding, salt, and pollution with organic pollutants due to a lack of arable land and a strong market demand for them. The adoption of appropriate rootstocks, which can lessen the influence of external pressures on the scion, is one strategy to substantiate or reduce these production losses **(Higgs, 1991)**. Orchard efficiency is primarily determined by rootstocks. They are responsible for water and mineral intake as well as tree anchoring. Tree size is determined by rootstocks. Rootstocks can be employed as inter stems in the propagation process, a short piece put between the rootstock and the scion "Three-piece" trees are what they're called. Inter stems can be used to overcome incompatibility barriers between the stock and the scion, as well as reducing the vigor of desirable rootstocks. Rootstocks may provide some tolerance to soils that are occasionally too wet or dry. It is impossible to have a single rootstock for India's large and diverse agro-climatic environments, which include frigid deserts, dry and wet temperate climates, flat and low, medium and high hills with varied degrees of slopes and topography, as well as varying soil and micro-climatic variables. Compatibility with various scion genotypes is one of the criteria for selecting an effective rootstock. Building a strong root system and transferring soil nutrients to the scion to maximize its genetic potential for increased fruit output and quality. Adaptable to a variety of soil types, pH, mineral content, nematode tolerance, salinity/alkalinity tolerance, limited water availability, pathogen load, and soil type should all be acceptable for adoption. Modern orchard systems require characteristics such as, uniformity in growth, size control, efficient canopy, yield efficiency, and adaptability to biotic and abiotic challenges. As a result, the rootstock is an extremely important component of a grafted fruit tree, and it often determines whether a commercial orchard succeeds or fails. As a result, rootstocks must be identified and developed to fit various situations.

Historical Perspective

The sixth century Chinese writings of Chia Shi-yi in the book Tsee Ming Yau Su contain records on the usage of a special seedling rootstock (*Pyrus betuelifolia*) for pear. With a few crops, the most notable of which are apple, pear, and grape. Careful selection of superior rootstock and grape is followed by clonal propagation, which is then maintained. Alexander the Great introduced a dwarf apple species named Spring Apple (*Malus pumila*) from Asia Minor to Greece, according to the Greek scholar Theophrastus (370 BC). He also discussed the practice of apple grafting. Researchers began using size-controlling rootstocks for apple trees in the late nineteenth and early twentieth century, and this need later expanded to other fruit tree species. Throughout the nineteenth century, Paradise was the most extensively used rootstock in Europe. Researchers at the UK's East Mailing Research Station were the first to classify rootstocks and produce new ones for specialized needs in 1912. They released M IX, or M 9, in 1917, which was produced from the Jaune de Metz Paradise rootstock. M 9 rootstocks produce compact, precocious apple trees, making them an ideal rootstock for commercial apple orchards even today. East Mailing Research Station, in partnership with Merton and Long Ashton Research Stations in the United Kingdom, created a series of virus-free rootstocks, of which M 27, M 9, M 26, MM 106, M 7, MM 111, and M 25 are still widely used today. In the

United States, resistance to fire blight disease and the capacity to withstand harsh cold were the primary goals. Work on these issues began at the New York Agricultural Experimental Station in Geneva, NY, and eventually led to the production of G-series rootstocks, which were quite popular in North America. Because of M 9's commercial significance, numerous researchers have created superior versions or substitutes.

Despite the introduction of clonal rootstocks of apple, pear, plum, citrus, grape, and other temperate fruits from different nations, most temperate fruits in India are still grown on seedling rootstocks. In addition, some native wild species have been recognized and commercialized. With the widespread use of traditional seedling rootstocks, genetic variability was observed in root system-specific characteristics such as root architecture (depth and distribution), tolerance to edaphic-environmental stresses (soil acidity, drought, salinity, and so on), and biotic stress tolerance (root-specific diseases and pests).

Effect of Rootstocks

Size and Growth Habit

Based on their effect on a scion cultivar, rootstocks have been classed as dwarf, semi-dwarf, vigorous, and very vigorous in apple. Rootstocks like M 27, M 9, M 26, and MM 106 are known to give scion cultivars dwarfing and semi-vigorous growth. Trifoliate oranges, troyer citrange, and other citrus fruits have a low vitality. Kurrakkan, Olour, Latra, Creeping, Vellaicolumban, and other rootstocks confer dwarfness in mango cultivars. Mango dwarfing interstock Anupam has been identified. Guava cultivars grafted onto *Psidium pumilum* are small.

Precocity in Flowering and Fruiting

Precocitiy is the period it takes from plating to fruiting and is regulated by rootstocks. Rootstocks influence fruiting in general, and dwarfing rootstocks in particular. Mandarins grown on *Citrus jambhiri* rootstock are more productive than those grown on sweet, sour, or acid lime rootstocks. In Kinnow mandarin, Troyer citrange causes precocity in bearing.

Fruit Set and Yield

Rootstock has a direct impact on bloom output and fruit set in many fruit crops. As grafted on *D. lotus* rootstock, Oriental persimmon (*Diospurous kaki* cv. Hachiya) produces more blossoms but fewer fruits when compared to *D. kaki* seedlings. The yield of acid limes budded on rough lemon is about 70 per cent higher than those budded on Troyer citrange, Rangpur lime, or the rootstock itself. Under south Indian conditions, sweet orange var. Sathgudi budded on Billikichilli rootstock yielded more than jambhiri or its own seedlings.

Fruit Size and Quality

Sathgudi sweet orange grafted on Gajanimma rootstock yields huge but poor-quality oranges, whereas it produces higher-quality fruits on its own roots. On the Cleopatra mandarin rootstock, the physiological problem granulation in delicious orange is quite low. Maximum granulation is achieved by budding rough lemon seedlings. The physiological condition when *Pyrus cummunis* was utilised as the rootstock, black end in Barlett pear was reduced.

Nutrient Uptake

Scion cultivars' nutrient intake and tissue state are influenced by rootstocks. When budded on Sathgudi sweet orange tree leaves had a better nutrient condition.

Winter Hardiness

Young grapefruit trees grown on Rangpur lime rootstock are more resistant to winter injury than rough lemon or sour orange rootstock. Cold resilience is good in sweet oranges and Mandarins on trifoliate oranges.

Diseases Resistance

There is a lot of variation in the responsiveness of rootstocks to diseases, pests, and nematodes in grapes, citrus, and other temperate crops. Rough lemon rootstock is resistant to Tristeza and Xyloporosis, as well as gummosis, although it is sensitive to Exocortis viral disease. Guava types grafted on Chinese Guava are also resistant to wilt disease and worms.

In numerous countries, the identification and use of appropriate rootstocks in crops such as apple, pear, citrus, mango, and grapes has been researched in relation to vigour management, nutrient uptake, soil salinity, moisture stress, and production efficiency. Most commercial fruit crops in India have yet to fully use rootstocks with enormous commercial potential. As a result, finding a rootstock with the most suitable qualities to satisfy the set of environmental conditions where the tree will be developed is a critical stage. It means that a root-stock that is ideal for one variety and set of environmental conditions may not be the best choice for another variety or set of conditions. This review article focuses on various studies that are primarily focused on the impacts of rootstocks on scion growth and cropping, which will aid in focusing and coordinating future research in this area.

Role of Rootstocks in Different Fruit Crops

Role of Rootstocks in Mango

Mango is one of the most significant tropical fruits in the world, and it is renowned in India as the "King of Fruits." There have been several initiatives to standardize rootstocks for various scion kinds, including the usage of polyembryonic cultivars for vigour management, salinity and drought tolerance, as well as to improve fruit yield and quality. In mango, vigour management is crucial, particularly for high-density planting and orchard management in terms of canopy management, harvesting, and plant protection. In comparison to other polyembryonic rootstocks, the growth and bearing habit of the 'Dashehari' mango on the seedling stock of 'Dashehari' itself demonstrated the most vigorous character and maximum yield, according to **(Jauhari, 1972)**. Similar findings from IIHR-Bangalore, based on a 21-year study of the performance of 'Alphonso,' showed that nucellar seedlings of 'Muvvandan,' 'Bappakai,' and 'Olour' were vigorous rootstocks in decreasing order of vigour, while 'Vellaikulumban' seedling imparted dwarfing in comparison to 'Alphonso' grafted on its own **(Reddy, 2003).** Some parameters, such as phenolic content, bark percentage, and chlorophyll fractions, have been found to be highly beneficial in assessing mango rootstock vigour at the nursery stage (Abirami). **(Murti, 2003)** stated that at the nursery stage, leaf water potential, total phenols, and possibly ABA can be used to choose size controlling rootstocks. Fruit output and productivity are critical in crops like mango, where productivity is low in the majority of mango-growing countries. **(Smith, 2008)** found that 'Kensington Pride' yield and yield efficiency were best on the 'Sg. Siput' rootstock and were poor on the Sabre rootstock, indicating the potential for influencing mango scion productivity through rootstock genotypes. **(Quijada, 2008)** discovered that the 'Sinamaica' rootstock is the most adaptable to the agro-ecological conditions of Venezuela's Maracaibo plain, producing higher yield and efficiency than other cultivars (Criollode Mara, Manzana and Sensation mango). The best rootstock for 'Dashehari' was discovered to be 'Bappakai,' followed by 'Muvvandan,' and 'EC 95862.' **(Chandan, 2006).** The maximum number of fruits per plant was recorded by 'Langra' grafted on 'Bappakai' rootstock, followed by 'Vellaikulumban' and 'Chandrakaran' **(Mahabir Singh, 2006).** Fruit quality is also a significant consideration, and various studies have been conducted in this area in various nations. A 21-year rootstock trial at IIHR in Bangalore found no significant effect of rootstocks (Vellaikulumban, Bappakai, Chandrakaran, Kurukan, Muvandan, Mylepelian, and Olour) on mango fruit quality **(Reddy, 2001)**. Saline testing of mango rootstocks revealed that the polyembryonic cultivars 'Olour' and 'Bappakai' could resist higher salinity levels **(Palaniappan, 2001).** In comparison to 'Kurukan,' 'Olour' was shown to be the best salt resistant rootstock **(Dubey, 2007)**. For calcareous soils or irrigation with saline water, mango cultivar '13-1' was chosen as a polyembryonic (3-6 embryos) rootstock. Mango trees on '13-1' rootstocks performed well in soil containing 20 per cent lime, but three other cultivars on the same rootstock performed well in sandy soil with 10-20 per cent lime **(Gazit, 1980)**. According to **(Zuazo Durán, 2004)**, the 'Gomera-1' rootstock is more resistant to saltwater environments. With increased salt concentrations of water, seedlings from the 'Kesar' variety were shown to be better, with significantly higher survival, germination, and growth indices **(Sharma, 1992).**

Role of Rootstocks in Citrus

In India, Rangpur lime (*C. jambhiri*) and Rangpur lime for kinnow, Rangpur lime and Cleopatra mandarin for Blood Red, and Jatti Khatti and Cleoptra Mandarin for Jaffa have shown potential **(Singh *et al.*, 1997)**. **(Dhatt and Singh, 1993)**. *Feronia limonia* has proven to be dwarfing and precocious, making it ideal for dense planting. 'Troyer Citrange,' 'Karna Khatta,' and 'Sohsarkar' were identified as dwarf, semi dwarf, and strong Kinnow rootstocks at IARI in New Delhi. Sathgudi and Mosambi tree volume was highest on Rangpur lime and rough lemon in Bangalore, followed by Kodakittuli and Cleopatra mandarin **(Haleem, 1984)**. In comparison to the other rootstocks, *Citrus volkameriana* identified superior rootstocks for Navel orange, Valencia orange, Ruby Red, and Marsh Grapefruit plants. The role of rootstocks in citrus is one of the most contentious and debated topics, and their selection is an important factor to consider when developing a citrus plantation. One thing is certain: selecting the appropriate rootstocks is critical to the orchard's success.

Role of Rootstocks in Grapes

Although grapevines can grow in a wide pH range (4.5-6.5), extremely acidic soils pose a concern. **(Himelrick, 1991)** discovered varied reactions of Vitis species or cultivars to low soil pH based on vine development. *V. labrusca* cv. 'Concord' and 'Catawba,' along with rootstock 'SO4' and '3309C,' and the hybrid cultivar 'Seyval,' were the cultivars most tolerant to strong acid soils; *V. vinifera* cv. 'White Riesling' and 'Chardonnay' were the most intolerant. It was strongly suggested that acid tolerant rootstocks such as 'SO4' and '3309C' be used. According to **(Leon, 1969)**, the grape rootstocks *V. champini* and *V. vinifera* are both salt tolerant. **(Walker, 2000)** investigated different grape rootstocks for salinity (0.4 or 2.3 dS/m) and found that grape cv. Shiraz grafted on 'Ramsey', '1103 Paulsen', and '140 Ru' had higher wine K+ pH, and colour hue than grape cv. Shiraz grafted on its own seedling. The same researchers discovered that grape cv. 'Sultana' grows well when grafted on rootstocks like 'Ramsey', '1103P', and 'R2'. **(Troncoso, 1999)** classified rootstocks as sensitive (41 B, R. Lot, 110 R, 140 R, and 161-49), moderately tolerant (13-5 and Ramsey), and tolerant (based on nature of tolerance to different salt concentrations (0, 50, 85, 120, 155 mM NaCI) (196-17, CH-1, CH-2 and Superior). (**Ezzahouani, 1995**) suggested that hybrids from the rootstocks *V. berlandieri* x *V. rupestris* ('110R', '140Ru', and '1103P') could be used in drought-prone areas where water is a limiting factor for grapevine productivity, and he also noted that drought resistance classification of rootstocks varies by country. Similarly, **(Kocsis, 1999)** claimed that *V. berlandieri* x *V. riparia* hybrids were drought tolerant. Only a few researches have been done on the impact of rootstock on scion cold hardiness. **Miller (1988)** discovered discrepancies in rootstock 'K5BB', '3309C', and 'SO4' cane and bud hardiness. Rootstock '3309C' had the coldest hardy canes and buds, and its fall acclimation was faster and spring declamation was slower than 'K5BB' and 'SO4'. The rootstock had little effect on the distribution of hardy canes within the canopy, according to **(Striegler, 1991)**; 'Seyval' grafted on '3309C' looked to be the cold hardy.

Role of Rootstocks in Guava

Guava is currently propagated via seedlings grown from open pollinated seeds rather than homogenous clonal rootstocks. *P. cattleianum, P. guinesee, P. molle,* and Philippine guava were shown to be appropriate as rootstocks **(Edwar, 1964).** The trees on *P. cattleianum* were the tallest and produced the most fruit. Although the *P. pumilum* rootstock had a dwarfing effect, the fruits on this rootstock had the most seeds, TSS, and total sugars. *P. cujavillis* trees produced the largest, highest-ascorbic-acid-content fruits; however, they were rough-skinned and uneven. Except for 'Allahabad Safeda,' all rootstocks were clear of wilt disease (*Fusarium solani/Macrophomina phaseoli*) **(Teaotia 1961).** In terms of plant height, plant spread, and tree volume, 'Allahabad Safeda' is dwarfed by rootstock aneuploid No. 82 **(Sharma, 1992).** This rootstock also produced a higher yield.

Role of Rootstocks in Apple

The materials utilised to raise rootstock seedlings in the apple industry vary greatly. The wild indigenous 'Crab apple' known as 'Trel' is used in Kashmir, while seedlings of 'Crab C' are used as rootstock in Himachal Pradesh **(Nauriyal 1965).** Crab C, M 2, M 13, Merton 779, Merton 793, and 'Local Selection' (seedling selection from 'Ribbistin Pippin') rootstocks were used in the first apple rootstock testing in 1937 at Chaubattia with the cultivars 'Red Delicious,' 'Jonathan,' and 'Rymer.' Following the introduction of the M and MM series of rootstocks from the East Malling Research Station in England in the early 1960s at Mashobra and Kotkhai in HP, Chaubattia in UP, and Shalimar in J&K, elaborate varietal-cum-rootstock trials were launched at Mashobra in 1967, Kotkhai in 1968, Chaubattia in 1969, and several locations in J&K in 1969. M 7, M 9, M 26, MM 106, and MM 111 for Himachal Pradesh; M 2, M 4, M 7, M 9 for J&K; and Merton 779, MM 106, M 13 for Uttarakhand **(Ananda 1993).** 2 rootstock response to prolific bearing and good yield in 'Red Delicious,' 'Jonathan,' and 'Rymer' at early stages was noted in Chaubattia (5th and 10th year) **(Om, 1983).** Following that, in later years (20th and 25th years), maximum yields were reported in 'Merton 779' and 'Crab C' rootstocks, while Merton 779 and M 13 rootstocks produced higher yields than other rootstocks after the 35th year. Similarly, apple trees grafted on MM 106 rootstock showed considerably higher fruit set, yield, and yield efficiency than those grafted on M 7. **(Chandal 1992)** found that 'Starking Delicious' on MM 111, MM 106, MM 104, and M 4 rootstocks had less reduction in growth, photosynthetic efficiency, and nutrient uptake, as well as higher stomatal resistance, lower transpiration rate, and more accumulation of proline, ABA, and carbohydrate under water stress (10 bar), making these rootstocks more drought tolerant than M 2, M 7, M 9, M 25, M 26.

Role of Rootstocks in Pear

For pears, there are fewer rootstock options than for apples. In terms of vigour, hardiness, and compatibility, domestic pear seedlings (*Pyrus communis*) are still the most suitable rootstocks for pear cultivars. All pear trees with seedling roots, however, are susceptible to fire blight. *Pyrus calleryana* seedlings are adaptable to a wide range of soil conditions and develop semi-vigorous growth. It's resistant to fire blight, but it's not winter hardy enough. *Pyrus betulaefolia* seedlings can be grown in a variety of soil conditions and are cold hardy. The trees are strong, larger than 'Bartlett seedling,' and fire resistant to a degree. *Pyrus ussurensis* is cold hardy, resistant to fire blight, and pear psylla.

Role of Rootstocks in Stone Fruits

Stone fruits such as peach, plum, apricot, and almond are typically propagated on their own seedlings in India, whereas cherry trees are grown from 'Paja' (*Prunus cerasoides*) seedlings. *P. mira*, often known as Behmi, is a popular almond rootstock **(Chadha 1998).** The relative contribution of organic and inorganic solutes to osmotic adjustment in three almond rootstocks exposed to soil salinity revealed that salinity affected leaf water and osmotic potentials in GF 677 and Bitter almond, but not so much in GN 15, implying a higher selectivity for K^+ and Ca^{2+} over Na^+ in this latter rootstock. Peach seedlings are used as rootstocks for peach plantations, as well as plum, apricot, and almond seedlings **(Dewan 1992; Singh 1973)**. Nematode susceptibility is common in peach seedlings; however, 'Nemaguard' and 'Okinawa' showed resilience to nematode attack **(Hansen 1956)**. Peach rootstock 'Behmi' (*P. mira*) is also used by some nurseries. Apricot can be utilized as a rootstock that is resistant to root knot nematode. Dwarfing rootstocks for peaches include Western sand cherry, Almond, and Nanking cherry. Other countries employ rootstocks such as 'Brompton' and 'Common plum' from Plum (*P. domestica*); St. Julien, 'Common Mussel' and 'Damson' from Plum (*P. institia*); Myrobalan from Plum (*P. cerasifera*); peach (*P. persica*), apricot (*P. armeniaca*), almond (*P. amygdalus*), and hybrids like Marianna plum Peach, apricot, behmi (natural hybrid of almond and wild peach), and plum seedlings are popular rootstocks in India **(Singh 1971)**. Hazard (*P. valium*) and Mahler (*P. mahaleb*) seedlings are the most common cherry rootstocks in other nations, with Stockton Morello (*P. cerasus*) being used on occasion. In Kashmir, Himachal Pradesh, and the Kumaon Hills of Uttar Pradesh, the most frequent rootstock is 'Paja' (*P. cerasoides*) **(Srivastava 1962)**, while old plantations are typically on Mazzard or Mahaleb rootstocks **(Singh 1971)**. Prunus besseyi seedlings of 'Russian sour cherry' could be employed for extraordinary hardiness **(Singh, 1971)**.

CONCLUSION

Fruit trees are propagated on rootstocks rather than their own roots because rootstocks offer various advantages. The review demonstrates the impact of rootstock on fruit quality in terms of physical characteristics and interior chemical composition. These factors have an impact on precocity/juvenile, yield, tree size control, disease resistance or tolerance, fruit respiratory behavior, crop load, and canopy. In this regard, rootstocks are an important component of contemporary agriculture's fruit crops and should be chosen based on the prevailing climatic conditions and soil type. Hence, the main prerequisites for large fruit production are a strong root system, its wide and deep dispersion in the soil, and a consistent and adequate yearly growth of absorbing roots.

REFERENCES

Abirami, K., Singh, R. and Baskaran, V. (2011). Studies on the influence of seedling physiological parameters with vigour in some polyembryonic and monoembryonic mango genotypes. *Indian Journal of Horticulture, 68*(1), 18-23.

Ananda, S. S. (1993). Rootstocks of stone fruits. Advances in Horticulture: Fruit Crops.

Bernstein, L., Ehlig, C. F., and Clark, R. A. (1969). Effect of grape rootstocks on chloride accumulation in leaves. *J. Amer. Soc. Hort. Sci.*

Chadha, K. L. and Singh, H. P. (1990). Citriculture scenario of India. In Citriculture in North-Western India. Procedings Citrus Show-cum-Seminar, Prospects and Problems of Kinnow Cultivation. *Punjab Agricultural University, Ludhiana* (21-64).

Dubey, A. K., Srivastav, M., Sharma, Y. K., Pandey, R. N. and Deshmukh, P. S. (2007). Dry mass production and distribution of nutrients in two mango rootstocks as affected by salinity. *Indian Journal of Horticulture, 64*(4), 385-390.

Edward, J. C., and Shankar, G. (1964). Rootstock trial for guava (*P. guajava* L.). *Allahabad Fmg, 38*, 521-527.

Ezzahouani, A. and Williams, L. E. (1995). Influence of rootstock on leaf water potential, yield, and berry composition of Ruby Seedless grapevines. *Amer. J. Enol.* Viticult., 46: 559-563.

Gazit, S. and Kadman, A. (1980). 13-1 mango rootstock selection. *HortScience, 15*(5).

Hansen, C., Lownrbery, B. and Hesse, C. (1956). Nematode resistance in peaches: Resistance to two widespread species of root-knot nematode ranged from almost immunity to none in peach seedling study. *California Agriculture, 10*(9), 5-11.

Hemstad, P. R. and Luby, J. J. (1998). Utilization of Vitis riparia for the development of new wine varieties with resistance to disease and extreme cold. In *VII International Symposium on Grapevine Genetics and Breeding 528* (487-496).

Higgs, K. H. and Jones, H. G. (1991). Water relations and cropping of apple cultivars on a dwarfing rootstock in response to imposed drought. *Journal of horticultural science, 66*(3), 367-379.

Himelrick, D. G. (1991). Growth and nutritional responses of nine grape cultivars to low soil pH. *HortScience, 26*(3), 269-271.

Jauhari, O. S., Teaotia, S. S. and Upadhyay, S. K. (1969, June). Rootstock studies in Mangifera indica L. In *Symposium on Mango and Mango Culture, 24* (107-109).

Kocsis, L., Granett, J., Walker, M. A., Lin, H. and Omer, A. D. (1999). Grape phylloxera populations adapted to *Vitis berlandieri* x *V. riparia* rootstocks. *American Journal of Enology and Viticulture, 50*(1), 101-106.

Miller, D. P., Howell, G. S. and Striegler, R. K. (1988). Cane and bud hardiness of own-rooted White Riesling and scions of White Riesling and Chardonnay grafted to selected rootstocks. *American journal of enology and viticulture, 39*(1), 60-66.

Murti, G. S. R. and Upreti, K. K. (2003). Endogenous hormones and phenols in rootstock seedlings of mango cultivars and their relationship with seedling vigour. *European Journal of Horticultural Science, 68*(1), 2-7.

OM, H. and Pathak, R. K. (1983). Influence of rootstocks on growth and productivity of some apple cultivars. *Indian Journal of Horticulture, 40*(3-4), 168-172.

Pongrácz, D. P. (1983). Rootstocks for grape-vines

Quijada, O., Matheus, M., Castellano, G., Camacho, R., González, C., Casanova, Á. and Noguero, N. (2008). Desarrollo y productividad de cultivares de mango injertados sobre el patrón sinamaica en la planicie de Maracaibo. *Agronomía Tropical, 58*(3), 207-213.

Reddy, Y. T. N., Kurian, R. M., Ramachander, P. R., Singh, G. and Kohli, R. R. (2003). Long-term effects of rootstocks on growth and fruit yielding patterns of 'Alphonso'mango (Mangifera indica L.). *Scientia Horticulturae, 97*(2), 95-108.

Rom, R. C. and Carlson, R. F. (1987). *Rootstocks for fruit crops* (No. 634.0432 R6).

Santhi, V. P., Nireshkumar, N., Vasugi, C., Parthiban, S. and Masilamani, P. (2020). Role of rootstocks to mitigate biotic and abiotic stresses in tropical and subtropical fruit crops: A. *IJCS, 8*(5), 499-510.

Sharma, D. D. and Chauhan, J. S. (1990). Effect of different rootstocks and training systems on growth and cropping of 'delicious' apples. *Indian Journal of Horticulture, 47*(4), 365-370.

Sharma, Y. K., Goswami, A. M. and Sharma, R. R. (1992). Effect of dwarfing aneuploid guava rootstock in high density orcharding. *Indian Journal of Horticulture, 49*(1), 31-36.

Singh, M. and Kanpure, R. N. (2006). Performance of mango cv. Langra on different rootstock. *Int. J. Agric. Sci, 2*, 564-565.

Singh, R. N. and Guptha, P. N. (1971). Rootstock problem in stone fruits and potentialities of wild species found in India. *Punjab Hort. J, 11*, 157-175.

Smith, M. W., Bright, J. D., Hoult, M. D., Renfree, R. A., Maddern, T. and Coombes, N. (2008). Field evaluation of 64 rootstocks for growth and yield of 'Kensington Pride'mango. *Hortscience, 43*(6), 1720-1725.

Srivastava, D. N. and Gustav, E. F. (1962). Propagation of temperate fruit plants in Kumaon hills. In fruit Nursery Practices in India. Directorate of Extension, Ministry of Food and Agriculture, New Delhi, 14-18.

Striegler, R. K. and Howell, G. S. (1991). The influence of rootshock on the cold hardiness of Seyval grapevines. I: Primary and secondary effects on growth, canopy development, yield, fruit quality and cold hardiness. *Vitis, 30*(1), 1-10.

Teaotia, S. S. and Pandey, I. C. (1961). Effect of growth substances on the rooting of guava stem cuttings. *Sci Cult, 27*, 442-444.

Troncoso de Arce, A., Matte, C., Cantos, M. and Lavee, S. (1999). Evaluation of salt tolerance of in vitro-grown grapevine rootstock varieties.

Varu, D. K. and Barad, A. V. (2010). Standardization of mango rootstock for mitigating salt stress. *Indian Journal of Horticulture, 67*(4), 79-83.

Walker, R. R., Read, P. E. and Blackmore, D. H. (2000). Rootstock and salinity effects on rates of berry maturation, ion accumulation and colour development in Shiraz grapes. *Australian Journal of Grape and Wine Research, 6*(3), 227-239.

Wolpert, J. A. (2005). Selection of rootstocks: Implications for quality. *Grapevine Rootstocks: Current Use, Research, and Application*, 25.

Zuazo, V. D., Raya, A. M., Ruiz, J. A. and Tarifa, D. F. (2004). Impact of salinity on macro-and micronutrients uptake in mango (*Mangifera indica* L. cv. Osteen) with different rootstocks. *Spanish Journal of Agricultural Research, 2*(1), 121-133.

Chapter 21

Techniques to Reduce the Postharvest Losses of Fruits and Vegetables

Pragya Uikey[1]*, Ankita Sharma[1], Amul Yadav[2], Reena Nair[3] and Rehan[1]

[1]Ph.D. Research Scholar, [2]M.Sc. Research Scholar, [3]Assistant Professor,

Department of Horticulture, College of Agriculture, Jawaharlal Nehru Krishi Vishwa Vidhyalaya, Jabalpur – 482 004, Madhya Pradesh

**e-mail: uikeypragya@gmail.com*

INTRODUCTION

Fruits and vegetables are essential food staples not just in India, but all throughout the world. Seasonal production of fruits and vegetables causes a glut during the season and a scarcity during the off-season. They are easily bruised or damaged during harvesting and other postharvest handling operations like packaging, transportation, and storage due to their soft texture. One of the issues with food security and world hunger in many nations is postharvest loss. Minimizing fruit and vegetable loss and waste is one of the key concerns for food security of world's population in the future since the amount of fruit and vegetable loss reaches up to 50 per cent. In order to feed the world's population sustainably in the future, postharvest losses must be reduced (Gustavsson *et al.*, 2011, Sharma *et al.*, 2021). If postharvest loss is significantly decreased, it may not be required to significantly increase production of fruits and vegetables in order to meet the rising demand. In general, it is less expensive to minimize harvest losses than to grow a comparable additional crop of high-quality fruits and vegetables.

Fruit and vegetable postharvest losses are primarily caused by handling and storage after harvest, processing steps, distribution, and consumption. Additionally, in developing and underdeveloped nations, this percentage is significantly higher.

Alao (2000) defined Postharvest loss of fruits and vegetables as "that weight of wholesome edible product (exclusive of moisture content) that is normally consumed by human and that has been separated from the medium and sites of its immediate growth and production by deliberate human action with the intention of using it for human feeding but which for any reasons fails to be consumed by human." Fruits and vegetables are affected not just in terms of quantity and quality but also in terms of appearance, which lowers their market value. Some diseases are not harmful because they affect non-edible components that may be removed before eating, such as leaf spots on cabbage's outer leaves. The main forms of deterioration that lead to disposal of fresh fruits and vegetables include moisture loss, bruising, and eventual decay (Ray and Ravi, 2005). While fruits and vegetables are being transported from the producer to the consumer, a number of factors (such as processing, storage, and transportation conditions) contribute to these deteriorative reactions. To identify probable causes and optimize the most effective postharvest techniques, it is essential to understand where losses occur in the food supply chain (Gustavsson *et al.*, 2011; Porat *et al.*, 2018). If these conditions are not appropriately regulated, large-scale postharvest losses will occur. While the changes that happen to fresh fruit cannot be controlled, they may be minimized with moderate care, including low storage temperatures, relative humidity management, appropriate packing, and safe transportation, *etc.* (Ahmad and Siddiqui, 2015). Due to high moisture content (65-95 per cent), insect attack, and damage from postharvest handling practices, postharvest losses of perishable food crops (vegetables and fruits) amounted to roughly 30 percent (packaging, storage and transportation). However, these losses can be reduced with the use of suitable packaging materials, suitable storage facilities, and appropriate transportation. In addition, the most important techniques to minimizing perishable food losses and maintaining product quality are recent methods of food processing and postharvest technology.

Development of value-added products with effective and efficient research programmes should be increased and promoted in developing countries. Improving postharvest management methods would minimize postharvest losses. Farmers frequently experience severe financial losses because they lack a thorough understanding of the nature and reasons of these losses, as well as appropriate preservation measures, transportation methods, and marketing strategies. However, this may be greatly decreased by applying appropriate cultural techniques, including cautious handling and packing. Furthermore, by shielding the fruit and vegetables from disease attacks and other environmental harm, the application of suitable chemicals throughout the pre- and postharvest stages may increase the shelf life of the food and make it available for a long time.

In recent years, the field of postharvest handling of fresh produce faced several technological advances. The combination of the online monitoring of the environment or produce state and active modulation of storage conditions with the emerging discipline of modeling based on big data and artificial intelligence (AI) tools introduces both opportunities and challenges to the field (Ziy and Fallic, 2021).

All fresh horticultural crops are metabolically active organisms, even after harvest, until they are either processed or consumed. Depending on storage conditions, various physiological processes, such as respiration, ripening, and senescence, can significantly impact the quality and shelf life of fruits and vegetables. Since fresh produce is not sterile, the storage time of fresh produce also largely depends on interactions with other organisms in its environment, namely pathogenic microorganisms, insects, and other pests. Consequently, two main processes cause fruit and vegetable deterioration: (1) physiological deterioration, which includes water loss, softening, ripening, and shedding of leaves; and (2) microbial decay, caused by pathogenic bacteria, yeasts, or molds. Both of these processes are highly dependent on environmental conditions and greatly affect one another (Ziy and Fallic, 2021).

The most prevalent approach to preserve fruit quality and to reduce produce loss after harvest is to delay ripening and senescence. Fruit ripening involves several metabolic processes that differ between 'climacteric' and 'non-climacteric' fruits. During the ripening of climacteric fruit, respiration increases until it reaches a peak, which is accompanied by an increase in ethylene production. In contrast, respiration of non-climacteric fruit does not increase during ripening, and ethylene is not required in order to complete the ripening process. Regardless of the type of ripening, this process, as well as other metabolic processes that lead to deterioration, are driven by respiration. After harvest, the fresh produce continues to respire–utilizing food reserves taking in oxygen, and releasing carbon dioxide and heat from stored carbohydrates. For that reason, postharvest treatments that reduce respiration will delay deterioration processes, prolong shelf life, and help to maintain produce quality (Ziy and Fallic, 2021).

Causative Factors of Postharvest Losses

Losses after harvest can be attributed to a number of factors, and they can happen at various postharvest handling phases.

1. Harvesting Methods

The right time should be chosen for crop harvesting since unripen fruits are more prone to mechanical damage and degradation and may develop undesirable traits such as high acidity and low sugar levels as they mature. Overripe fruits, on the other hand, have a short shelf life. Fruits are more prone to physiological disorders in both situations (overripening and under ripening).

Crops lose value economically and nutritionally when they are harvested too early. Whole items may sometimes be abandoned because they are unfit for consumption (Azabaðaoðlu, 2018).

Postharvest losses may result from improper harvesting methods. The loss in the harvesting period is between 4 - 12 per cent due to incorrect calculation of harvesting time, inappropriate application of harvesting method, and failure to apply pre-cooling to fruits like cherry. At the time of harvest or following harvest, farmers could not have storage container. This causes mechanical damage to fruit, vegetable, root, and tuber plants during harvest, which causes losses of fresh produce.

2. Packaging

Fresh produce needed to be packaged after harvest, throughout transportation, and during marketing. Packaging has a significant role in lowering losses and prolonging the shelf life of fresh produce. Therefore, improper packing and the use of inappropriate packaging materials are two of the main reasons why fruits and vegetables are lost during postharvest stages. Farmers offer their goods for sale at wholesale or fresh marketplaces. Fresh produce is either offered in an unpackaged form or in bundles at the retail level. If fresh product is handled in this way at the market and is not soon sold, its shelf life is significantly reduced (Kiaya, 2014).

Poorly constructed packaging materials are unable to fully protect fresh goods against harm and may even hasten its deterioration. Unfortunately, because they are inexpensive, low-quality packing materials are often utilized in many regions of the world. Particularly, the usage of mediocre packing materials is more widespread in developing and underdeveloped nations. Even some of the delicate fruits and vegetables are packaged in polybags, seriously harming the produce. The majority of packaging don't give the product the complete protection it needs. Some nylon-sack structures do not enable cold air to circulate because of lack of aeration, which raises interior temperatures and causes a buildup of heat from respiration. Additionally, containers with sharp edges and surfaces can cause mechanical injuries to the produce (Arah I. K., 2015).

3. Storage

According to a study by the International Refrigeration Institute (IIR), 23 per cent of perishable commodities in developing nations went bad because they weren't kept in coolers (IIR, 2009). Postharvest losses in developing nations are considered to be mostly caused by inadequate storage facilities (FAO, 2013). Postharvest products may have to be stored for a period of time ranging from a few hours to a few months after undergoing certain procedures including cleaning, sorting, and packing. Storage of Commodities, allowing for more leisurely marketing and consumption of fresh product. Storage is consistently provided in developed nations from the production, along the whole supply chain. Perishable goods have a much longer shelf life when cold storage is paired with postharvest technologies such a controlled environment.

4. Transportation Facility

The majority of developing nations, including India, have marketing centers that are both distant from the source of production and inaccessible by road. Products are harvested and transported from the point of production to the local market via trolleys, buses, motorcycles, or by manual. Bus, trucks, and other vehicles are used for long distance travel. On the other side, wealthy nations have excellent infrastructure, including refrigerated facilities. Poor road conditions have resulted in improper transportation of harvested goods, which has become a significant challenge for both production and consumption levels. Because transportation takes time between production and consumption, it can be one of the major reasons for losses, especially for fresh goods.

Transportation of perishable commodities in refrigerated vehicles is a common practice in developed nations. Poor roads, a lack of suitable transportation, and ineffective logistics management make it difficult for underdeveloped nations to effectively preserve perishable crops. Additionally, in these nations, untrained, uneducated employees who typically handle goods carelessly do loading and unloading activities. Agricultural products are mechanically harmed as a result (Azabaðaoðlu, 2018). Food losses during transport result from bad roads in the rural areas where the majority of production takes place.

5. Waste During Consumption Stage

The period from purchase of fruit and vegetables and consumption is known as the consumption stage. At the consumption stage of the food supply chain, fruit and vegetable losses are significant. Fruit and vegetables account for 39 per cent of all household waste, according FAO (2011). Another study by WRAP (2008) found that this proportion can reach up to 40 per cent of all house waste. The major reasons why customers waste so much are because they buy too much, don't plan in advance, handle their home storage poorly, *etc.* The extent of consumer waste is influenced by both sociocultural and material aspects, including gender, way of life, money, and available home storage (Porat *et al.*, 2018). According to studies, significant food losses occur in developing nations as a result of technical and managerial shortcomings in harvesting methods, storage, transportation, and processing operations, as well as a lack of adequate infrastructure, packaging, and marketing systems. On the other hand, in developed countries, postharvest loss is primarily related to consumer behaviour, strict safety regulations, and criteria for fruit and vegetable quality (FAO, 2011, 2015).

The losses so generated by the waste produced by Fruit and vegetable not only represent the wasting of food commodities, but also indirectly include wasting of critical resources such as land, water, fertilizers, chemicals, energy, and labour. (Venkat, 2011; Vilari˜no and others 2017). In due course, household garbage, fruit and vegetable processing units commonly produce the highest wastes into the environment (Gowe, 2015). The waste so generated by the horticultural commodity is rich in potentially valuable bioactive compounds. The scenario now a days is changing as the horticulture byproducts have not been taken very seriously in the past as very valuable materials, but now can be used to recover highly valuable biomolecules. Horticultural by-products are exclusive sources of different pigments, phenolic compounds, dietary fibers, sugar derivatives, organic acids, and minerals, among other components. Several bioactive compounds possess beneficial health attributes: antibacterial, antitumor, antiviral, antimutagenic, and cardio protective activities (Yahia, 2017). Vegetables like potatoes, carrots, green peas, onions, artichokes, and asparagus can be used to prepare juice or pulp extraction, jams, and frozen pulp, producing significant amounts of waste (Rodriguez and others, 2006 and Sharma *et al.,* 2021). FVWs (Fruit Vegetable waste) can are utilized to extract and isolate potential bioactive compounds. Therefore, although some of the waste can be considered unavoidable one, the proper use of waste materials acquired from horticultural commodities may establish an initiative for sustainable development to mitigate environmental problems and to improve human health through foods enriched with health-enhancing substances (phenols, carotenoids, and other pigments, vitamins, dietary fibers, among several others). Vegetable processing industries are incriminated for a high quantity of pollution, hence, it is becoming very crucial to solve this problem by developing optimized systems for the treatment of vegetable wastes (Sharma *et al.,* 2021). Among the several processes that are being used nowadays, the ones described are the following: thermal processes, evaporation, membrane processes, anaerobic digestion, anaerobic co-digestion, biodiesel spraying, combustion, transesterification, coagulation and composting. Waste can contain many reusable substances of high value (Sharma *et al.,* 2021). Depending on there being an adequate technology, this residual matter can be converted into commercial products either as raw material for secondary processes, as operating supplies or as ingredients of new products. Numerous valuable substances in food production are suitable for separation and recycling at the end of their life cycle, even though present separation and recycling processes are not absolutely cost efficient (Sharma *et al.,* 2021).

6. Cultural Practices

The fundamental management techniques as cultural practices extend the shelf life of perishable horticultural crops. Inconsistent irrigation before and after harvesting reduces the shelf life and sensory quality of fruits and vegetables. Practices like pruning and thinning enhance fruit size and reduce TSS and acidity of fruits. These activities can also add up to postharvest losses of fruits and vegetables. Problems like lack of improved fruit and vegetable cultivars, nutrient deficiencies, a lack of tools and infrastructure, and diseases and pests like fruit ball worm in guavas, borer in tomato and okra, premature fruit drop in citrus (which reduced the crops' quality and shelf life) causes their prices to drop, this also contributes to postharvest losses of fruits and vegetables.

7. Sanitation

Before being transported, fruits and vegetables require careful handling. Produce like root and tuber crops are frequently cleaned to get rid of the soil that has adhered to it. Before being packaged, fruits and vegetables that have been treated with toxic chemicals need to be thoroughly washed. Washing in clean water prevent wilting in fruits and vegetables. Before marketing, in leafy vegetables such as lettuce, spinach, and cabbage, trimming is essential. Ultraviolet or gamma radiation, hydrogen peroxide liquid or vapour, or both are used for disinfection.

8. Harvesting Factors

Crop produce that are harvested too early or too late lose quality and have a lower shelf life. Fruits and vegetables are selected for harvest based on maturity index, including size, taste, total soluble solid (TSS), and skin colour. The crucial phases that primarily govern the shelf life of fruits and vegetables include maturation, ripening, and senescence. The fruit's maturity stage indicates when it is ready to be harvested. Harvesting method comprises manual method and mechanical method, it can also cause postharvest loss in fruits and vegetables if farmer select in appropriate harvesting method.

9. Marketing Facility

Compared to other durable agricultural products, the marketing of fresh perishables is more problematic. In comparison to cereal crops, horticulture crops are more challenging to market because of their perishable nature, bulkiness, seasonal produce, high level of care, and need for prompt disposal. As a result, there are numerous issues with the supply of horticulture commodities, including significant price fluctuations. Fruits and vegetables also have a distinct pricing range from producer to consumer, which is the result of numerous intermediaries at various levels in the marketing system (Neeraj, 2017). Loss of products can occur at intermediaries.

10. Environmental Factors

Environmental factors including temperature and humidity are significant contributors to fruit and vegetable postharvest loss. The produce sustains severe damage as a result of the microorganisms that thrive in the high temperatures and relative humidity. High temperatures also speed up the rate at which fruits and vegetables respire, which causes the interior tissues to break down.

Additionally, high temperatures and high relative humidity speed up the decay of fruits and vegetables, whereas low temperatures, especially below 5 °C, slow down the rate of microbial infection on many crops. Tropical and subtropical fruits and vegetables are most commonly observed to suffer from chilling injury, which is caused by low temperatures. Relative Humidity (RH) plays a similar impact in the postharvest environment as temperature. The capacity of air to store moisture fluctuates with temperature, making the effects of temperature and relative humidity mostly comparable and related. The aeration in storage containers or in retail spaces affects RH and, in turn, indirectly influences the onset of disease.

Table 21.1: Postharvest Loss (Per cent) and their Causes

Sl.No.	*Cause of Postharvest Loss*	*Postharvest Loss of Fruit and Vegetables (per cent)*
1	Harvesting factors	20 per cent
2	Postharvest handling and storage	8 per cent
3	Packaging and processing of product	10 per cent
4	Marketing	10 per cent
5.	Consumer waste	5 per cent

Source: Elik *et al.*, 2019.

Postharvest Handling of Fruits and Vegetables

Vegetables and fruits keep respiring even after they have been harvested. Harvested commodities alter in appearance, texture, taste, and flavour, rendering them unsuitable for the market. By using the right postharvest management techniques, the quality of harvested goods can be preserved until consumption. Poor produce handling practices have caused significant postharvest losses, as well as time and money lost. The risk of loss can be reduced, though, by putting produce through the necessary grading, sorting, storage, transportation, and marketing processes. By protecting products from dangerous infections and contaminants, pre- and postharvest chemical treatments also prolong the shelf life of commodities.

1. Precooling

The purpose of precooling is to quickly cool a commodity to the necessary transport or storage temperature. The most crucial action following harvest is immediate cooling because increased field heat causes unwanted changes in certain metabolism and leads to quality degradation. The field heat of the harvested vegetables must be removed, especially if the harvesting is done in hot weather. Therefore, cooling preserves the fresh produce's weight, giving an added benefit during the prolonged storage period. Pre-cooling during storage can lower physiological weight loss from 6 to 2.9 percent in fruits like tomatoes. While effective cooling can lower storage losses in vegetables that decay quickly. Precooling lessens the impact of microbiological activity, metabolic activity, respiration rate, and ethylene generation while slowing the rate of ripening, water loss, and decay, conserving quality and extending shelf life of harvested tomatoes (Shahi, 2012). Precooling can be accomplished using a variety of techniques, including ice cooling, vacuum cooling, hydro-cooling, forced air cooling, and room cooling. However, the pace of cooling is influenced by the cooling medium, the produce's thermal characteristics, and the way the produce is stacked. some fruits, such as grapes and leafy vegetables, which are typically very difficult to cool using water or air, they can be quickly and uniformly cooled by vacuum cooling.

2. Cleaning

Cleaning is required before fresh produce is marketed in order to get rid of dust, debris, and spray residue. All farmers, middlemen, and consumers are extremely concerned about the cleanliness and hygienic conditions of the produce in order to prevent both postharvest infections and food-borne illness of consumer (Arah IK, 2015). Suitable cleaning methods and equipment should be used for fruits and vegetables. It has been discovered that washing solution is most effective between at pH 6.5 and 7.5. Common chemicals for cleaning fresh vegetables include thiabendazole, sodium hypochlorite solution, and chlorinated water.

3. Sorting and Grading

Sorting is the process of separating healthy, clean fruits that are fit for marketing and storage from bruised, damaged, and rotten fruits. When sorting, only sensory quality factors are taken into account. The most crucial marketing tactic that must be taken into account while handling is sorting. Fruits and vegetables that have been damaged are separated because they are unfit for the market. Then, based on shape, colour, and size, products are classified into grades. Sorting and grading are two of the most crucial steps in the packing and marketing of fruits and vegetables (Arjenaki, 2013). By eliminating contaminated and damaged products, further contamination can be reduced. Grading is the way of categorizing fruits and vegetables according to their shape, size, colour, and internal properties. A, B, C, 1, 2, 3, fine, coarse, large, medium, small, *etc.* are all valid grade.

4. Waxing

Reduced evaporation and respiration are the main advantage of wax coating. Wax skin coating is another technique used to extend the shelf life of fresh fruit in situations when refrigeration storage facilities are either unavailable or inadequate. Many fruits and vegetables, including tomatoes, okra, mangoes, and other produce, have longer shelf lives after they have been processed through water emulsified with mustard oil.

5. Packaging

Produce is enclosed in packaging to protect against exterior contamination (physical, chemical, and biological) and mechanical injury (Kochhar, 2014). Most developing nations employ nylon sacks, plastic or wooden crates, woven palm baskets, cardboard boxes, and polythene bags as their standard packing materials. Specific sizes of plastic trays or cartons are used for tomatoes and corrugated cardboard boxes for apples, and locally accessible wrapping/cushion materials are advised which are suitable for local condition (Bhattarai, 2018). Fruits, root and tuber crops were also packaged using local sacks, wooden boxes, polyethylene bags, and other packing materials. Leafy vegetables including cabbage, spinach, and lettuce were packed utilizing local materials such as plastic materials, wooden baskets, and nets. The packaging must meet the requirements for handling and marketing, and have enough mechanical strength, be chemically free, and allow rapid cooling of the contents.

6. Storage

The practice of maintaining agricultural products' quality and preventing degradation for a set amount of time after their usual shelf life is known as storage (Kiaya, 2014). According to their rate of deterioration, market demands, and relative prices, some commodities are kept in storage for a short period of time while others are kept for a longer period of time. Ripe tomatoes can be stored at a temperature of 10-15°C and 85-95 per cent relative humidity for long-term preservation. Most enzymes function best up to a temperature of 30°C and are inactivated above 40 °C (Castro, 2005).

7. Marketing

Commodities are transported from the site of production to the point of consumption through the marketing process. The final and concluding step in the postharvest system is marketing. Pre-harvest contract is a widely used marketing strategy in perennial and uncontrolled volume production. A few months before the production season, contractors buy the orchard for lump sum.

D. Dhakal (2005) identified the following four categories of marketing channels for acid lime.

Channel 1: Producers-Retailers- Consumers

Channel 2: Producers-Wholesalers-Retailers-Consumers

Channel 3: Producers-Commission Agent-Wholesaler- Retailers-Consumers

Channel 4: Producers-Collectors-wholesalers-Retailers- Consumers

While customers overpaying, farmers receive a lower return on their investment. When fruits and vegetables are overstocked on the market, there is a greater loss. This makes farmers less likely to grow fruits and vegetables. Thus, efforts should be taken to prevent glutting and the loss should be kept to a minimum (Mardiyya SY, 2019).

Table 21.2: Postharvest Losses in some Fruits and Vegetables

Vegetables or Fruits	*Postharvest Losses (per cent)*
Citrus	8-31 per cent
Grapes	23-30 per cent
Mango	17-37 per cent
Banana	12-24 per cent
Pea and Beans	7-12 per cent
Cauliflower	10-15 per cent
Brinjal	10-13 per cent
Garlic	1-3 per cent
Potato	15-20 per cent
Onion	15-30 per cent
Tomato	10-20 per cent

Ways to Reduce Postharvest Losses and Waste of Fruits and Vegetables

Proper cultural operations, harvesting, transportation, storage, and pre- and postharvest treatments can significantly reduce the amount of postharvest loss in fruits and vegetables.

1. Cultural Practices

A good cultural operation is desired to ensure normal development and extend the shelf life of fruits and vegetables. To prevent root forking, it is important to prepare the soil to a fine tilth of porous nature, for root crops like carrot and radish. It is crucial to maintain consistent irrigation during the growth stage of fruits, vegetables, bulbs, and tubers. Onions' outer scales split and carrot, radish, and tomato skins crack due to insufficient water availability. Tomatoes and watermelons, particularly at late maturity after a prolonged dry season, might crack when there is excessive or intense irrigation. These perishables have a short shelf life and little visual appeal. Three weeks prior to harvest, irrigation should be discontinued to promote improved keeping quality for onions and garlic.

When nitrogenous fertilisers are applied heavily, fruits and vegetables typically experience faster tissue deterioration; nevertheless, a regular supply of potassium fertilisers increases the quality of fruits and vegetables' storage. Physiological disorder like greening of the almond and heart rot of the cabbage are caused by molybdenum and boron deficiencies. Such fruits often have a relatively short shelf life and may noticeably deteriorate in quality. Therefore, it is advised to fulfil the optimal need for these micronutrients at the appropriate moment.

The supply of water, nutrients, and sunlight in orchards and vegetable fields was improved as well as natural growth of fruits and vegetables by proper control of weeds. Fruit trees need adequate sunshine to develop their colour and size, therefore proper training, pruning, and removal of dead or diseased branches ensure that the trees receive sufficient sunlight. However, mulching should be used when growing vegetables because it improves the quality of the produce while conserving nutrients, water, and weeds. Staking in pea, tomato, and other vine crops helps to keep vegetables from coming into touch with the soil, which prevents rotting. While using specific mechanical devices along with adequate and attentive supervision, the damage caused by birds in orchards can be decreased. Some strategies for reducing fruit damage by birds include creating noise with a drum or placing bird-scarers over trees.

2. Harvesting Factors

The quality and storage life of commodities are determined by harvesting, which is an important basic activity that also helps to minimize significant losses of fruits and vegetables. Consumers consistently prefer fresh, well matured, insect and disease-free fruits and vegetables with a pleasing appearance. Vegetables should therefore be harvested as soon as they reach their ideal and maximum size and are still delicate. Maximizing crop yield, minimizing crop losses and quality deterioration, and maintaining the harvested produce in good condition until it is consumed or sold are the objectives of a successful harvest. Fruits that are harvested too early or too late lose quality and have a lower shelf life. Therefore, for optimum quality and better storage, the fruits should be harvested at the appropriate stage of maturity. Important variables that affect postharvest losses and product quality include maturity criteria, harvesting procedures, harvesting instruments, field packing, and transportation conditions to the packinghouse.

Harvesting factors that can help to minimize fruit and vegetable losses.

- ☆ harvesting at the best possible time (Set an optimum maturity date).
- ☆ The collection of mature fruits is also crucial (the produce has to be harvested and handled gently).
- ☆ the right harvesting equipment and technology
- ☆ Containers used for harvesting could cause a loss (Kiaya, 2014).

Some fruit and vegetable maturity indices include size, shape, hardness, juice content, textural qualities, tenderness, colour, and nutrient composition, including sugar, starch, oil, dry matter content, soluble solids, and titratable acidity. After harvest, temperature control is crucial for minimizing postharvest losses and maximizing shelf life.

The technique of harvesting has an effect on the fruit and vegetable quality as well. Crop products can be damaged as a result of improper harvesting techniques. Therefore, in order to reduce damage at the time of harvest, harvesting methods must be standardized for each and every fruit and vegetable. It is best not to delay harvesting root crops like carrots and radishes because doing so leads to pithiness and sponginess. The reduced shelf life of onions and garlic, as well as nearly all other crops, is a result of delayed harvesting. It should be emphasized that in order to reduce mechanical damage, it is crucial to avoid unnecessarily wounding, bruising, crushing, or damaging product. Harvesting techniques include mechanical and manual methods alone or in combination. Numerous mechanical devices are employed in mechanical harvesting to harvest products on a large scale. Harvesting containers include picking baskets, bags, sacks, carts, and buckets constructed of a variety of materials, including plastic, wood, natural fibres, and synthetic fibres. Avocados and citrus fruits have firm skins, therefore harvesting bags with waist or shoulder slings can be employed. Sacks are frequently used for crops including pumpkins, potatoes, and onions. Harvesting fruits that are readily crushed, like tomatoes, requires for the use of smooth plastic buckets without any sharp edges. The type of fruit or vegetable should determine the appropriate harvesting container.

Table 21.3: Maturity Indices of some Fruits and Vegetables

Crop	*Maturity Index*
Mango	Tapka, Fullness of cheeks, Bulging of shoulder
Banana	Angularity (Disappearance of angle)
Grapes	14-17 per cent SSC
Citrus	Juice content 50 per cent
Apple	T stage, 11-12 SSC
Pineapple	Flattening of eyes
Potato	Yellowing and drying of haulm
Onion and Garlic	Neck fall
Okra	4-7 days after flowering
Pea	Change in pod colour
Watermelon and Jackfruit	Tapping

3. Transportation

Distance, perishability, and product value largely determine the best method of transport for fresh produce (Harris, 1988). Fresh food must be transported through several stages from the place of origin to the point of consumption.

During the transportation, produce should be kept at the best possible temperatures. The most crucial methods for ensuring the quality preservation of perishables are adequate temperature control systems and air circulation systems. Since temperature has a significant impact on the rate of biological reactions, it is the primary environmental factor affecting the quality of product.

Ambient humidity and water loss, air gas concentration, mixed loading, physical damage and stress, and transportation condition are other elements that affect product quality (surface road conditions, time of the day, *etc.*) (Vigneault *et al.*, 2009). During transport, the produce is vulnerable to microbiological infection, physical or chemical damage, and both. Fresh fruit and vegetables must be transported with special precautions to ensure their safe storage.

Some guidelines to be follow to ensure safe transportation

- ☆ Transportation periods should be minimum
- ☆ The best possible care should be taken during loading and unloading
- ☆ The product should be adequately protected in terms of its fitness for physical harm
- ☆ Maintaining the proper temperature throughout the transportation duration
- ☆ Provide ventilation and aeration and maintain proper relative humidity
- ☆ Providing shade and cover from sun and rain at sites for loading and unloading

4. Storage

The practice of maintaining agricultural products' quality and preventing degradation for a specific amount of time after their usual shelf life is known as storage. The primary goal of storage is to reduce the rate of ageing caused by respiration, moisture loss, and disease deterioration. Throughout the world a diverse variety of storage structure are employed to successfully store horticultural produce. General terms, the structure must be maintained cool (refrigerated, or at the very least aired and shaded), and the produce placed in storage must be of excellent initial quality (Kiaya, 2014). According to Alao (2000), the lowest temperature that doesn't harm fresh produce by chilling is the best temperature for storing fruits and vegetables.

Controlled Atmosphere Storage (CAS), which entails regulating the gaseous atmosphere in a storage facility in addition to temperature and humidity, has been found to increase the maintenance of postharvest quality. Given that increased CO_2 levels typically have a negative effect on the development and growth of microorganisms, one major advantage of controlled environment storage is the reduction of food deterioration due to pests and diseases. The amount of postharvest chemicals necessary to preserve fruits and vegetables from insects and microbes may therefore be greatly reduced by storing them in a controlled environment (Thompson, 1998).

Cold storage is the most frequently used practice to decrease metabolic activity. However, low-temperature (LT) storage conditions must be adjusted and optimized for each specific type of produce. Additionally, other storage conditions such as humidity (Adeeko *et al.*, 2020) and air composition should be adjusted for the best results. Thus, it is necessary to determine the optimal storage conditions to prolong fruit and vegetables' storage time. Specifically, controlled atmosphere (CA) (Paulauskienë *et al.*, 2020) for kiwifruit, as well as dynamic controlled atmosphere (DCA) or ultra-low oxygen (ULO) (Koriæanac *et al.*, 2020) for apples, are increasingly being evaluated as chemical-free applications to prolong storage time (Dällenbach *et al.*, 2020).

Table 21.4: Storage Conditions for some Fruits and Vegetables

Crop	*Temperature (°C)*	*Relative Humidity (Per cent)*	*Storage Time*
Tomato (ripe)	8-10	90	1 week
Brinjal	8-12	90-95	1-2 week
Potato	4-5	90-95	4-9 months
Onion	0-1	70-80	6-8 months
Pea	0-1	95-100	1-3 weeks
Broccoli	0-1	95-100	1-2week
Capsicum	7-10	90-95	1-3 week
Carrot	0-1	95-100	4-8 months
Cucumber	8-11	90-95	1-2 week
Spinach	0-1	95-100	1-2 week
Mango	11.7-12.8	85-90	2-3 week
Papaya	7.2	85-90	1-2 weeks
Strawberry	0.0	90-95	5-7 days
Lemon	8.9-10.0	85-90	1-6 month
Apple	-1 to 4	85-90	4-8 month
Guava	7.2-10.0	90	2-3 week
Grape	-1	90-95	3- 6 weeks
Pomegranate	0.0	90	2-4 weeks
Litchi	2.1	90-95	3-5 weeks

5. Packaging in Reduction of Postharvest Losses and Wastes

Depending on their physical and physiological characteristics (primarily transpiration, respiration, and ethylene production rate), as well as susceptibility to microbial decay, various fresh produce requires different forms of packaging. The fresh produce's ability to last after harvest mainly depends on temperature, relative humidity, and ventilation. The requirements for fresh produce packaging can be summed up as protection against physical damage and bruising, protection against microbial contamination and deterioration, protection against moisture/weight loss, delaying ripening and increasing storage life, controlling ethylene concentrations in the package (Ahmad and Siddiqui, 2016; Ramaswamy, 2014).

Packaging is important step to reduce postharvest losses during postharvest losses. It is effective for enhancing the fresh produce's shelf life and protecting it during the postharvest process. Fresh fruit and vegetable packaging reduces food waste by using better and smarter design to keep food fresher for longer. As a result, effective packaging techniques should be created to minimize food losses. Fruits should be packaged carefully to prevent bruising and damage during transportation. Between the layers of fruits, dry grasses or paddy straw can be utilized successfully for this purpose. However, packaging is a very inventive technology, and better methods and materials are always being added, both for processed goods and fresh produce.

Fresh fruits and vegetables have some degree of protection from external elements by packaging materials used in conventional packaging procedures. To decrease postharvest losses and lengthen the shelf life of fresh fruit and vegetables, improved packaging technologies and appropriate packaging materials must be developed. In fact, there are a number of cutting-edge packaging technologies for the packing of fresh fruits and vegetables. The most significant advanced packaging technologies are regarded as modified atmosphere packaging (MAP), active packaging, and smart (intelligent) packaging. For instance, by employing well-designed MAP, the shelf life of packed goods can be increased by 50 per cent to 200 per cent. (Siddiqui, 2016).

6. Use of Edible Coating

In addition, the use of natural components for edible coating applications is referenced as a safer option for extending the shelf life of perishable produce and improving food appearance (Fawole *et al.*, 2020). Edible coatings (ECs) of fresh produce with semipermeable film can prolong postharvest fruit life through reducing moisture loss, respiration, gas exchange, and additionally delaying changes related to ripening such as fruit softening, color changes, loss of organic acids, and the breakdown of starches into sugars (Tzortzakis *et al.*, 2019). Thus, ECs are increasingly gaining attention as alternative storage methods for fresh agricultural produce to extend storage time and shelf life. For example, the application of polysaccharide-based edible coatings for plums was shown to prevent shriveling–an economically important postharvest disorder (Fawole *et al.*, 2020). In some cases, edible coatings are applied in combination with additional natural compounds that serve as antimicrobials to increase the efficiency of a treatment (Tzortzakis *et al.*, 2019). Nevertheless, more research is required to fully understand the specific effects of different coatings on fruit physiology and postharvest performance.

Essential oils (EO) from the aromatic plant species are an economic source of unsaturated fatty acids, posing antimicrobial and antioxidant activities that can aid in postharvest preservation. Such natural extracts as citrus (Shehata *et al.*, 2020) and sage essential oil (Tzortzakis *et al.*, 2019) were found to delay microbial activity and to extend the shelf lives of strawberries and tomatoes, respectively. EO can be applied alone or as an additive for the edible coating of fresh produce. In tomatoes, the combined application of aloe vera gel coating with sage EO decreased fruit ethylene emission and decay symptoms, maintaining fruit firmness (Tzortzakis *et al.*, 2019).

7. Inhibition of Ethylene Synthesis

Ethylene is the plant hormone best-known for its effect on fruit ripening. The inhibition of ethylene synthesis is advantageous because inhibiting ethylene production delays processes related to fruit ripening, *i.e.*, softening, chlorophyll degradation, and the breakdown of complex sugars, *e.g.*, starch, into simple sugars. Reducing ethylene synthesis by LT/CA storage and/or by application of the competitive ethylene antagonist, 1-methylcyclopropene (1-MCP), which blocks the response of plants to ethylene, can greatly delay the ripening of climacteric fruits, and additionally alleviate physiological disorders such as scald in apples. However, treatment of 1-MCP for late-harvested apples stored in CA increased flesh browning (Jung and Choi, 2020), indicating that more work is required to optimize performance of this treatment. Interestingly, co-regulation of methionine biosynthesis, ethylene production, and respiration in 1-MCP-treated tomatoes has shed new light on the effect of this treatment on fruit physiology (Gamrasni *et al.*, 2020) and may facilitate the improvement of its implementation. While manipulating ethylene production is a common practice, the increase in applications of other growth regulators, such as salicylic acid and the polyamine putrescine, holds great promise to extend the repertoire of postharvest treatments and resultantly prolong the storage time of non-climacteric fruits, such as mandarins (Ennab *et al.*, 2020 and Ziy and Fallic, 2021).

8. Chemical Treatment

Ripening, senescence, and mechanical injuries not only affect fruit quality but also significantly increase the susceptibility of fresh produce to microbial attacks that cause decay. Chemical treatments with synthetic or natural pesticides are used for fresh produce preservation, as they are highly efficient practices to inhibit microbial activity and to reduce losses. However, this practice may pose negative implications on human and environmental health, due to possibly harmful by-products and residues. For that reason, there is a growing public demand to reduce the use of these chemicals. Furthermore, several health and environmental concerns for hazardous food contamination due to preharvest application of pesticide, as shown for cherry tomatoes (Shimshoni *et al.*, 2019) also render this practice questionable. (Ziy and Fallic, 2021)

Other than fungicides, a number of chemicals extend the shelf life of fruits and vegetables by postponing the ripening and senescence processes. Vegetables' shelf life has been proven to be increased by the use of postharvest plant hormones. Although cytokinin has a noticeable effect, particularly on green vegetables, the ripening of tomatoes is slowed down by the application of gibberellins after harvest. Gibberellic acid postharvest treatment significantly slowed down the ripening of bananas. It has been discovered that the maleic hydrazide is particularly effective ripening of several fruits. Mango ripening is postponed by maleic hydrazide at 1000–2000 ppm (parts per million). Application of a solution of indole acetic acid aids in preserving the green colour of many vegetable pods. Additionally, maleic hydrazide (MH) sprays applied prior to harvest prevent potato and onion sprouting during storage.

Fruits and vegetables are partially dehydrated to keep them succulent and soft, and then the moisture content is reduced using in-package desiccants. The shelf life of dried fruits and vegetables can be extended by using in-package desiccants like calcium oxide or fumed silicas to reduce the moisture content of the storage environment to 1 per cent or less. The infections in the harvested produce can be significantly reduced by pre-harvest fungicide application. Antibiotics and fungicides work best when combined in wax emulsions, which is the ideal way to apply them. The use of sulphur dioxide (SO2) in fumigating materials that are stored helps to stop the spread of fungus like botrytis. For packaged spinach and lettuce, dipping in 1000 ppm streptomycin (50–500 ppm) can stop bacterial soft rot.

REFERENCES

Adeeko, A., Yudelevich, F., Raphael, G., Avraham, L., Alon, H., Presman, M. Z., Alkalai-Tuvia, S., Paris, H. S., Fallik, E., Ziv, C. (2020). Quality and Storability of Trellised Greenhouse-Grown, Winter-Harvested, New Sweet Acorn Squash Hybrids. *Agronomy, 10*, 1443.

Ahmad, M. S. and Siddiqui, M. W. (2015). Factors affecting postharvest quality of fresh fruits. Postharvest quality assurance of fruits, Springer, p 7-32.

Ahmad, M. S. and Siddiqui, M. W. (2016). Postharvest quality assurance of fruits. Chapter, 2, Springer, p7-12.

Alao, S. L. (2000). The importance of postharvest loss prevention. Graduation ceremony of School of food storage technology. Nigeria: Nigerian Stored Products Research Institute, Kano, 1-10.

Arah, I. K., Harrison, A., Ernest, K. K. and Hayford, O. (2015). Preharvest and postharvest factors affecting the quality and shelf life of harvested tomatoes: a minireview. *International Journal of Agronomy.*

Arjenaki, O, OM. (2013). Online tomato sorting based on shape, maturity, size and surface defects using machine version. *Turkish Journal of Agriculture and Forestry*.37(1): 62-68.

Azabaðaoðlu, M. Ö. (2018). Investigating fresh fruit and vegetables losses at contemporary food retailers. *Sosyal Bilimler Araþtýrma Dergisi*.7(4): 55-62.

Bhattarai, D. R. (2018). Postharvest Horticulture in Nepal. *Horticulture International Journal*. 2(6): 458-460.

Castro, L. V. (2015). Effect of cooling delay and cold-chain breakage on 'santa Clara' tomato. *Journal of Food, Agriculture and Environment*.3(1): 49-54.

Dällenbach, L. J., Eppler, T., Bühlmann-Schütz, S., Kellerhals, M. and Bühlmann, A. (2020). Pre- and Postharvest Factors Control the Disease Incidence of Superficial Scald in the New Fire Blight Tolerant Apple Variety "Ladina". *Agronomy, 10,* 464.

Dhakal, D. K. T. (2005). Marketing survey of acid lime and hill lemon in Nepal. *Journal of Institute of Agriculture and Animal Science*.26: 107-116.

Elik, A., Yanik, D. K., Istanbullu, Y., Guzelsoy, N. A., Yavuz, A. and Gogus, F. (2019). Strategies to reduce postharvest losses for fruits and vegetables. *International Journal of Scientific and Technological Research*. 5(3): 2422-8702.

Ennab, H. A., El-Shemy, M. A. and Alam-Eldein, S. M. (2020). Salicylic Acid and Putrescine to Reduce Postharvest Storage Problems and Maintain Quality of Murcott Mandarin Fruit. *Agronomy, 10,* 115.

Fawole, O. A., Riva, S. C. and Opara, U. L. (2020). Efficacy of Edible Coatings in Alleviating Shrivel and Maintaining Quality of Japanese Plum (*Prunus salicina* Lindl.) during Export and Shelf-Life Conditions. *Agronomy, 10,* 1023.

Food and Agriculture Organization of the United Nations (FAO). (2013). Food Losses and Waste in Turkey, Country Report, 48. Available at: http: // www.fao.org/3/a-au824e.pdf

Food and Agriculture Organization of the United Nations (FAO). (2011). Global food losses and food waste – extent, causes and prevention. Available at: http: //www.fao.org/docrep/014/mb060e/mb060e.pdf.

Food and Agriculture Organization of the United Nations (FAO). (2015). Global Initiative on Food Loss and Waste Reduction. Available at: http: //www. fao.org/3/a-i4068e.pdf.

Gamrasni, D., Feldmesser, E., Ben-Arie, R., Raz, A., Tabatznik Asiag, A., Glikman, M., Aharoni, A. and Goldway, M. (2020). Gene Expression in 1-Methylcyclopropene (1-MCP) Treated Tomatoes during Pre-Climacteric Ripening Suggests Shared Regulation of Methionine Biosynthesis, Ethylene Production and Respiration. *Agronomy, 10,* 1669.

Gowe, C. (2015). Review on potential use of fruit and vegetables by-products as a valuable source of natural food additives. Food Sci Qual Manag 45: 47–61.

Gustavsson, J., Cederberg, C., Sonesson, U., Otterdijk, R. and Meybeck, A. (2011). Global food losses and food waste, food and agriculture organization of the united nation (FAO). Swedish Institute for Food and Biotechnology, Düsseldorf, Interpack.

Harris. S. R. (1988). Production is only half the battle - A training manual in fresh produce marketing for the Eastern Caribbean, Food and Agriculture Organisation of The United Nations, Bridgetown, Barbados.

International Refrigeration Institute (IIR). (2009). The role of refrigeration in worldwide nutrition. 5th Informatory Note on Refrigeration and Food. Available at: http: //www.iifiir.org/userfiles/file/publications/notes/NoteFood_05_EN.pdf

Jung, S.-K., Choi, H.-S. (2020). Browning of Early and Late-Harvested 'Empire' Apples Affected by Cold Storage and 1-MCP. *Agronomy, 10,* 1050.

Kiaya, V. (2014). Postharvest losses and strategies to reduce them. *Technical Paper on Postharvest Losses, Action Contre la Faim* (*ACF*).25: 1-25.

Kochhar, P. P. (2014). Active packaging in food industry; a review. *Environmental Science, Toxicology and Food Technology*. 8(5): 1-7.

Koriæanac, A., Miletiæ, N., Popoviæ, B., Mitroviæ, O., Lukiæ, M., Pešakoviæ, M., Tomiæ, J. (2020). The Effect of ULO and NA Storage on Changes in the Quality of Apple Fruit (Malus domestica Borkh.) during Shelf Life. *Agronomy, 10,* 25.

Mardiyya, S. Y. (2019). Review of Postharvest Losses of Fruits and Vegetables. *Biomedical Journal of Scientific and Technical Research*. p10192-10200.

Neeraj, A. C. (2017). Marketing and production of fruits and vegetables in India. *International Journal of Current Microbiology and Applied Sciences*.6(9): 2896-2907.

Paulauskienë, A., Tarasevièienë, Þ., Þebrauskienë, A. and Pranckietienë, I. (2020). Effect of Controlled Atmosphere Storage Conditions on the Chemical Composition of Super Hardy Kiwifruit. *Agronomy, 10,* 822.

Porat, R., Lichter, A., Terry, L.A., Harker, R. and Buzby, J. (2018). Postharvest losses of fruit and vegetables during retail and in consumers' homes: Quantifications, causes, and means of prevention. *Postharvest biology and technology*.139: 135-149.

Ramaswamy, H. S. (2014). Postharvest technologies of fruits and vegetables: DEStech Publications, Inc., Lancaster, Pennsylvania, USA.

Ray, R. and Ravi, V. (2005). Postharvest spoilage of sweet potato in tropics and control measures. *Critical reviews in food science and nutrition*. 45(7-8): 623-644.

Rodriguez, R., Jimenez, A., Fernandez-Bola, ´ nos, J., Guillen, R. and Heredia, A. (2006). ´ Dietary fibre from vegetable products as source of functional ingredients. Trends Food Sci Technol 17: 3–15.

Shahi, N. C. (2012). Effects of precooling treatments on shelf life of tomato in ambient condition. *International Journal of Food, Agriculture and Veterinary Science*.114: 50-56.

Sharma, A., Banerjee, J., Nishant and Nair, R. (2021). Vegetable Waste Management: an assessment of Sustainability in Environmental Chemistry for Sustainable world. Cape Comorin Publisher, Kuzhiodial Vilai, Kanyakumari, pp170-179.

Shehata, S. A., Abdeldaym, E. A., Ali, M. R., Mohamed, R. M., Bob, R. I., Abdelgawad, K. F. (2020). Effect of Some Citrus Essential Oils on Postharvest Shelf Life and Physicochemical Quality of Strawberries during Cold Storage. *Agronomy, 10,* 1466

Shimshoni, J. A., Bommuraj, V., Chena, Y., Sperling, R., Barel, S., Kaye, Y. and Fallik, E. (2019). Residual Distribution Kinetics of Pesticides in Cherry Tomato Peel, Pulp, and Fruit as a Function of Irrigation Water Salinity, Household Rinsing, and Storage Regimen. *Agronomy, 9,* 800.

Siddiqui, M. W. (2016). Eco-friendly technology for postharvest produce quality: Academic Press.

Thompson, A. K. (1998). Controlled atmosphere storage of fruits and vegetables. CAB International, London.

Tzortzakis, N., Xylia, P., Chrysargyris, A. (2019). Sage Essential Oil Improves the Effectiveness of Aloe vera Gel on Postharvest Quality of Tomato Fruit. *Agronomy, 9,* 635.

Venkat, K. (2011). The climate change and economic impacts of food waste in the United States. Intl J Food Sys Dyn 2: 431–46.

Vigneault, C., Thompson, J., Wu, S., Hui, K. C. and LeBlanc, D. I. (2009). Transportation of fresh horticultural produce. *Postharvest technologies for horticultural crops*.2: 1-24.

WRAP. (2008). Helping Consumers Reduce Fruit and Vegetable Waste: Final Report.

Yahia, E. M. (2017). Fruit and vegetable phytochemicals: chemistry and human health. Ames, Iowa: Wiley-Blackwell Publishing.

Ziv, C. and Fallik, E. (2021). Postharvest Storage Techniques and Quality Evaluation of Fruits and Vegetables for Reducing Food Loss. *Agronomy*. 11(6): 1133.

Chapter 22

Technique to Reduces the Postharvest Losses of Fresh Produces

Ajit Kumar Singh, Pooja B.K. and Suman Dutta

Ph.D. Research Scholar, Division of Postharvest Technology, ICAR- Indian Agriculture Research Institute, New Delhi

**e-mail: 718713ajit@gmail.com*

INTRODUCTION

Fruits and Vegetables are perishable in nature because of their high metabolic activity, water loss, and susceptibility to various pathogenic organisms, as a result, huge losses in horticultural products occur each year during handling and transportation and it nearly about 25-30 per cent. That means one-quarter of what is produced never reaches the consumer to whom it was grown, as well as the effort and money expended in producing it are lost forever. Fruit crops are much less hardy and mostly perishable, and if they are not harvested at proper time, with care less handling duringf transports, will cause to decay and make the produce unfit for human consumption.

Factor for the Losses

All the fresh produces are living part of the plant and contains around 65 to 95 per cent water and continue their transpiration and respiration even after harvest. Their postharvest life is determined by the rate at which they deplete their stored food material (CHO, Fat and Protein) and the rate at which they losses water. Produce will start to shrivel and decays when water reserves are depleted. Anything that speeds up the moisture loss process may render the produce inedible. The primary causes of loss are discussed further below.

Physiological Deterioration

Abnormal environmental condition that accelerates natural deterioration, such as high temperature, low atmospheric humidity, physical injury along with pre harvest (from growing to harvesting) and postharvest factor (precooling, handling, storage, temperature relative humidity, disease and disorder, gaseous composition, postharvest treatment *etc.*) which lead to an increase in the rate of quality loss of produce. When fresh produce is subjected to temperature extremes, atmospheric modification, or contamination, that cause the abnormal physiological degradation and it might be result in unpalatable flavors, failure to ripen and finally it makes the produce unfit for human consumption.

Mechanical Damage

Fresh produce is vulnerable to mechanical injury due to their high moisture content and soft texture, which can occur at any stage from production to marketing due to the following factors:

- Poor harvesting methods
- Unsuitable packaging material with splintered wood, sharp edges, or poor nailing or stapling happen during the field or marketing.
- Harvested fresh produce are filled up either over or under-packaging
- Mishandling, such as dropping, throwing, or walking on packed produce
- Improper packaging of fresh produce in containers during grading, transport, or marketing processes
- Injuries can occur through many ways, such as: splitting of fruits due to impact when they are dropped; internal bruising caused by impact that is not visible from the outside;
- Surface level grazing or abrasions that damage the skin of the outer layer of cells
- Leafy vegetables and other soft produce often crushed during the loading
- Rupture of cell due to improper cutting or scraping during the harvesting of fruit and vegetable that create entry points for molds and bacteria to lead the decay.
- Increases the losses of higher amount of water from the damaged area.
- Increase the speed of respiration coincide with heat production during the abnormal condition.
- Internal discoloration of damaged tissues due to the enzymatic activity such as peroxidase, polyphenol oxidase
- Off-flavors (lipid peroxidation) occur in produce during anaerobic respiration under the abnormal storage condition.

Losses Caused by Biotic

Disease-causing agents found in the air, soil, and water can infect fresh produce before or after harvest. Pathogens can infected the any living material when if they get congenial condition. However disease-causing agents are found in the air, soil, and water that can infect fresh produce before or after harvest. Some pathogen is capable to damage healthy tissue due to secretions of proteolytic enzyme, that it leads to the rotting of skin *viz.*; Anthracnose of papaya and gray mold of grapes.

Management of Postharvest Losses

Harvesting at the Appropriate Maturity Stage

Generally, any type of fresh produce is ready for harvest when it has reached the ideal size. However, the decision on when and how to harvest a crop is a critical time for growers, and such condition is commonly referred to as harvest maturity. Confusion may arise due to the word maturity, which in the botanical sense refers to when the plant has finished its active growth (vegetative growth) and has it arrived at flowering and seed production stage that is called physiological maturity. As a result, harvest maturity refers to the time when the "fruit" attained their aesthetic value with emission of volatile flavor that indicated the produce is ready to be harvest.

When are the Conditions Favorable for Harvesting of the Crop?

The basic rules to follow are:

- ☆ Harvest the produce during the coolest part of the day: early morning or late afternoon.
- ☆ Do not harvest fresh produce that is wet from dew or rain. Wet produce will overheat if not properly ventilated, and it will have caused to decay faster.
- ☆ Some produce, such as citrus fruits, may be more susceptible to rind damage when wet that is called 'Oil spot disorder' (Kinnow mandarin more prone to it).
- ☆ When transport is not immediately available, protect harvested produce in the field by placing it under open-sided shelter. Produce left in direct sunlight will become extremely hot that may lead to bleaching of the colour.

Method of Harvesting

In most cases, hand harvesting has been common because of less damage to produce than machine harvesting if done correctly. Hand harvesting is common in areas where fruit or vegetables or other produce is grown at various stages of maturity within the crop, indicating a need for repeated harvesting visits over a period of time. However, machine harvesting is typically feasible only when the entire crop is harvested all at once.

Special Postharvest Operation

Sorting

Before the produce reaches to the next stage of processing, it must be sorted to remove unmarketable pieces and foreign matter such as damaged plant part, debris, soil, or stones *etc.* All discarded material should be removed away from the packing house as soon as possible or placed in closeable bins for later removal. This is due to the fact that accumulations of decaying or infested waste in or near the packing house will contaminate marketable produce.

Cleaning and Washing

The exclusion of the soil stones and other plant debris can be accomplished by hand-picking or sieving. Some fruits and vegetables can be washed, brushed, or cleaned with a soft cloth. Hand-polishing or machine-brushing produce, especially fruit, can remove light soil contamination or dust. This should be done with caution because damaging the skin of fresh produce promotes early decay. Washing is also required to clean produce that has acquired latex stains from harvesting injuries, most notably mangoes, sapota, and bananas.

It's indeed important to remember that washing should only be done when absolutely necessary. When washing required use only clean, running water. Washing produce in recirculated or stagnant water should be avoided because it can quickly become heavily contaminated with decay organisms, and that it leads to heavy rotting of the washed produce. If disinfection is required a fungicide should be applied immediately afterwards.

To decrease the risk of residual affect, dilute the wash water with the fungicide below effective concentration and washed produce that is treated with fungicide should be drained after washing. When washing is not followed by fungicide treatment, washed produce should be spread out in a single layer on raised mesh or slat racks in the shelter but subjected to proper ventilation to assist in rapid drying.

Fungicide Treatment

Fungicide could be applied on a small scale in packing operations by:

Dipping- Treatment is done by hand, with a hand-agitated fungicide suspension; wire-mesh baskets may be used to soak several small pieces at once; after dipping, produce should be drained and dried in a shaded, shiny location.

Spraying - While produce is still in trays or racks after washing for drying, then a hand-operated knapsack sprayer should be used to spray wholly and to the point of runoff. Larger spraying operations may necessitate a simple mechanized spray or drenching arrangement with a mechanical fungicide mixer.

Produce is sprayed or drenched in perforated trays while moving on a belt or roller conveyor. Numerous different application methods, such as smokes, dusts, or vapour, can be used in large-scale operations where produce is to be stored. For example, Fungicidal (Carbendazim) application can be used to effectively extend/improve the shelf life of fresh produce. Litchi fruit treated with 0.25 per cent SOPP (Sodium-ortho-phenyl-phenate) in 6 per cent wax emulsion is effective against a wide range of fungal diseases.

Chemical Treatment

Chlorination

To control microbial spores' buildup in produce during packing operations, chlorine treatment (100-150 ppm available chlorine) can be used in washing water because of it action as anti-germicidal. For best results, keep the pH of the wash water between 6.5 and 7.5.

Use of Plant Growth Regulator

The use of cytokinins or GA3 (20 ppm) could really help to extend the shelf life of citrus fruits, especially Baramasi lime.

Use of Nutrient Treatment

Calcium

Postharvest treatment with of CaCl2 or Ca (NO3)2 help to extend the storage and marketability life of fruits by preserving their firmness and quality.

Use of CaCl2 (2-4 per cent) for a 5-10-minute dip after harvest increases the storage period of pear by up to two months whereas in the case of apples by up to 7 months, and plums by up to 4 weeks at 0-2°C. will maintain excellent color and quality. Mangoes treated with 2-8 per cent $CaCl_2$ solution are less softened for 8–12 days longer when stored at 20°C compared to untreated fruits.

Thermal Treatment

Fruits could be dipped in hot water before marketing or storing to control postharvest diseases and improve or maintain quality attributes. for instance, in harvested mangoes, a hot water treatment at 50-52 ° C. for 5 minutes is recommended to prevent fungal infection through elicited phytoalexin, as well as the expression of certain heat shock protein that led to the delaying the ripening of fruit by suppression of ethylene action. Cool water showers or forced air should be provided to assist in returning the fruit to their optimum temperature as soon as possible after treatment.

Delay the ripening metabolic activity of treated fruits by heat inactivation of degradative enzyme (Polygalacturonase and exo- and endo- cellulase activities were significantly lower, PME).

Hot water treatment has proven to be extremely beneficial in a variety of other fruits (Table 22.1).

Table 22.1: Hot Water Treatments are Recommended for Controlling Various Pathogens in Fruit Crops

Commodities	*Inhibited Pathogen*	*Temp. (°C)*	*Time (min)*	*Injuries that Could Occur*
Apple	*Penicillium expansum* *Gloeosporium* sp.	45	10	Reduce storage life
Papaya	*Collectotrichum gloeosporioides*	48	20	
Peach	*Monolinia fructicola* *Rhizopus stolonifer*	52	2.5	Motile skin
Grapefruit	*Phytopthora citropthora*	48	3	
Mango	*Collectotrichum gloeosporioides*	52	5	No stem rot control
Orange	*Diplodia* sp. *Phomopsis* sp. *Phytophthora* sp.	53	5	Poor degreenin

Vapour Heat Treatment

For this treatment, the fruit boxes are stacked in a room that is heated and humidified by the injection of steam. The temperature and duration of exposure are adjusted to kill all stages of insects (egg, larva, pupa, and adult), but fruit should not be adversely affected. Citrus, mangoes, papaya, and pineapple should have been treated at 43°C in saturated air for 8 hours and then held at that temperature for another 6 hours. It is mandatory for export of mango to subject to VHT.

Fumigation

SO_2 fumigation has been employed effectively to control grape postharvest maladies. It is accomplished by placing the fruit boxes in a gas tight room and introducing the appropriate concentration of gas from a cylinder. There are special sodium metabisulphite pads available that can be packed into individual boxes of grapes and it slowly release SO_2. The primary goal of treatment is to inhibited the growth of *Botrytis cinerea* to minimize the postharvest rotting. SO_2 fumigation is also employed to protect the discoloration of litchi peel. The fumigation of litchi fruits of sulfur dioxide (600 g/tonne of fruits), after a two-minute dip in 1M HCl, aids in maintaining the red color of litchi fruits and maintain market quality for three weeks at 2°C with 90-95 per cent RH. Citrus fruits are frequently treated with paper pads or wraps impregnated with biphenyl fungicides. The chemical slowly vaporizes, protecting the fruits from fungal infection.

Irradiation (Cold Pasteurization)

Ionizing radiation has the potential to control microorganisms by inhibiting or preventing cell reproduction, and cause some chemical changes in fresh fruits. It can be used by exposing the crop to radiation from radioisotopes (typically gamma-rays, but X-rays can also be used) and machines that produce a high-energy electron beam. Grays are the units of measurement for radiation doses (Gy). And 1 Gray equal to 100 rads.

Irradiation can be used as a postharvest treatment to prevent the infestation of fruit from insect. While 300 Gy has been shown to control mango stone weevil, meanwhile 150 Gy has been shown to prevent adults which are hibernated in fruit. Radiation exposure doses greater than 600 Gy caused lenticel spotting, surface discoloration, and ripening retardation in Kensington Pride mangoes, but only a minor improvement in disease control. However, irradiation followed immediately by hot benomyl treatment to prevented anthracnose and stem-end rot during 15 days of storage at 20°C.

Irradiation at a minimum dose of 250 Gy was recommended in Hawaii for commercial control of fruit fly in papaya, in addition, along with combination of hot water treatment. Dosages of 100 Gy were reported to prevent the most resistant stage (third in star larva) from developing into adults in papayas. Mango and banana fruits should be given a dose of 25 to 50 Krad to slow the ripening process.

Table 22.2: Doses of Irradiation for Various Fruits and Vegetables for Postharvest Disease Control

Fruit and Vegetable	*Minimum Dose Required (Gy)*	*Maximum Dose Tolerant (Gy)*
Apple	100	100-150
Apricot, peach and nectarine	200	50-100
Orange	150	200
Grape	—	25-50
Potato	8-15	20
Tomato	150	300

Curing of Root Crops and Onions

The use of surface drying and curing processes is one of the most important methods of reducing postharvest losses in many root crops such as potatoes, colocasia, yam, and sweet potato, as well as onions. Curing is done by keeping the produce at a high temperature and high relative humidity for several days while harvesting wounds heal and a new, protective layer of cells forms. While curing can be initial expensive at first, its long enhanced the storage life and makes it commercially valuable. Curing is a natural wound healing process that replaces and strengthens damaged areas in sweet potatoes by forming a corky layer that protects against water loss and infection by decay organisms. on the other hand, curing in onions, is primarily a drying process in which excess moisture is removed from the onion's outer skin and neck. Simultaneously, by exposing the onion to higher temperatures, the skin color darkens and natural fungicidal compounds accumulate in the skin. Both processes work together to form a protective layer, which reduces water loss and acts as a physical and chemical barrier against infection.

Commodity	*Temperature (°C)*	*Relative Humidity (Per cent)*	*Days*
Cassava	30-40	95-95	2-5
Yam	30-42	90-104	1-4
Sweet Potato	30-32	86-90	4-7
Potato	15-20	90-95	5-10
Onion	37.8	–	6

REFERENCES

Maxie, E. C., Sommer, N. F. and Mitchell, F.G. (1971). infeasibility of irradiating fresh fruit and vegetables. *HortScience* 6, pp.292-94.

Lurie, S. (1998). Postharvest heat treatments. *Postharvest Biol. Technol.* ***14***: 257–269.

Mahajan, P. V., Caleb, O. J., Singh, Z., Watkins, C. B. and Geyer, M. (2014). Postharvest treatments of fresh produce. Philos Trans A Math Phys Eng Sci. doi: 10.1098/rsta.2013.0309.

Palou, L., Serrano, M., Martinez-Romero, D. and Valero, D. (2010). New approaches for postharvest quality retention of table grapes. Fresh Produce ***4:*** 103–110.

Wilson, J. (2008). Careful Storage of Yams: Some Basic Principles to Reduce Losses.

Yadav, P. K., Narolia, R. K., Rathore, S. R. and Sharma, N. K. (2020). Manual on postharvest and value addition in fruit and vegetables ***20:*** 1-65

Chapter 23

Underexploited Vegetables: Importance, Constraints and Improvements

Sanjive Kumar Singh[1], Nirankar[2], Rajat Singh[2] & Shefali Sachan[3]*

[1]Assistant Professor, [2]Research Scholar,

[2]Department of Vegetable Science, Chandra Shekar Azad University of Agriculture and Technology, Kanpur, Uttar Pradesh, India, 208002.

[3]Senior Research Fellow, Organic College, Punjab Agriculture University, Ludhiana, Punjab, India, 141004.

**e-mail: Sanjive.csau@gmail.com*

INTRODUCTION

India has very diverse agro-climatic conditions and about 175 vegetables (82 leafy, 41 tubers and 52 others) are grown in country. Among these, only some vegetables are grown conventionally, commercially and at mass scale whereas remaining are grown at very limited scale. The vegetables which are grown in limited scale referred as underexploited vegetables. These underexploited vegetables are very rich in nutraceuticals, minerals and nutrients and have good marketing and income generation values.

Important Features of Underexploited Vegetables

- ☆ No special attention is paid from growers, researchers, extension workers, decision and policy makers on these vegetables.
- ☆ These vegetables are cultivated in smaller area other than conventionally grown vegetables.
- ☆ These crops are recognised to have indigenous uses.
- ☆ These vegetables are adapted in specific agroecologically zones.
- ☆ These vegetables have poor seed supply system.
- ☆ Underexploited vegetables are highly nutritious and have therapeutic properties.

Difference between Underexploited and Underutilized Vegetables

Underexploited Vegetables

Underexploited vegetables are those vegetables which are neither grown commercially on large scale nor traded widely. Ex. Lettuce (*Lactuca sativa* L.), Bathua (*Chenopodium album* L.), Amaranth. (*Amaranth spp.*) *etc.*

Underutilised Vegetables

Underutilised vegetables are defined those vegetables which are less available, less utilized of rarely used. Leek (*Allium porrum* L.), Pointed gourd (*Trichosanthes dioeca* L.), Asparagus (*Asparagus officinale*)

Major Constraints for Research and Development of Underexploited

- ✰ Lack of knowledge among farmers about the nutritional and therapeutic quality of underexploited vegetables.
- ✰ Poor recognition of these vegetables.
- ✰ Lack of appropriate planting materials.
- ✰ Lack of research and development and negligible application of novel and innovative technologies such as, biofortification, biotechnology and plasticulture.

Other Names of Underexploited Vegetables

Minor vegetables, lesser- known vegetables, rare vegetables, neglected vegetables and orphan vegetables.

Classification of Underexploited Vegetables

1. Underexploited Leafy Vegetables
2. Underexploited Legumes vegetables
3. Underexploited rhizome and corn
4. Underexploited tree vegetables
5. Underexploited cucurbits vegetables

1. Underexploited Leafy Vegetables

There are more than 22 underexploited plant spp. which provide leaves, petioles and twigs as green. Major vegetables of this group belong to family *Convolvulaceae.* These are highly beneficial and have nutraceuticals properties. Therefore these greens should be popularised among consumer for further cultivation and improvement.

Table 23.1: Underexploited Leafy Vegetables

Sl. No.	Hindi/Vernacular Name	English Name	Scientific Name	Family	Plant Part Used	Other Uses	Growing Area
1	Bak Phool and Agastoya	Agathi	*Sesbania grandiflora* L.	Papilionaceae	Flowers, fruits and tender leaves	1. Plant is used as support and betel vine and vine 2. Used as shade and wind break 3. Leaves are used as fodder	Assam, West Bengal, Punjab, A.P. and Rajasthan
2	Makhana, Kakwagi and Thargigiojdi	Grogan nut	*Euyale ferix*	Euyalaceae	Young leaves, petioles, fruits and seeds	1. Seeds are roasted and eaten 2. Used in religious ceremonies.	Kashmir, UP, Bihar, Rajasthan, Assam and Tripura
3	Patsan and Ambari	Bimlipatan jute, Kenaf and Mesta	*Hibiscus cannabinus* L.	Malvaceae	Tender leaves, shoots and fruits	1. Source of fiber 2. Used as paper pulp 3. Seed oil is used for burning 4. Seed cake is used as cattle feed making	A.P., Assam, U.P., M.P., Bihar, T.N. and West Bengal.
4	Dudhia kalmi	Moonflower	*Ipomea alba*	Convolvulaceae	Leaves and fleshy calyxes	Fragrant night blooming flower	Throughout India
5	Ghiabato Boota	-	*Ipomea eriocarpa* L.	Convolvulaceae	Leaves and stem	—	Throughout India
6	Nari	-	*Ipomea aquatica*	Convolvulaceae	Young terminal	—	Throughout India in aquatic Habitat
7	Kalmi Sag, Sarmali, Gandhian	Swamp cabbage, water spinach	*Ipomea replance* L.	Convolvulaceae	Young leaves and shoots	Vines are used as cattle feed	Maharashtra, Odisha, Bihar and South India.
8	Ban kalmi	-	*Ipomea sepiaria* L.	Convolvulaceae	Whole plant	—	Throughout India

Sl. No.	Hindi/Vernacular Name	English Name	Scientific Name	Family	Plant Part Used	Other Uses	Growing Area
9	Michai	-	*Ipomeaturbinate* L.	Convolvulaceae	Flower, pedicel	—	Throughout India
10	Khesari	Grass-pea, chickling pea, Jarosse	*Lathyrus sativus* L.	Paplionaceae	Foliage twigs	1. Foliage is used as cattle feed 2. Grains are used as pulse	Bihar, West Bengal, Odisha, Chhattisgarh
11	Kasari	Grass pea	*Lathyrus nudicandis*	Compositae	Leaves are used in curries	1. Plant is used as fodder 2. It is used for makings	
12	Poi	Indianspinach	*Basella alba*	Basellaceae	Leaves	It is rich in calcium, phosphorus and oxalate	A.P., Assam, U.P., M.P., Bihar, T.N. and West Bengal
13	Khersa Kulfa	Purslane	*Portulaca oleracia* L.	Portulacaceae	Leaves and twigs	1. Rich in protein, vitamin 2. Used for treatment of liver diseases 3. Blood purifier	Throughout India during Summer and rainy season
14	Bathua	Chenopo d, pig weed, limbsquarters	*Chinopodium album* L.	Chinopo-diaceae	Leaves and tender twigs	1. Used as fodder 2. Rich in iron and protein	Throughout Northern India
15	Sabji jute patand tilapat nercha	Vegetabl e jute, white jute	*Corchorus capsularis* L.	Tilliaceae	Leaves	1. Rich in protein (18- 22 carotene 2. It is rich source of fibre	W.B., Assam, Bihar, U.P., Odisha and Maharashtra.
16	Har Farauri	Star gooseberry	*Phyllanthus acidua* L.	Euphorbiaceae	Leaves and fruits	1. Bark is used as tan 2. Used as fodder 3. It has medicinal value	West Bengal And South India

Sl. No.	Hindi/Vernacular Name	English Name	Scientific Name	Family	Plant Part Used	Other Uses	Growing Area
17	Kusum	Safflower	*Carthamus tinctorious* L.	Compositae	Tender leaves and twigs	1. Flowers are used as dye. 2. Seeds yield edible oil. 3. Oil is used in soap, paint, varnish.4.Oil cake is used as cattle feed	A.P., M.P. Karnataka, Maharashtra, Rajasthan, Haryana.
18	Kuttu	Buch wheat	*Fagopyrum esculentum*	Polygonaceae	Tender leaves and shoots	1. Grains are mixed with millets. The product is fasting 2. The husks are used for pillows	Hilly region of North India and Nilgiri.
19	Jangali palak, Laln bibi, Khatti palak, amruleand ambarch	-	*Rumex dentatus*	Polygonaceae	Tender leaves	1. Roots are source of red dye. Rich in calcium, carotene and Vitamin C	A.P., W.B., Maharashtra, Karnataka
20	Amloo, lemon hoo	Mountai n sorrel	*Oxyriya dygina*	Polygonaceae	Leaves	Good source of Calcium, Iron and minerals.	Throughout India
21	Chakwat	Orach	*Atriplex hortensis*	Chinopo-diaceae	Leaves and branches	1. Leaves are rich in protein and minerals. 2. It has medicinal qualities	J&K, H.P. Laddakh, alpine region of Himalaya
22	Makoy	Black Night shade	*Solanum nigrum* L.	Solanaceae	Tender twigs and leaves	1. It cures dropsy disease 2. Ripe fruits are edible	Throughout India

Figure 23.1: Underexploited Leafy Vegetables

1. Agathi, 2. Makhana, 3. Patsan, 4. Dudhiya Kalami, 5. Giabato Boota. 6. Nari, 7. Khesari, 8. Basella, 9. Kersa Khulfa.

Nutritional, Medicinal and Socioeconomic Importance of Underexplored Leafy Vegetables

1. Bathua is rich in rich in iron, mineral, and protein provides nutritious fodder also for cattle.
2. Jangli Palak and mountain sorrel are rich in Vitamin C.
3. Makhana is a typical source vegetable where young leaves, petioles,

fruits and seeds are used as greens. The seeds are popped and are used in religious ceremonies.

4. Patsan and jute are useful source of fibre, besides, leaves and fruits being used as vegetable.

2. Underutilised Legumes Vegetables

All legume vegetables belong to family leguminaceae. There are many indigenous food legumes whose potential is under exploited and untapped. Many of these indigenous food legumes play a vital role in protein nutrition to poor farm families, especially to women and children, in the region. Looking at the total area cultivated and production, legumes such as soybean, groundnut, chickpea, lentil, common bean, field peas, chickpea, and pigeonpea can be considered as major legume crops. Other legumes that are indigenous and under-exploited are: adzuki bean, bambara groundnut, blackgram, broadbean (faba bean), horsegram, lablab bean, lathyrus, moth bean, rice bean, and winged bean **(Gowda *et al.*,2007)**

Nutritional, Medicinal and Socioeconomic Importance of Underexplored Legumes Vegetables

1. Fix the atmospheric nitrogen and increase productivity of soil.
2. Serves as green manuring crop
3. Some of them serve as cover crop and check soil erosion
4. Kiwach is very source of minerals, vitamins and protein, vitamins, protein. It also contains aldopa which cures Parkinson disease.
5. Sunhemp besides being used source of vegetable, provide stem fibre.

3. Underexploited Rhizome and Corm

1. In Lotus, rhizomes, seeds and young leaves are used as vegetable.
2. The juice of lotus leaves are used for curing dysentery and diarrohea.
3. Petioles of arvi leaves is used in burn, sore and wounds.

4. Underexploited Tree Vegetables

Nutritional, Medicinal and Socioeconomic Importance of Underexplored Trees

1. These trees are hardy in nature and thus require least care.
2. Besides the use as vegetable, the fruits are also used in pickling for example. Kathal, Karil and Sahjan *etc.*
3. The bark of these trees provide tannin (Badhal, Goolar, Lasoda), Dye (Kathal) and Gum (Chhonkar).

Table 23.2: Underutilised Legumes Vegetables

Sl.No.	*Hindi/ Local Name*	*English Name*	*Botanical Name*	*Plant Part Used*	*Other Uses/Qualities*
1	Chaudhari phalli, Charkori sem	Goa bean	*Psopocarpus tetragonolubus* L.	Tender pods and tuberous roots, flowers, sprouts	Grown as green manure, cover and fodder crops
2	Sanai, Jhunjhuna, Sum	Sunnhemp	*Crotalaria juncea*	Flowers	1. Stem fiber is used for various purpose 2. Stem fiber is used for cigarette and tissue paper 3. Green stem and leaves make green manure 4. Seed contain about 15 per cent gum 5. It fixes nitrogen in the soil
3	Bakla	Broad bean, Windsor bean	*Vicia fava* L.	Fruits and seeds	1. Grown for fodder and hey 2. It fixes nitrogen in the soil
4	Kiwach Alkusi	Velvet bean and Horse eye bean	*Mucuna puriens* L.	Fruits	1. Pods, seeds and roots have medicinal value. 2. Rich in vitamin A and protein 3. It contains aldopa which is used in curing Parkinson disease. 4. It fixes nitrogen in the soil.
5	Makkhan sem	Sword bean	*Canavalia*. gladiata L.	Pods (fruits), green seeds, flower, leaves	1. Used as green manure and cover crop 1. Fixes nitrogen in the soil
6.	Bara Sem	Horse bean, Jack Bean, Gotani Bean	*Canavalia ensiformis* L.	Pods and Immature seeds	1. Used as forage purpose 2. Acts as cover crop 3. Flowers and young leaves are used as condiment 4. Used as green manure crop 5. Roasted seeds are used assubstitute of coffee

Figure 23.2: Underutilised Legumes Vegetables

1. Goa Bean, 2. Sunhemp 3. Broad Bean 4. Velvet Bean 5. Sword Bean 6. Horse Bean.

Figure 23.3: Underexploited Rhizome and Corm.

1. Kamal, 2. Arvi.

4. Cluster Fig acts as host for lac insect.
5. These trees can be also grown as avenue and wind break tree Kathal, Kachnar and Imli *etc.*)
6. Mahua flowers are resource of alcohol.

Table 23.3: Underexploited Rhizome and Corm

Sl.No.	Hindi/ Local Name	English Name	Botanical Name	Family	Plant Part Used as Vegetable	Other Uses/ Qualities
1	Kamal, Kanwal	East Indian lotus, Sacredlotus	*Nelumbo nucifera, Nymphoea nelumbo*	Nelumbonaceae	Rhizomes, seed and young leaves	Fragrant flowers are mu garlands and used
2	Arvi	Arum lily commonballa	*Zantedeschia aethiopica*	Araeceae	Young leaves and petioles	Poultice of leaves is use wounds and sores

Table 23.4: Underexploited Tree Vegetables

Sl. No.	Hindi/ Local Name	English Name	Botanical Name	Family	Plant Part Used	Other Uses/ Special Quality
1	Badhal	Bread fruit	*Artocarpus lakoocha* L.	Moraceae	Fruits	1. Wood is used as timber 2. Bark is source of tannin 3. Seeds are purgative
2	Kathal, Kanthal, Panas	Jack fruit, Jack tree	*Artocarpus hetrophyllus, A. integrifolia*	Moraceae	Fruits and seeds	1. Heart wood is valuable timber 2. Wood yields yellow dye 3. Leaves are used as cattle fodder 4. Bark is used for tanning
3	Imli and Tamarind	Tamarind	*Taramindus indica*	Caesalpinia-ceae	Fruit pulp is used for	1. It is rich in vit C, tartaric acid and citric acid. 2. Grown as avenue tree. 3. Fruit pulp has medicinal values.
5	Goolar/Domoor	Cluster fig	*Ficus glomerata*	Moraceae	Fruit	1. Grown along road side 2. Leaves are used as fodder 3. Bark is used as tannin host for lac insect

Sl. No.	Hindi/ Local Name	English Name	Botanical Name	Family	Plant Part Used	Other Uses/ Special Quality
6	Mahua, Moha	Mahua tree	*Madhuca indica*	Sapotaceae	Flowers	1. Flowers are source of Alcohol. 2. Seed oil is used for cooking and soap making. 3. Oilcakes make good Manure.
7	Chokargond, Lasoora and Labherah	Sebesten, clammy cherry and Indian cherry	*Cordia dichotoma*	Euphorbiaceae	Fruits	1. Fruits are eaten raw or pickle 2. They have medicinal value 3. Bark is rich source of tannin 4. Trees are hardy
8	Ker, Karar, Karira and Kareel	-	*Capparis aphylla, C. decidua*	Capparidaceae	Flower buds, green and mature fruits	1. Top shoots and young leaves are used as fodder 2. Bark is useful in Coughing and asthma 3. Wood is hard and heavy 4. It is termite resistant 5. Suitable for wind break and shelter belt
9	Khejri, Chhokar, Jand, Sami Sangari	-	*Prosopis cineraria, P. spicigera*	Mimosaceae	Young pods	1. Grown as hedge plant 2. Plant produces gum 3. Drought tolerant 4. Fruits are rich in protein and calcium
10	Lal Kachnar, Kaliar	Camel's foot tree, Pink Bahaunia	*Bauhinia purpurea*	Caesalpiniaceae	Flower buds and fruits	1. Leaves are used as fodder 2. Tannin is extracted from bark 3. Wood is useful for agricultural implements
11	Kurkurj wah	-	*Leea indica*	Leeaceae	Fresh shoots	1. Roots have medicinal quality in diarrhoea
12	Meethi neem	Curryleaf	*Murraya koenigii,*	Rutaceae	Leave for flavouring the curries	1. Rich in vitamin C, calcium and phosphorus. 2. High in medicinal values 3. Oil is used in perfumery

Figure 23.4: Underexploited Tree Vegetables

1. Bread fruit, 2. Jack fruit 3. Tamarind, 4. Cluster Fig, 5. Mahua tree, 6. Sebesten, 7. Blue cherries 8. 9.Kachnar, 10. 11. Curry leaf.

5. Underexploited Cucurbits

The cucurbits vegetables play a very significant role in the household food, nutritional security (particularly through vitamin and mineral supplementations) and income generation of the small and marginal farmers and landless labourers.

Nutritional, Medicinal and Socioeconomic Importance of Underexplored Leafy Vegetables

1. Kakora and Kundru are useful in curing diabetes.
2. The sponge gourd fruits are used in making mat, hand fan and sponge for cleaning utensils.
3. The seed of snap melon are used in confectionaries.
4. The unripe fruits of kachri are cut into pieces and dried and stored for off season use.
5. Most of these cucurbits are rich in vitamin C

Role of Underexploited Vegetables in Food Security

All underexploited vegetables are very nutritionally rich and are adapted to low input agriculture. The erosion of these species, whether wild, managed or cultivated can have immediate consequences on the food security and well-being of the poor. The enhance use underexploited vegetables can bring better nutrition. Focusing attention on these vegetables help maintain adverse and healthy diet and to combat micronutrient deficiency, so called hidden hunger and other dietary deficiency particular in rural and more vulnerable social group in developing countries.

Conservation of Underexploited Vegetables

After discussing the above points, it is clear that underexploited vegetables are fulfilled with nutritional and medicinal properties and farmers, scientists need to give focus on conservation of these vegetables. There are four methods described in this chapter for conservation of underexploited vegetables.

1. Diversity Distribution Assessment

The diversity of underexploited vegetables is not well assessed and if national programmes are not taking care of this species, these may under threat. Safe gaurding/conserving such diversity in national seed/field gene bank is prime importance.

2. Biotechnology Application

Biotechnology is a rapidly developing field The wider application of biotechnology can provide sufficient scope to realize the full potential of underexploited vegetables in achieving increased production, quality (nutritional, shelf life, storage *etc.*) and meeting market needs. Initial major role

Table 23.5: Underexploited Cucurbits

Sl. No.	*Hindi/ Local Name*	*English Name*	*Botanical Name*	*Plant Part Used*	*Other Uses/ Special Quality*
1	Ban karela, kakrol, kakora	Spine gourd	*Momordica cochinachinesis* (Roxb.)	Fruit	Cures diabetes
2	Kundru, kanduri	Kovai fruit and Ivy gourd	*Coccinia indica, C. grandish, C. cordifolia*	Fruit, tender leaves and twigs	1. Juice from leaves and roots is used in diabetes 2. Rich in iron, phosphorus, minerals and carotene
3	Amantmul, kundri, Bam	Melothria	*Melothria hetrophylla, Solena hetrophylla*	Leaves, twigs and tuberous roots	1. It is rich in calcium and other minerals.
4	Launku	Chayote, chow-chow and christoph ine	*Sachium edule*	Fruits, tender leaves, twigs and tuberous roots	1. Leaves are rich in vit A, B, C, calcium, iron 2. Leaves are used as fodder 3. Stem fiber is used in basket making
5	Thumba	Bitter cucumber	*Citrullus colocynthis*	Bitter fruits are used as salad and pickle	1. Dry seeds are made in to floor 2. Seed contains 16 per cent oil
6	Phoot and kachra	Snap melon	*Cucumis melo var. momordica*	Immature fruit	1. Ripe fruits as dessert 2. Seeds are useful in confectionery
7	Kacharia, petha, sendha	Kachri	*Cucumis cellosus*	Fruits	1. Used in chantey, salad, pickle 2. Fruits are cut in pieces and dried which are used during off season
8	Meetha karela	Sleeper gourd	*Cyclathera pedata* L.	Fruits, tender twigs and leaves	1. Used as salad and ripe fruits are useful in pudding
9	Jugali karela	Balsom apple	*Momordia balsamina*	Fruits, leaves,and tuberous roots	1. Fruits are used for petha 2. Leaves are good source of vitamin C.
10	Kumra, villaiti kaddu	Pumpkinand marron	*Cucubita pepo*	Fruits, flowers,growing tender twigs	—
11	Ghia torai	Vegetable sponge	*Luffa aegyptiaea*	Immature fruits and unopened flower buds	1. Dried fruits yield fibrous structure which is widely used in fabrication of articles 1. Fibrous sponge is used in scrubbing the utensils

Figure 23.5: Underexploited Cucurbits

1. Spine gourd, 2. Ivy gourd, 3. Melothria, 4. Chow-Chow, 5. Bitter cucumber, 6. Snap melon, 7. Kachri 8. Balsam apple, 9. Summer Squash, 10. Sleeper gourd, 11. Vegetable Sponge.

of biotechnology are increase availability through micropropagation and tissue culture and to characterize genetic diversity through molecular marker studies.

i) Micropropagation and Tissue Culture

This study assumes importance in above context in addressing the application of biotechnology. Activities have been undertaken on a number of underexploited vegetables including *in vitro* propagation (via micro cutting and somatic embryogenesis) of Hyacinth bean and taro *etc.* Virus free planting material distribution has been achieved through adopting these biotechnological techniques.

ii) Through Molecular Marker

Another important application of biotechnology that would need focus is genetic diversity studies. Some examples are of *Artocarpus hetrophyllus* (Jackfruit; isozyme study, among accessions), and sago palm; AFLP among accessions and related species).

3. Documenting Indigenous Knowledge

In centres of diversity of crop plants in particular, semi domesticated and gathered diversity from natural habitats still helps support household food security and also healthcare through use of indigenous/herbal medicines. The role of ethnobotany/application of ethnobotany in conservation of native plant genetic diversity and use and in community development has been highlighted by **Martin,1995** by focusing on the ways the local culture classify, manage and use plant resources or locating new and more valued use of these resources.

4. Habitat Protection

Human destruction of natural habitats weather exploited for commercial growing needs or subsistence cultivation, is presently one of the major ecological threats. As a result, there has been pressure on natural/wild vegetation sites and due to ecological imbalance, the original habitants are lost, and so also the native plant diversity occurring here. In each of the several underexploited, minor, less known species occurs and these need to be collected and conserved and studied for their use to meet social and economic need of the rural people.

Institution, Programmes and Schemes for Farmers to Encourage the Area of Underexploited Vegetables

1. **International Bureau of Plant Genetic Resource (IBPGR):** IBPGR has done very specific activities at international level for better conservation and use of underutilised and underexploited vegetables crops. IBPGR is very active global media to raise awareness on underexploited as well as underutilised crops and development of systems for share of benefits arising from their use.

2. **National Bureau of Plant Genetic Resource (NBPGR):** Presently, ICAR-NBPGR, New Delhi is coordinating the All India Coordinated Research Network on Potential Crops (AICRN-PC) and conducting research on 17 crops of food, fodder and industrial value through 14 main, nine cooperating and 15 voluntary centers located in diverse agro-climatic zones of India. The main objective is to generate improved technologies in selected crops of the minor economic importance for food, fodder and industrial use. Presently, ICAR-NBPGR, New Delhi is coordinating the network as a coordinating unit and conducting research on 17 crops of food, fodder and industrial value through 14 funded, nine cooperating and 15 voluntary centers located in diverse agro-climatic zones of India.
3. **Mission for Integrated Development of Horticulture (MIDH):** MIDH provides, technical advice administrative support to state governments for underutilised and underexploited crops and other horticulture related activities.
4. **International Centre for Underutilized Crops (ICUC):** ICUC is international autonomous, research and training centre. It was established in 1988 to address ways of increasing the use of underutilized and underexploited vegetables for food, medicinal and for environmental development **(Madhvi, *et al.*, 2020).**
5. **AICRP on Underutilized Crops (AICRP – UUC):** Indian Council of Agricultural Research (ICAR), with a view to exploit the potential of UUC plants initiated the organized work in 1982 with the establishment of the All India Coordinated Project on Under-utilized Crops (AICRP-UUC) with emphasis on crops such as amaranth, buckwheat, chenopods, cluster bean, winged bean *etc.*

REFERENCES

Gowda, C. L. L., Upadhyaya, H. D. and Ghaffar, M. A. (2007). Importance of under-utilized indigenous legumes in Asia-pacific region. *Acta Horticulture.* 752, 95-98.

Madhvi, Sharma, D., Chauhan, Banyal, S. K. and Divya. Underutilized vegetable crops: A lost tresasure. *International journal current microbiology applied science.*

Martin, G. J. (1995). Ethnobotany: A Conservation Manual. Chapman and Hall, London. *Scientific research.* 2(1): 26-45.

Singh. S. K. and Prakash H. G. (2021). Introduction of unexploited/underutilized vegetables. Practical manual on vegetables crops. pp-61-41.

www.midh.gov.in

Chapter 24

Underutilized Vegetables: The Treasure of Nutraceuticals

*Ankita Sharma[1] *, Rehan[1], Reena Nair[2] and Akhilesh Tiwari[3]*

[1]Research Scholar, [2]Assistant Professor, [3]Professor,

Department of Horticulture, College of Agriculture, Jawaharlal Nehru Krishi Vishwa Vidhyalaya, Jabalpur, Madhya Pradesh

**e-mail: ankitasharma199511.as@gmail.com*

INTRODUCTION

In today's globalized world, a greater need for antioxidant-rich, calorie- and protein-rich, healthful, and nutrient-rich sources of food to alleviate the problems of malnutrition and to combat ageing diseases like cancer. Underutilized vegetable crops are ascertained as 'undervalued' or 'Minor-crops' that are being cultivated regionally and globally but still neglected because of their still-relatively low global output and market value. This incredible increase in vegetable production has grown as a result of the introduction of new varieties, hybrids, production, and protection technologies attributable to meticulous research and widespread farmer implementation. However, just a modest number of significant crops contributed towards this amazing yield. Though diverse agroclimatic conditions in India facilitate for the growth of around 60 cultivated and approximately 30 wild species, not much attention has been paid to lesser-known vegetable crops or underutilised vegetables. These vegetable cops are a significant source of valuable and nutritional medicinal and nutraceutical components which can be utilized for human consumption coupled with their use in pharmaceutical industries which many staple crops including field crops, vegetables and fruits are devoid of. There are intrinsic agronomic, ecological, nutritional, and economic concerns associated with the current overreliance on a small number of important staple crops, which is likely unsustainable in the long term. A wider use of today's underutilised minor crops would increase resistance to biotic and abiotic stress and provide farmers more alternatives for incorporating temporal and geographical variability into uniform cropping systems. A vital source of vitamins, minerals, and protein, many traditional vegetables and underused legume crops are an important part of achieving nutritional security. Underutilized vegetable crops are of appreciable commercial value and consequently essential source of household-income. Lettuce, Asparagus, Celery, Amarnath, Globe artichoke, Leeks, Chekurmanis, and Chinese cabbage are a few examples of neglected vegetable crops. Because of their high vitamin, micronutrient, and protein contents, these crops are regarded as essential for achieving nutritional security. The majority of underused vegetable crops can withstand challenging agro-climatic conditions. Lack of planting material, ignorance of the importance of nutrition and medicine, and ignorance of production methods are probable causes of the slow growth and poor condition of underutilised vegetable crops.

Significant research, breeding and development efforts are needed for a range of promising crops to convert existing local landraces into competitive varieties with wide adaptation and promising commercial potential. Access to genetic diversity of these selected crops is a pre-condition for success. Three underutilized minor crops–amaranth, drumstick tree, and mungbean–are highlighted and briefly described. All three crops are well-represented in AVRDC's genebank with substantial inter- and intra-specific genetic diversity, and already have demonstrated their potential for wider adoption and commercial exploitation Underutilized vegetable crops are those which are neither grown commercially on large scale nor traded widely. Some underutilized vegetable crops are Lettuce, Asparagus, Celery, Amarnathus, Globe artichoke, Leek, Chekurmanis, and Chinese cabbage *etc.* These crops are considered as valuable component to attain nutritional security because of their high content of vitamins, micronutrients and proteins. Most of underutilized vegetable crops are tolerant to harsh agro-climatic conditions. The possible reason for slow development and poor status of underutilized vegetable crops are lack of availability of planting material, lack of awareness on nutritional and medicinal importance and lack of information on production techniques. The government of India has been taking some steps to do research on underutilized crops like MIDH (Mission for Integrated Development of Horticulture), MEIS (Merchandize Export from India Scheme) and a national coordinated project has been also launched by Ministry of Agriculture. Basically, these underutilized vegetable crops have great potential for food security, income generation and environmental services and ultimately it makes Indian economy sounder.

However, presently only one fourth is utilized as a major vegetable crops and rest are named as minor, underutilized, rare vegetables, wild edible vegetables or so on. Underutilized or wild vegetables may not be widely known outside of a specific area or region, and there is a perception that they are grown mostly in rural areas. In some places consumption of these plants is not socially acceptable by some community sectors because they are considered to be food for the poor (Jaenicke and Hoeschle, 2006). Although they are important locally or regionally, underutilised vegetables seldom receive national notice or admiration. The plant species that are historically employed for their food, fibre, fodder, oil, or medicinal characteristics are the underutilised vegetable crops. However, those species have untapped potential to provide environmental benefits, nutrition, health, income, and food security (Jena *et al.*, 2018).

Significance of Underutilized Vegetables for Enhancing Nutritional Aspects

In the world literature, the term "underutilised crop" has been defined in a variety of ways, with the majority of these emphasising features such as linkages to local cultural heritage, multiple uses, traditional crops in localised areas, and being overlooked by agricultural research and development agencies (Singh *et al.*, 2019). Wild edible plant species, which can be gathered from cultivated production technologies or from wild or semi-natural environments, nonetheless play a significant role in diet diversifying, especially in local food systems. Utilizing underutilised vegetables more frequently and successfully is helping to combat economic prosperity, poverty, and hunger. They are essential biological resources for the rural poor and can help the majority of tribal people dwell more healthily. Vitamins, minerals, and other health-promoting ingredients, such as high levels of antioxidants are abundant in underutilised vegetables. These species, as well as the genetic richness they possess, have enormous potential for food and nutritional security, in conjunction with countering "hidden hunger" caused by micronutrient (vitamin and mineral) inadequacies (Padulosi *et al.*, 2013). They have a significant impact on diet diversification, which equates to more balanced sources of micronutrients. Nevertheless, underutilised vegetables are resilient to a variety of biotic and abiotic stresses, and their yields can help feed the hungry by satisfying the nutrient needs of vulnerable populations. Numerous veggies, especially those that are undervalued, offer a natural and abundant source of vitamins, minerals, and antioxidants.

Underutilised vegetables play an important role in the life of rural people; they form an important part of food and nutrition of local population as many of them are traditionally been esteemed for their utilization in terms of medicinal, therapeutic and nutritional values since time immemorial and are consumed either as raw or as cooked vegetables as traditional delicacies and the sales from the surplus of these vegetables add to the income of many rural families Malnutrition and subsequent food shortage among the poor rural population are conspicuous. They have potential to contribute to poverty elimination through employment opportunities and income generation and also through improved efficiency and profitability of farm household labour use in both rural and urban environments. (Jena *et al*,2018). It is possible to lessen the danger of being overly dependent on a small number of key crops by using underutilised vegetable crops. By increasing the variety of edible foods, they can promote sustainable lives by ensuring the food security of households. For low-income city dwellers, they sometimes serve as convenience food and contribute nutrients to the diet. They are able to thrive in delicate conditions and support the stability of agroecosystems, especially in dry and semi-arid regions, mountains, steppes, and tropical woods. They offer a wide range of crops to boost output, safeguard global food security, and fulfill new market demands. These underutilised and underappreciated resources provide numerous benefits. They are strongly linked to the biocultural heritage of their origins they are highly adapted to marginal, complex, and complicated environments and have made a significant contribution to the diversification and resilience of agroecological niches; they can be gathered from the wild or cultivated in conventional production systems with little or no external inputs (Padulosi *et al.*, 2011); and they could contribute to enhanced earnings (Moraza *et al.*, 2018).

In general, nutrition science considers diet distinctiveness only in the context of inter-species diversity, whilst also food composition data has principally been confined to an aggregate level, frequently ignoring substantial compositional complexities associated to agroecological zone, seasonality, and, most importantly, genetic diversity. Improving expertise and evidence acknowledges the necessity for increased food composition evidence on the world's Neglected and underutilized species (NUS). Compositional data can and should be employed in nationwide nutritional planning, suggestions, food processing, and breeding programmes. This element also emphasizes the necessity to employ locally fitting approaches to priorities the significant concentrations of variety of NUS that exist in order to identify and promote a more manageable quantity. A further phase in this prioritizing process will need refining this list to a smaller collection of species with more immediate or medium-term potential for cultivation, promotion, and commercialization, which we shall refer to as "target" species in this study (Hunter *et al.*, 2019). Sustainable and healthy food sourcing, institutionalizing high-quality diets through public food procurement, including food distribution in schools, public food procurement to promote local agroecological produce, establishment of short supply chains, development of food-based dietary guidelines related to local foods and biocultural heritage, and providing policy incentives for diversification and agroecology are all examples of such approaches. However, previous NUS research has been restricted in breadth, failing to provide a suitable path for enhanced NUS incorporation into value chains and diets (Mabhaudhi *et al.*, 2017).

Causes of Un-utilization of the Vegetable

The majority of these green vegetables have been discovered to be tough, adaptable, and tolerant of unfavorable climatic circumstances, despite the fact that they may be grown at reduced management costs even on poor marginal soils. The underutilization of these leafy vegetables, particularly in Eastern India, is owing to a variety of factors such as geographical, social, and economic considerations, among others. The un-assessable location, geographical remoteness, completely undiscovered status of production agronomic approaches, and ultimately too little institutional interest in these plants are all factors contributing to the poor production standards of these veggies (Sing and Munda, 2008). Despite the presence of variety in plant kinds, morphological and physiological variances, adaptations, and responses to diseases and pests, they have remained neglected owing to a lack of knowledge and advertising. The lack of a popularization programme, mobilization of farmers and consumers of all classes toward better utilization of these underutilized leafy vegetables, insufficient facilities for processing and value addition, bottlenecks in research, weak policy support from various organizations, insufficient marketing facility(ies), and supply chain are the main reasons why these crops remain underutilized.

Though domestication and cultivation of these species are quite limited, current awareness among conscientious people about the great nutritional and therapeutic benefits of these neglected green vegetables is encouraging farmers to plant them, although the scope of this cultivation is yet unknown.

Role of Underutilized Veggies

A. Proper Nutrition and Food Security

Neglected and underutilized species (NUS) provide valuable macronutrients such as carbohydrates, proteins and fats, micronutrients such as vitamins and minerals, as well as bioactive non-nutrients that contribute to dietary health (Fanzo *et al.*, 2013; Dulloo *et al.*, 2014; WHO/CBD 2015; Bioversity International 2017). This food diversity represents a natural wealth for many countries yet most, if not all, fail to use them adequately for this purpose. Many underutilized vegetables are nutritious and well-suited to low-input production. The extinction of these species, whether wild, managed, or cultivated, can have substantial ramifications for impoverished people's food security and well-being. Their increased use may lead to enhanced nutrition. Many underused vegetables, for example, have more vitamin C and pro-vitamin A than commonly accessible commercial species and cultivars. Focusing on neglected and underused vegetables is an effective strategy to promote a diversified and balanced diet and battle micronutrient deficiencies, so-called "hidden hunger," and other nutritional inadequacies, particularly among the rural poor and vulnerable socioeconomic groups in developing nations.

B) New Market Niche and Increased Income for Rural Poor

The increased consumer desire for variety and innovation in meals in both developed and developing nations is generating new market niches for neglected and underexploited veggies. These market options can help impoverished farmers earn more money in less-favorable areas if these crops have comparative benefits over major staples or commercial crops. Furthermore, current technology' capacity to turn crops and other plants into varied goods and extend their shelf-life has offered new chances to find new applications and therefore commercialize these species and their products.

C) Stability of Ecosystem

Climate change and the deterioration of land and water resources have increased interest in crops and species suited to harsh habitats such as desert borders, those with low soil or degraded vegetation, or those vulnerable to drought.

D) Cultural Biodiversity

Plants have always been an integral element of local cultures and traditions. Many neglected and underused vegetables contribute to the preservation of cultural variety through culinary habits, health practices, religious rites, and social interchange.

Underutilized Vegetables and their Nutritional Significance

Leafy Vegetables

Many species of these neglected green vegetables, both cultivated and wild, are available fresh and dried at these local marketplaces. These veggies' fresh leaves, young shoots, or blooms are cooked like vegetables with or without seasonings. During peak season, the locals collect and dry a considerable amount of these green vegetables. The collecting or harvesting of leafy vegetables from agricultural fields, domestic environs, or woods is followed by washing and sorting to eliminate dirt and other undesired particles. The leaves are then stretched out in the sun to dry on a clean towel. Depending on the weather and the type of green vegetable, complete drying normally takes 2-5 days. The dried leaves are next crushed by hand or mashed in a pestle and mortar before sifting to achieve a fine powder. These powdered, dried leaves are typically served as a soup with boiled rice water to be consumed during the lean season, when other vegetables are scarce and costs are high.

Local leafy vegetables are among the most nutritious since they are high in minerals like as calcium, magnesium, iron, and potassium, as well as vitamins. They are also high in fibre, very low in fat and carbs, and a good source of protein. Because of their strong nutritional properties, these crops may be useful in overcoming nutritional deficits prevalent in many rural regions of the nation, as well as in improving the socioeconomic status of the community.

The nutritional importance of underutilized leafy vegetables:

1. Sessile Joyweed (Sanchi saag) (*Alternanthera sessilis*)

It is a perennial herb, often found near ponds and water buries, prefers places with constant or periodically high humidity. A much-branched prostrate herb, branches often purplish, frequently rooting at the lower nodes; leaves simple, opposite, somewhat fleshy, lanceolate, oblanceolate or linear-oblong, obtuse or subacute, sometimes obscurely denticulate, glabrous, shortly petiolate; flowers small, white, in axillary clusters; fruits compressed obcordate utricles, seeds suborbicular.

Nutritional value: The plant is commonly used as a cooked leafy vegetable in India and as a salad in Sri Lanka. It is high in protein and dietary fibre. It also contains alpha spinasterol, beta sitosterol, beta sitosterol glucoside, campesterol, oleanolic acid, saponins, stigmasterol, and other nutrients.

Medicinal value: It is particularly beneficial in skin illnesses, but it is also useful in acne, asthma, debility, diarrhoea, dizziness, and digestive disorders. It has antiulcer, cholegogue, diuretic, febrifuge, galactagogue, memory enhancer, and antihyperglycemic properties (Hossain *et al.*, 2014 and Walter *et al.*, 2014). Stems and leaves can help with eye problems. To prevent blood vomiting, a decoction is consumed with a little salt. Shoot with additional virility-restoring chemicals (Gupta *et al.*, 2012). Poultice used to treat boils.

2. Amarnathus (*Amaranthus cruentus*)

Amarnathus is an annual or short lived perennial plant. Amarnathus serves as a cheap rich source of protein and dietary fibre (Shukla *et al.*, 2010).

Nutritional value: It is a good source of vitamins, particularly pro-vitamin A. Amarnathus leaves are high in carotene, micronutrients, and phytochemicals such as isothiocyanates and phenolic compounds, which have been shown to assist in the prevention and suppression of illnesses such as cancer, ageing, and arteriosclerosis (Madhvi *et al.*, 2020).

Medicinal value: Due to the inclusion of several phytochemicals, it is beneficial to young children, breastfeeding mothers, and people suffering from constipation, fever, haemorrhage, anaemia, or renal ailments (Mathur *et al.*, 2010 and Choudhury, 2012). It has a diuretic effect. As an infant laxative, the roots are cooked with honey. The ash from the stems is used to wounds, while the heated leaves are applied to tumours.

3. Purslane (Nunia saag) (*Portulaca oleracea*)

The leaves of the plant are succulent, alternate, clustered at stem joints and ends.

Nutritional value: It comprises substantial amounts of Fe, Ca, Zn, Vit-A, riboflavin, folic acid, flavonoids, and other nutrients (Srinivasa *et al.*, 2008). For hundreds of years, it has been used in salads, cooked vegetables, and as a medicinal herb (for humans). Purslane is a tasty food that is popular throughout Europe and Asia. Because of its mucilaginous nature, it is ideal for soups and stews.

Medicinal value: It is claimed to be used as a diuretic, sedative, analgesic, and cardiotonic. It is also used to treat rheumatism, gynaecological disorders, renal failure, and colorectal cancer.

4. Water Cress (Helencha) (*Enhydra fluctuans*)

A trailing marsh herb that floats on water; the stem is 30-60 cm long and roots at the nodes. Sessile, 2.5-7.5 cm long, linearoblong leaves that are acute or obtuse, whole or subcrenate. Heads are axillary and terminal, sessile, and the flowers are white.

Nutritional value: This plant is high in protein and a strong source of -carotene. Saponins, myricyl alcohol, kaurol, cholesterol, sitosterol, glucoside, sesquiterpene lactones such as germacranolide, enhydrin, fluctuanin, and fluctuandin, a number of diterpenoid acids and their isovalerate and angelate derivatives, stigmasterol, cholesterol, sitosterol, glucoside, and other steroids are also present.

Medicinal value: The plant is thought to have antidiabetic, antibacterial, antiinflammatory, antioxidant, anticancer, haepatoprotective, neuroprotective, analgesic, and other medicinal properties. The leaves are laxative and antibilious; they treat inflammation, leucoderma, bronchitis, and biliousness; they are beneficial in skin and neurological disorders; and they are particularly useful in liver torpidity (Sarma *et al.*, 2014).

5) Lettuce (*Lactuca sativa*)

Lettuce occupying the maximum area among salad vegetables. The tender leaves and head are consumed as salad. It is initially cultivated for its edible oil which is extracted from its seeds. It is rich source of vitamins A, calcium and iron. There are 4 types of botanical varieties of cultivated Lettuce 1. Head Type 2. Cos Type 3. Leafy Type 4. Stem Type. It is herbaceous annual and produces seeds in plain.

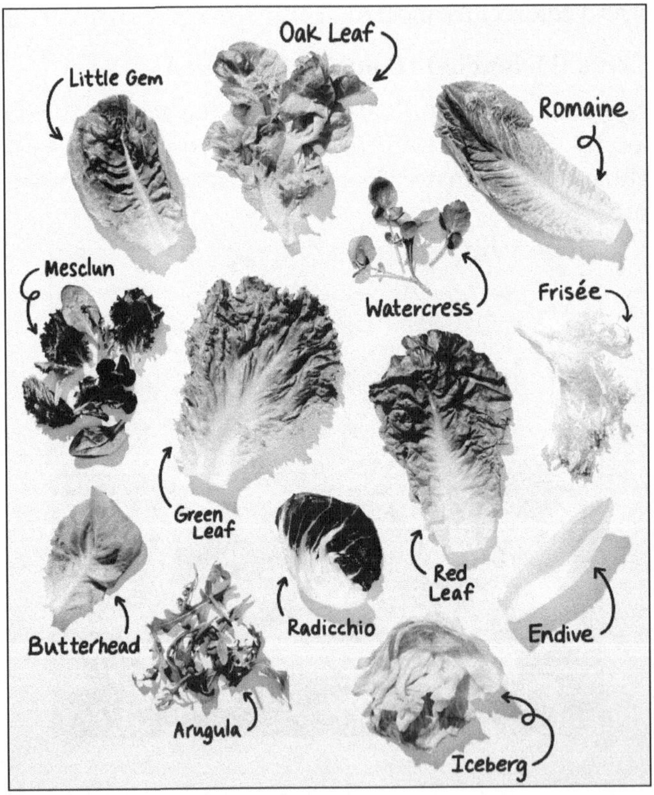

Nutritional Value: It helps to prevent cancer, insomnia, heart diseases, control anxiety and it also has antimicrobial properties. As lettuce is generally eaten raw, more nutrients are preserved than in other cooked or processed vegetables, such as potatoes. Red lettuce is highlighted for its lipophilic antioxidant activity and ascorbic acid and phenolic contents compared to other leafy vegetables (chicory, green lettuce, lamb's lettuce, mizuna, red chard, red lettuce, rocket, spinach, Swiss chard, tatsoi), especially when exposed to low photosynthetically active radiation (PAR) light intensity. Among plant types, the mineral content was higher in butterhead, romaine, and leaf lettuces than in crisphead (iceberg).

6. Cleome

Cleome is known for its nutritional and medicinal value.

Nutritional Value: Increased soil fertility results in increased crude protein, but a decrease in beta-carotene, ascorbic acid and iron content of the leaves. Increased soil fertility does not affect phenolic compounds or the calcium and sodium contents of the leaves. In *C. gynandra* the amount of ascorbic acid lost increases with cooking time, with up to 80 per cent destroyed if cooked for 15min in 8 volumes of water (Van den Heever and Coertze 1996b).

Medicinal Value: It is used to help mothers recuperate after giving birth and during breast-feeding; also for stomach ailments (Chweya and Mnzava 1997, Schippers 2000, Schippers *et al.*, 2002a, Van den Heever and Coertze 1996b).

Other Underutilized and under-exploited vegetables

7. Asparagus (*Asparagus officinalis*)

Asparagus is an herbaceous perennial plant and known as Shatawar in Hindi. It is mainly grown for its tender succulent shoots known as 'Spears' which are used as vegetable and in the preparation of soups. It continues to produce yield for 10-15 years.

Nutritional value: Asparagus juice contains a white crystalline substance known as 'Asparagine' which has diuretic properties. It has great importance in diet because of its valuable salts, vitamins and for its large amount of cellulose contents. Large quantities of asparagus spears are canned and frozen. It contains bioactive metabolites such as fructo-oligosaccharides, polysaccharides, asparosides, shatavarins, sapogenins, racemosols, isoflavones, glycosides, mucilage, and fatty acids, while saponin is one of the main active constituents of asparagus roots.

Medicinal value: Asparagus helps in fertility promotion, stress management, and hormone modulation. It also treats stomach ulcers, kidney disorders, and Alzheimer's disease. Substitution of asparagus powder or extract for value addition of the food products (such as beverages, bakery, and milk) enhances the nutritional and functional properties (Kohli *et al.*, 2022).

8. Elephant Foot Yam (*Amorphophallus paeoniifolius*)

Suran or Zimikand gained its popularity due to its yield potential and nutritional compositions. It is a tropical tuber crop belonging to family Aracea. The tubers are rich source of nutrient and a delicacy as a food.

Elephant foot yam is also known as white spot giant arum, stink lily. It is perennial herbaceous plant. It is used as vegetable and also used in preparation of pickle, chips and others. It is also grown for its medicinal properties.

Nutritional value: It contains betulinic acid, stigma sterol, triacotane and lupeol (Ramalingam *et al.*, 2010). A rich source of trace elements like potassium, magnesium, selenium, zinc, phosphorous and calcium, that help in improving concentration and memory. It is helpful in improving immunity in the body. It is a good anti-inflammatory agent as well as a detoxifier.

Medicinal value: It also contains blood purifier properties and used for treatment of piles, tumours, asthma and rheumatism (Kirtikar and Basu, 1989), vomiting, abdominal pain and blood purifiers. The corms of Elephant foot contain calcium oxalate which imparts acridity and cause irritation in mouth and throat.Tubers are used in medicine preparation (Ayurvedic).

9. Chow-Chow (*Sechium edule*)

Chow Chow a cucurbitaceous vegetable is a perennial vine which is propagated through sprouted fruits (viviparous germination).

Nutritional value: Chow chow, also known as chocho or chayote squash is the richest gourd in nutritive value particularly in Carbohydrates (6.3 per cent) and Calorific value (29 calories/100g). how chow contains essential vitamins and minerals. It is especially rich in vitamin C. It also contains folate (which is great for cell division and for pregnant women), vitamin B6 and vitamin K. Minerals like potassium, manganese, zinc and copper are also found in this squash.

Medicinal value: Chow chow is rich in antioxidants, keeping inflammation at bay and ailments like cancers and joint pains. Quercetin, myricetin, morin and kaempferol are the antioxidants found. Not only do these prevent cell-related damage, but they also prevent the onset of type 2 diabetes by controlling blood sugar levels. This squash is also fibre-rich, keeping your gut healthy. It can prevent fatty liver disease since it is closely linked with liver health.

10. Pointed Gourd (*Trichosanthes dioica* Roxb.)

Pointed gourd is known as 'King of gourd'. It is also known as Parwal, Pravar in Hindi. In India, it is often called Green potato. Its leaves are very nutritive and used leafy as vegetable. It is dioecious climber with perennial root stock.

Nutritional value: It is rich source of carbohydrates, Vitamin A, vitamin A, vitamin B1, vitamin B2, and Vitamin C and minerals. Its juice is used as tonic, alopecia and subacute cases of enlargement of liver (Nadkarnier, 1996). Pointed Gourd is also a great source of calcium which is necessary for the health of the bones. The best part of Pointed Gourd vegetable is that it is very low in calories and hence helps in controlling cholesterol levels in the body.

Medicinal value: Consumption of this gourd helps to purify blood, Good for maintaining a healthy digestive tract as it help cure gastric ulcers. They are also rich in dietary fibre, beneficial for a smooth digestive process. Helps to cure diabetes- by blood sugar regulation. Improves cardiovascular health and their consumption boosts immunity and fights common infections and ageing of body cells.

11. Ivy Gourd (*Coccinia indica*)

Ivy Gourd is nutrient rich fast growing perennial vine plant of Cucurbitaceae family. It is a dioecious, perennial and herbaceous climber or trailing vine with glabrous stems and tuberous roots. It is known for its medicinal and nutritional qualities

Nutritional value: It is rich in vitamin C, which is straightening the bones. It also provided vitamin B1 and B2 and also small amount of vitamin A and tendril of Ivy gourd is also rich in mineral like potassium, calcium and iron. All part of plant are used for treating various disease, leaves extraction of *Coccinia grandis* for antibacterial activities (Bhattacharya 2010). Khatun *et al.*, 2012 evaluated the different nutrient Components of *Coccinia grandis* are carbohydrate-12.62 per cent, total protein- 15 per cent, water soluble protein-11.25 per cent, lipid-4.0 per cent, total phenol-61.92mg/100g, vitamin C-25.55 mg/100g, βcarotene-70.05mg/100g, potassium-3.3 mg/100g, phosphorous- 1.15 mg/100g, sodium-

0.95mg/100g, iron-2.23 mg/100g and calcium-3.79 mg/100 g. Fruits of Coccinia grandis contains steroids, saponin, ellagic acid, terpenoids, lignin, other compound like alkaloids, tannins, flavonoids, glycosides, phenols, B-amyrin acetate, lupeol, taraxerol, β, Carotene, lycopene, cryptoxanthin, xyloglucen, carotenoids and β-sitosterol.

Medicinal value: All plant parts like roots, leaves and fruits are used in local medicinal purposes like jaundice, diabetes, wound healing, ulcers, stomach ache, skin disease, fever, asthma, cough. It is a wonderful vegetable plant species having various pharmacological properties like analgesic, antipyretic, anti-inflammatory, antimicrobial, antiulcer, antidiabetic, antioxidant, hypoglycemic, hepatoprotective, antimalarial, antidyslipidemic, anticancer, antitussive, mutagenic.

12. Coleus (*Solenostemon rotundifolius*)

A minor tuber crop grown for its edible tuber possessing special flavour and taste belonging to Labiatea family. It is popularly known as 'Hausa potato' or 'Chinese potato'.

Nutritional value: Tubers used in curry preparation due to its aromatic flavour and sweetness. Tubers are rich in starch (73 per cent on dry weight basis) but poor in protein content. They are also rich in minerals like calcium and iron and contains certain vitamins including thiamine, riboflavin, niacin and ascorbic acid.

Medicinal value: Coleus contains forskolin, a substance that may help dilate blood vessels and improve the forcefulness with which the heart pumps blood.

13. Tannia (*Xanthomonas saggitifolius*)

Cocoyam belonging to family Araceae is an herbaceous edible species grown in India. It is usually cultivated by small farmers in Asia, Africa and Latin America.

Nutritional value: Starchy cormels occupy a vital place in the diet of many tropical countries. Tubers have nutritional value comparable to that of potato and are digestible than potato. Corms, Cormels and leaves of cocoyam are importance source of carbohydrates, vitamins and minerals for human use and animal feed. Cooked tannia tubers are more nutritious than taro but rather difficult to digest because of its larger starch grains. Cormels can be eaten as boiled, baked, paraboiled and as oil fried to get rid of the toxins and oxalic acid in the tubers.

14. Winged Bean (*Psophocarpus tetragonolobus*)

Winged beans are one of the versatile edible legumes of tropical origin. The plant is a climber in the Fabaceae family and closely related to the pole beans. Its name is ascribing to its fleshy four-angled fruit pod. All the plant parts, *viz.*, seeds, immature pods, leaves, flowers and tubers are edible. Winged bean, a high protein crop, is one of the important underexploited legumes of the tropics.

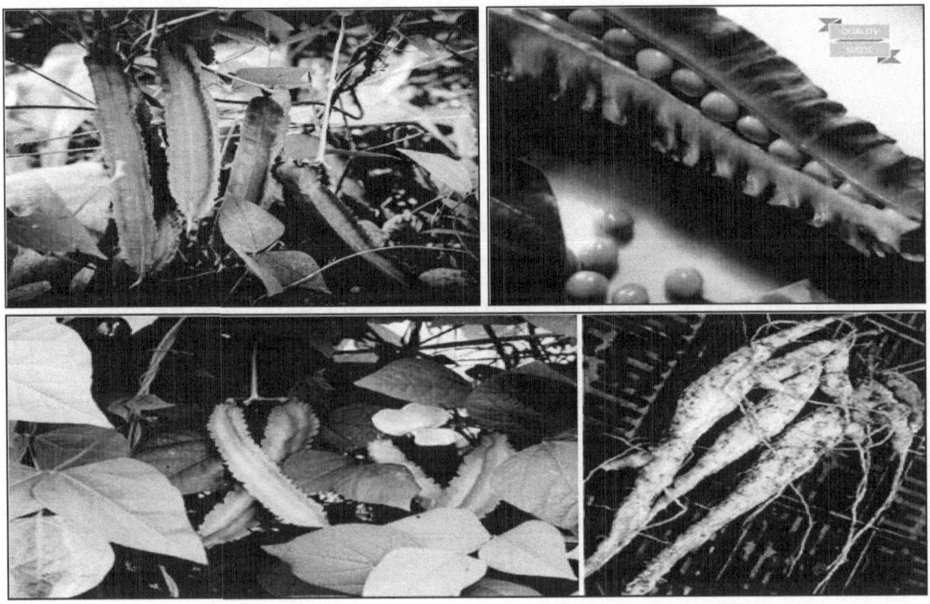

Nutritional value: Mature seeds contain between 29 and 37 per cent protein and 15 and 18 per cent oil. It contains adequate levels of phosphate, iron, and vitamins B and C. Winged bean's essential amino acid content is quite close to that of soybean. Tender, immature pods of winged beans are a low-calorie vegetable; 100 g beans contain just 49 calories. Mature winged bean seeds, on the other hand, provide 409 calories per 100 g and a comparatively high protein level comparable to that of soybean protein. Fresh, young bean pods are an excellent source of folates. Furthermore, winged beans are high in minerals and vitamins. They contain vital minerals such as iron, copper, manganese, calcium, phosphorus, and magnesium. The fatty acid makeup is extremely similar to that of peanut. It has a high concentration of behenic acid and parinaric acid. Heat resistance has been demonstrated for the trypsin inhibitor in winged bean. Other hazardous components have been identified, including hemagglutinins and cyanide. Cooking winged bean seeds is difficult.

CONCLUSION

Underutilized vegetable crops have the ability to alleviate nutritional issues, provide job possibilities, and ultimately contribute to the national economy. As a result, these crops require specific care and must be marketed in order to be used to cure a variety of lifestyle-related diseases. The cost-intensive nature of their production, farmers' reliance on traditional crops, and the present high market price are all probable explanations for their poor output.

Underutilized vegetables with strong nutritious potentials and the capacity to withstand unfavorable climatic conditions may prove beneficial to all concerned - producers, consumers, and environmentalists - if properly controlled. The probable causes for the poor usage of underused vegetables, despite their acknowledged relevance, include a shortage of planting material, a lack of awareness about nutritional and therapeutic value, and a lack of information on crop production techniques. In this environment, there is an urgent need to initiate a programme of genetic resource investigation, management, exploitation, and development of underutilized vegetable crops in order to assure future food and nutritional security. Underutilized veggies are vital to India's national economy.

REFERENCES

Bhattacharya Bolay. (2010) *In vitro* evaluation of antifungal and Antibacterial activities of the plant *Coccinia grandis* (L.) Voigt. (Family- Cucurbitaceae). Journal of Phytology. 2(11): 52-57.

Bioversity International (2017) Mainstreaming agrobiodiversity in sustainable food systems: scientific foundations for an agrobiodiversity index. Biodiversity International, Rome

Choudhury, A. (2012). Evaluation of Physicochemical and Phytochemical Parameters of Amaranthus spinosus Leaves, J. Trop. Med. Plants. Vol. 7 No. 2 (Dec'2006): 210-211.

Dulloo, E., Hunter, D. and Leaman, D. L. (2014) Plant diversity in addressing food, nutrition and medicinal needs. In: Garib Fakim A (ed) Novel plant bioresources: application in food, medicine and cosmetics. Wiley-Blackwell, Chichester, pp 1–21

Fanzo, J. (2019) Biodiversity: an Essential Natural Resource for Improving Diets and Nutrition. In: Yosef S, Pandya-Lorch R (eds) Fan S. Agriculture for improved nutrition: Seizing the Momentum, CAB Int, pp 26–46

Gupta, H. C., Raj, Rathi, J., Sundaram, A. E. N., Kumar, S. and Manchanda, R. K. (2012). Morpho-anatomy of leaf, stem and root of *Alternanthera sessilis* (L.) R. Br. ex DC and *Alternanthera pungens* Kunth (Amaranthaceae) and its significance in drug identification, Ind. J. Res. Homeopathy, 6(4): 1-6

Hunter, D., Borelli, T. and Beltrame, D. M. O. (2019). The potential of neglected and underutilized species for improving diets and nutrition. *Planta* **250**, 709–729.

Jaenicke, H. and Hoeschle-Zeledon, I. (2006). Strategic Framework for Underutilized Plant Species Research and Development, with Special Reference to Asia and the Pacific, and to Sub-Saharan Africa. ICUC, Colombo and GFU, Rome. 33p.

Jena, A. K., Deuri, R., Sharma, P. and Singh, S. P. (2018). Underutilized vegetable crops and their importance. Journal of Pharmacognosy and Phytochemistry. 7(5): 402-407

Khatun, S., Pervin, F., Karim, M. R., Ashraduzzaman, M. and Rosma, A. (2012). Phytochemical screening and antimicrobial activity of *Coccinia cordifolia* L. Plant. *Pakistan Journal of Pharm. Sci.* 25: 757-761.

Kirtikar, K. R. and Basu, B. D. (1989). Indian Medicinal Plants, Published by Lalit Mohan Basu, Allahabad, India, 4, 2609-2610.

Kohli, D., Champawat, P. S. and Mudgal, V. (2022). Asparagus (*Asparagus Racemosus* L.) Roots: Nutritional Profile, Medicinal Profile, Preservation and Value Addition. Journal of the science of food and agriculture. 10.1002/jsfa.12358.

Mabhaudhi, T., Chimonyo, V. G. P., Chibarabada, T. P. and Modi, A. T. (2017). Developing a roadmap for improving neglected and underutilized crops: a case study of south africa. Front. Plant Sci.

Madhvi, Sharma, D., Chauhan, A., Banyal, S. K. and Divya (2020). Underutilized Vegetable Crops: A Lost Treasure. *Int. J. Curr. Microbiol. App. Sci.* 9(04): 2172-2178.

Mathur, J., Khatri, P., Samanta, K. C., Sharma, A. and Mandal, S. (2010). Pharmacognostic and Preliminary Phytochemical Investigations of *Amaranthus Spinosus* (Linn.) Leaves, *International Journal of Pharmacy and Pharmaceutical Sciences,* 2(4): 121-124

Moraza, C., Stöber, S., Ferenczi, Z. and Bokelmann, W. (2018). Remembering forgotten crops developing new value chains. Rural 21: 40–43

Padulosi, S., Thompson, J. and Rudebjer, P. (2013). Fighting poverty, hunger and malnutrition with neglected and underutilized species (NUS): needs, challenges and the way forward. Bioversity International, Rome.

Ramalingam, R., Hima, B. K., Bindu, M. B., Ravinder, N. A. and Banji, D. (2010). Phytochemical and anthelmintic evaluation of corm of *Amorphophallus campanulatus. International Journal of Pharma and Bio Sciences,* 4, 1-9.

Pal R, Abrol G, Singh AK, Punetha S, Sharma P and Pandey AK. (2019) "Nutritional and Medicinal Value of Underutilized Fruits". Acta Scientific Agriculture 3.1: 16-22

Sarma, Upasana, Borah, Vedant V., Saikia, Kandarpa, K. and Hazarika, N. K. (2014). Enhydra Fluctuans: A Review on Its Pharmacological Importance as A Medicinal Plant and Prevalence and Use In North–East India, *International Journal of Pharmacy and Pharmaceutical Sciences,* 6 (2): 48-50.

Shukla, S., Bhargava, A., Chatterjee, A., Srivastava, A. and Singh, S. P. (2010). Genotypic variability in vegetable amaranth (*Amaranthus tricolor* L) for foliage yield and its contributing traits over successive cuttings and years. Euphytica, 151(1), 103-110.

Singh, R. D. and Munda, G. C. (2008). Status of Natural Resources and Suitable Crops and Cropping System for North Eastranran Region. B. ARTICLES, pp: 3-9.

WHO/CBD, World Health Organization and Secretariat of the Convention on Biological Diversity, (2015) Connecting global priorities: biodiversity and human health: a state of knowledge review

Image Sources

https: //www.thekitchn.com/types-of-lettuce-explainer-23312912

https: //foragersyear.wordpress.com/2015/01/30/purslane-portulaca-oleracea-saag-plus-other-weeds-and-hunted-meat/

https: //en.wikipedia.org/wiki/Coccinia_grandis

https: //www.naturehomeopathy.com/11-health-benefits-of-alternanthera-sessilis-joy-weed-ponnanganni.html

Chapter 25

Use of Drip Irrigation in Vegetable Crops

Vibhootee Garg[1], Simran Choudhary[2], Vedant Gautam[3] and Deepak Singh[4]

[1]Ph.D. Research Scholar (Vegetable Science), Department of Horticulture, College of Agriculture, Jawaharlal Nehru Krishi Vishwa Vidyalaya, Jabalpur, (M.P.)

[2]Ph.D. Research Scholar (Vegetable Science), Department of Horticulture, College of Agriculture, Jawaharlal Nehru Krishi Vishwa Vidyalaya, Jabalpur, (M.P.)

[3]Ph.D. Research Scholar, Department of Mycology and Plant Pathology, Institute of Agricultural Sciences, Banaras Hindu University, Varanasi (U.P.)

[4]Ph.D. Research Scholar (Fruit Science), Department of Horticulture, College of Agriculture, Jawaharlal Nehru Krishi Vishwa Vidyalaya, Jabalpur, (M.P.)

**e-mail: vibhootee007garg@gmail.com*

INTRODUCTION

The most effective method for delivering nutrients and water to crops is drip irrigation. Each plant receives the precise amount of water and nutrients it requires at the precise time for optimum growth by being delivered straight to the root zone of the plant. Drip irrigation allows farmers to increase yields while using less water, fertilizer, energy, and even crop protection goods.

Drip irrigation is a regulated irrigation technique in which water is supplied gradually to the root systems of numerous plants. The root zone of your plant receives water directly through a drip irrigation system. In this place, it is permitted to seep into the earth one drop at a time.

It drips water into the soil at very low rates (2-20 liters/hour) using a system of small diameter plastic pipes equipped with outlets known as emitters or drippers. Drip irrigation is sometimes referred to as trickling irrigation. In contrast to surface and sprinkler irrigation, which wets the entire soil profile, water is supplied near to plants so that just the portion of the soil where the roots develop is moistened. When compared to other approaches, drip irrigation uses water applications more frequently (typically every 1-3 days), which creates a very favourable high moisture level in the soil where plants can thrive.

Advantages of Drip Irrigation

- Door-to-door delivery of the resources your plants require, at precisely the right time and in the right amount, is what drip irrigation is likened to.
- Because there isn't any extra water in the soil, weed development is inhibited.
- Drip irrigation helps those who lead extremely busy lives save time and effort.
- Because plants grown with drip irrigation aren't vying for resources, they develop significantly more swiftly and are lot healthier.
- Additionally, these plants are immune to certain ailments like powdery mildew because they don't develop in wet foliage.
- Because trickle irrigation may utilise less water than sprinkler irrigation does, lower volume water sources can be employed.
- Lower operational pressures result in lower pumping energy costs.
- Because plants may get more accurate water volumes, high levels of water-use efficiency can be attained.
- Labor and operating expenses are often lower.
- Extensive automation is conceivable.
- Plant root zones receive direct water applications. Between rows or in other unproductive regions, no treatments are performed, which improves weed control and saves a lot of water.
- Field operations, including as harvesting, can continue while irrigation is running since the spaces between the rows stay dry.
- Fertilizers can be supplied effectively via the drip system.
- Compared to spray irrigation, irrigation can reduce soil erosion and nutrient leaching under a variety of field situations.
- A variety of field conditions can be used for irrigation.

Limitations and Drawbacks of Drip Irrigation

- Management requirements are somewhat higher
- Initial investment expenses per acre may be higher than those of alternative irrigation solutions. Delaying important operating choices could harm crops permanently.
- Frost protection cannot be achieved with drip systems; sprinkler systems must be used instead.

- ☆ Possible sources of leaks include rodent, insect, and human damage to drip lines.
- ☆ To prevent the tiny emitter holes from clogging, water filtering is required.
- ☆ Soil water distribution is more constrained with drip irrigation than with spray irrigation.

Requirement of Drip Irrigation

How does it Operate

- ☆ Drip tape with prepunched emitter holes is used to moisten a continuous strip along the row because vegetables are typically planted in rows. Thin-walled disposable tape (8 to 10 mm thick) is often used for just one season because most vegetables are cultivated for just one season. To enable the system to be disassembled and transferred from season to season, less emphasis is placed on underground mainlines and sub-mainlines. You should create a workable system that enables maximum production at a minimum cost because costs could be expensive. You can either adapt your own components or buy the full system from a drip irrigation provider. The proper system design will assist you in preventing future issues.
- ☆ Water used for irrigation can be obtained from wells, ponds, lakes, rivers, streams, or public water sources. Given its relative cleanliness, groundwater might just need a screen or disc filter to remove debris that can block emitters. Before a drip system is constructed, though, a water quality test should be done to look for precipitates or other impurities. The presence of bacteria, algae, and other aquatic life in surface water from ponds and streams makes more expensive sand filters absolutely necessary. Results of water quality tests are typically provided by municipal water providers, making it simpler to identify any issues. However, this water will cost you a lot of money.

Components of Drip Irrigation

This system has usually six major components:

A. Delivery System

1. **Primary distribution to the field:** Water is transported from its source (pump, filtration system, *etc.*) to the sub-mainline via underground polyvinyl chloride (PVC) pipe or above-ground aluminium pipe (header line). irrigation mainline connected to sub-main with screen filter, pressure regulator, pressure gauge, and water metre

2. **Sub-mainline (heading):** Polyethylene pipe known as vinyl "laid flat" hose is frequently used as the sub-mainline (header line). This hose is strong and long-lasting, and when not in use, it lays flat so that machinery can be driven over it. After each growing season, the lay flat hose, connectors, and feeding tubes are withdrawn and put away until the following year. Due to its relative rigidity, polyethylene pipe is difficult to roll up at the end of the season.
3. **Connectors and couplings:** To join the drip line to the sub-main, plastic connectors and couplings are utilised.

B. Drip Lines/Drip Tapes

☆ Commercial vegetable production uses two primary types of drip lines, with turbulent flow drip tape being the most popular. This polyethylene product has thin walls, collapses in the absence of pressure, and emitters that were moulded into the seam during production. Pressures of 6 to 15 psi are used to operate drip cassettes. An alternative to turbulent flow drip tapes are drip tubes with inbuilt emitters. Despite being more expensive, products with in-line or internally attached emitters frequently have higher water distribution uniformity and clogging resistance. The distance between emitters along the drip line is referred to as emitter spacing. Emitter spacings of 8 to 16 inches are typical for vegetables.

☆ A closer spacing could be necessary on particularly sandy soils to guarantee proper water distribution. The ability of a drip line to sustain a specific emission rate throughout a variety of pressures is referred to as pressure compensation. Over a variety of pressures, a pressure compensating line emits water at the same flow rate. Water is released from a non-pressure compensating line at a rate that rises linearly with pressure. Partially pressure compensating drip lines are those that fall in the middle and are frequently employed.

C. Filters

☆ A drip system cannot function without filters. There are numerous tools and management strategies available for cleaning irrigation water. Depending on the water source, drip irrigation systems may also include settling ponds, self-cleaning suction equipment, sand separators, media filters, screen filters, and disc filters. A drip system must be kept clear of debris at all times because most clogs render a system permanently inoperable. The size of the holes that the water travels through in the filter element distinguishes media, screen, and disc filters from one another. The mesh size of the filter determines the size of the holes. The relationship between mesh size and filter opening size is inverse. A 200-mesh filter, for instance, will be able to catch smaller particles than a 100-mesh filter.

- Media, disc, or screen filters may occasionally be placed in front of sand separators. The most typical filters used in industrial vegetable production are media filters. They have a diameter of 14 to 48 inches and are typically installed in pairs. Despite being costly, bulky, and massive, media filters may efficiently clean low-quality water at high flow rates. In a media filter, the top inch or two of the media (often made of sand or crushed rock) function as a three-dimensional filtering agent, trapping particles. If the water source is generally pure, screen filters are the most popular irrigation filter employed by small businesses and are commonly used in commercial vegetable cultivation.
- Like a media filter, screen filters may effectively remove dirt; however, they cannot remove as much particles before needing to be cleaned. Devices called disc filters combine elements of media and screen filters. A disc filter's screening component is made up of stacks of thin, doughnut-shaped, grooved discs. Water flows from the exterior of the stack of discs into the centre of the cylinder.

D. Pressure Regulators

- The water pressure in the irrigation system manifold (the pipeline supplying the drip lines) is lowered by pressure regulators to the drip lines' operating pressure. For a large variety of flow rates, devices with fixed outlet and adjustable outlet pressure are both offered. Globe valves limit the water flow path in order to control pressure. They are not advised, though, as any modification to operational pressure or system flow rate also influences downstream pressure.
- This can take place if water is sent to another zone or if a system starts to clog. The risk of a faulty pressure regulator is that it could cause the system to over pressurize. Pressures as low as 30 psi can cause drip tape to bend or break.
- Using automatic or manually operated valves to open and close various zones can water multiple fields or sections of fields from a single water supply. To control irrigation zones, one can use either manually operated (gate or ball type) or automatically operated (electric solenoid valves using a timer, water need sensor, or automatic computer controller box) valves. A water metre should be placed to track the system's flow rate and overall water usage. If you use a well or municipal water supply or when adding fertiliser or chemicals to the system, a backflow/anti-siphon valve is also required.

E. Chemical Injectors

- Chemigation is the process of using a drip irrigation system to inject and apply fertilisers, insecticides, and anti-clogging chemicals. Fertilizers are frequently injected; the capacity to "spoon-feed" nutrients is part of what makes drip irrigation boost yields. In order to keep insects under control and safeguard plants from disease, systemic insecticides are commonly administered into drip irrigation systems. Additionally, chemicals that prevent or fix clogging issues can be administered. Acids are used to change the pH of water to dissolve some precipitate blockages, whereas chlorine is used to kill algae.
- Choosing the right chemical injector requires careful consideration of the sort of chemical being injected. Unless fertiliser is continuously injected, maintaining an exact injection rate is not essential for fertilisers. The ability of a fertiliser injector to finish the injection cycle in a fair amount of time is its most crucial feature. For injecting fertiliser into irrigation zones smaller than 10 acres, a 1 gpm injector is likely to be adequate. On the other hand, injecting chemicals accurately at a very low injection rate is necessary to prevent clogging. A separate injector is frequently employed since these compounds are typically constantly injected at concentration rates of 1 to 10 ppm. Injection of pesticides is similar to that of fertiliser, but the amount of material needed is typically less than that of fertiliser. Due of this, the majority of pesticides can employ injectors designed for either clogging prevention (low injection rate/high accuracy) or fertilisers (high injection rate/poor precision).
- They can be used to inject anti-clogging chemicals or fertilisers. Injectors that run on water are propelled by the irrigation system's pressure. Therefore, their main benefit is that they don't need an external power source. There are piston and diaphragm types. Their injection rate is related to the injector's flow rate or the system pressure. Chemicals are added via proportional injectors in accordance with the flow rate. They are especially helpful when a specific chemical concentration is required and chemicals are injected to prevent blockage. The concentration of material injected using proportional injectors is unaffected by changing the system flow rate (for instance, by moving from one zone to another).

Drip Line Arrangement

- The placement of the drip lines within the planting is one factor to take into account when designing a drip system. One drip line can be present every two rows (every other row), one line, or two lines. With spacings of 6, 12, and 24 inches being the most typical, the distance between the drip line and the plant row can also differ.

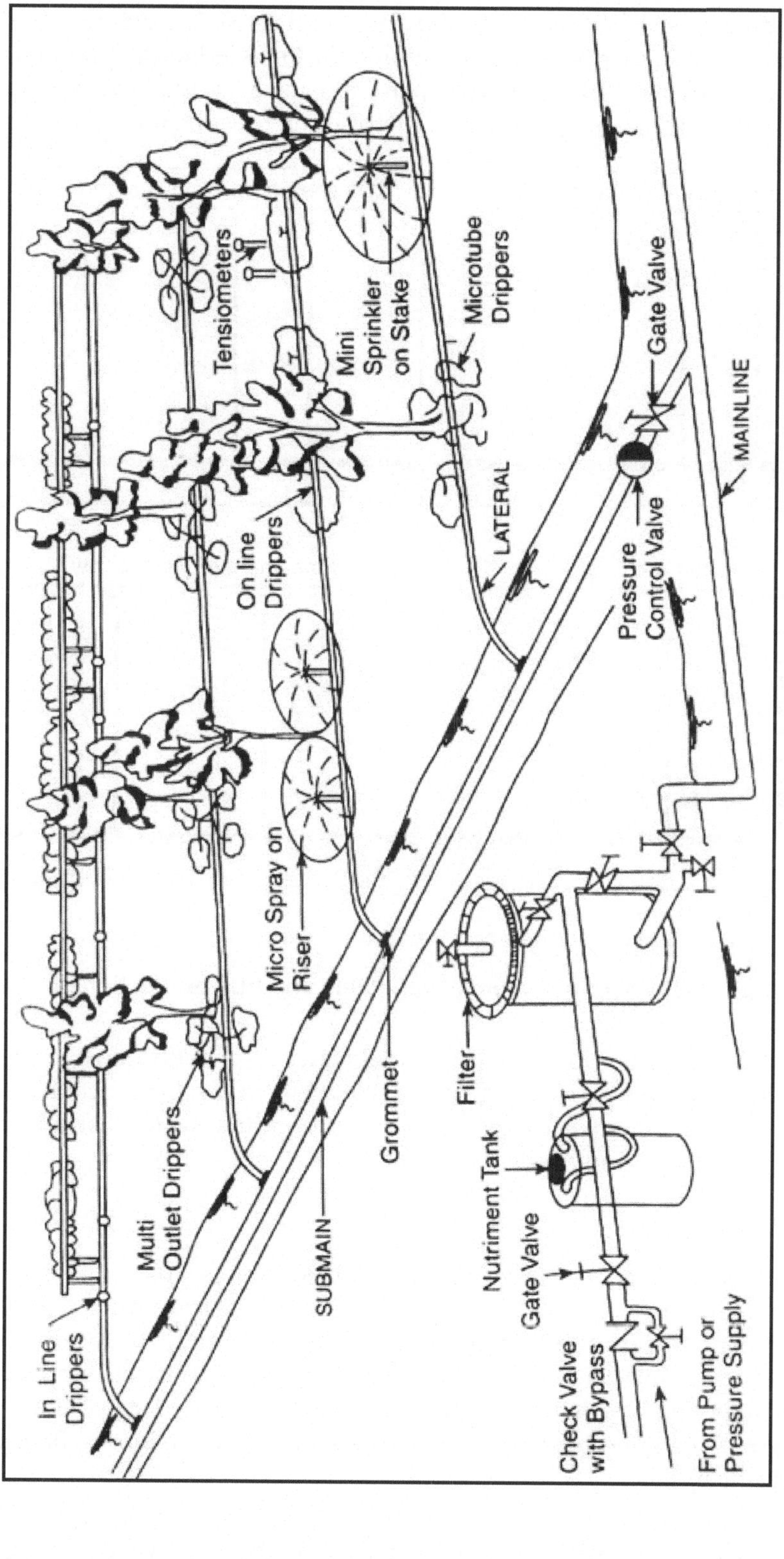

Figure 25.1

[*Source*: Phytochemistry of fruits and vegetables (2019)].

☆ The number of lines utilised depends on the crop's water needs, the soil's characteristics, the economy, and the knowledge and preferences of the growers. Additionally, when choosing a layout, considerations for system cost and ideal water distribution must be made.

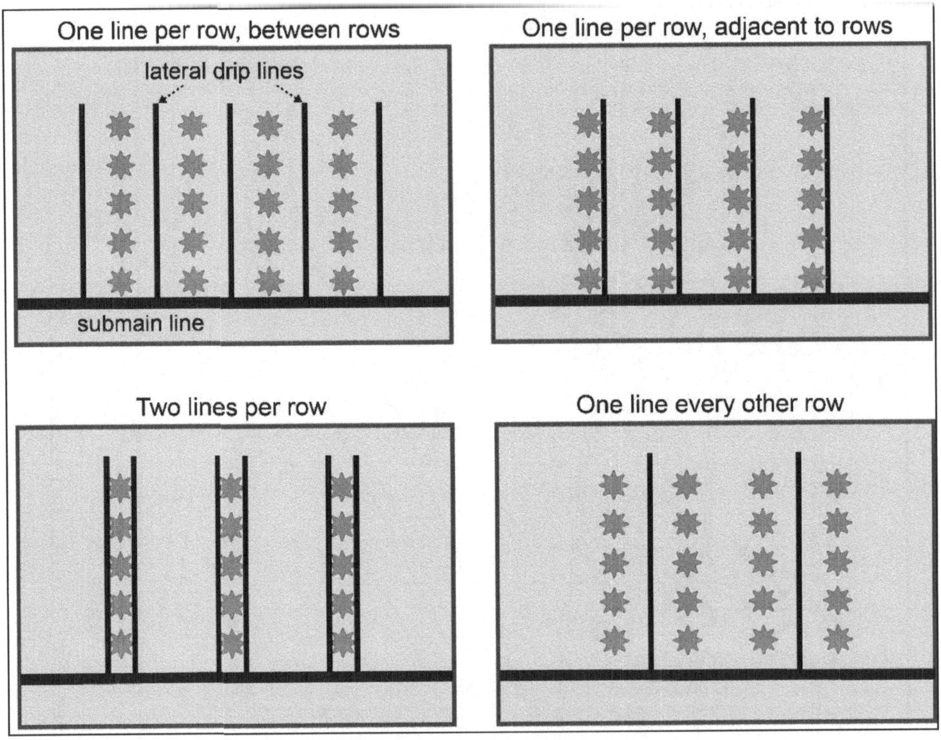

Figure 25.2: Layout for Lateral Drip Lines Relative to Crop Rows.

[*Source*: Drip irrigation for commercial vegetable and fruit production (2017)].

☆ In heavier soils with good horizontal water dispersion, one line every other row can give enough water delivery at a reduced cost while at least one line per row is necessary for optimal water distribution, especially in sandy soils.

System Maintenance

☆ The greatest danger to a drip irrigation system is clogging, which is brought on by chemical, biological, and physical impurities. Physical impurities can be eliminated or removed through filtration, and biological and chemical contaminants must frequently be eliminated or removed by chemical water treatment. Tapes are far less likely to become clogged by mineral deposits when they are buried beneath plastic mulch.

- Chlorine or commercial bacterial control chemicals can be pumped through the fertiliser injection system to remove bacteria, algae, and slime from irrigation lines. If slime starts to be an issue in the system, a 2-ppm chlorine daily rinse at the end of the irrigation cycle or a 30-ppm "shock treatment" might be utilised. For information on dilution rates for commercial cleaning solutions, contact your irrigation system dealer.
- A great maintenance procedure is routine flushing of the mainline, sub-mainline, and drip tape. Each drip tape has an adapter that may be used to flush lines at the end of each irrigation cycle automatically, or the adapter can be manually opened to let a few gallons of water flush from the end. This will stop any accumulation of dirt or slime at the drip line's end.

Use of Drip Method of Irrigation in Vegetable Crops

Brinjal

- Brinjal is grown on 0.73 million hectares in India, or 7.12 per cent of the nation's total area dedicated to vegetable crops.
- Brinjal has a 23 '000 hectare area, produces 430 '000 MT, and has a productivity of 19.12 MT/HA. A major contributing factor to low productivity is the insufficient and inconsistent supply of irrigation water.
- Vegetables need irrigation at regular intervals and during their crucial growth phases since they are fleshy and sensitive to water stress. Water stress may cause crop maturation to be delayed, as well as poor quality and low yield.
- But due to severe water constraint farmers are unable to irrigate crops on a regular schedule.
- Even in extremely dry situations, drip irrigation has the capacity to entirely alleviate water stress for crops.
- The water is provided directly to the root zone using a system of pipes and emitters. This approach is completely different from the traditional approach, which supplies water to the entire cropland.
- Water that is supplied in excess using traditional methods could be lost, which could occasionally have a negative impact on crop yield.
- DMI allows for the direct delivery of water to the crop's root zone, which significantly reduces water losses that would otherwise occur during transportation, distribution, and application in the field.
- Because of variations in the horsepower of the pump sets used to raise the water, groundwater depth, the capacity and quality of water delivery pipes, conveyance distance, field terrain, *etc.*

- It is challenging to determine exact estimations of the water utilised for irrigation under field settings.

Tomato

- Application of water by drip irrigation increased the yields tomato crop.
- Drip irrigation produced tomatoes with the maximum fruit yield of roughly 55 per cent. In terms of vegetables grown with drip irrigation, tomato had the biggest percentage improvement in yields over furrow irrigation (58.70 per cent).
- The appropriate soil moisture regime in the crop root zone and lower nutrient losses are responsible for the higher yields produced by the drip irrigation method.
- Drip irrigation systems that are used frequently lowered the water stress on the plants, improving yields.
- In comparison to the furrow irrigation mode of water application, drip irrigation increased the water productivity of all the vegetable crops. The maximum water productivity was found in tomatoes the drip irrigation system.
- According to the study, tomato growing with drip irrigation produced the highest benefit cost ratio. This suggests that tomato farming may be able to produce higher net benefits per unit of expenditure. The findings imply that tomato was the crop that benefited from drip irrigation the most.
- When using the drip method, the tomato is in the high economic water productivity and high water productivity zone as compare to furrow technique. This is mainly due to the fact that using this method results in higher tomato yields.

Summer Squash

- In an open field, growing vegetables is solely dependent on the thermal-rainfall conditions present during the growing season.
- The very best soil moisture is the significant contributor to high and high-quality yield. Therefore, the irrigation systems should be linked to the cucurbit plant output.
- Fruits' chemical characteristics were affected by drip irrigation. The concentration of both monosaccharides and total sugars, as well as vitamin C, were both markedly elevated by drip irrigation.
- The fruits that were collected from plants grown with drip irrigation also had increased levels of dry matter and carotenoids in them.

Okra

- ☆ Okra is a significant vegetable crop in India. Okra yields of 7 to 12 Mg ha^{-1} are regarded as good, however reports suggest that they have a high potential output of 30 to 40 Mg ha^{-1}.
- ☆ To maximise okra's potential yields in India, the effect of surface drip irrigation on the plant was investigated. Plant growth and water use efficiency was more pronounced when drip irrigation was used. As opposed to surface irrigation, it can be seen that moisture content varies less and is more uniformly distributed along depth with surface drip irrigation.

Chilli

- ☆ India is the foremost nation in terms of producing chillies, making it the most dominant participant in the global chilli industry. India provides one fourth-largest producer of chillies in the world, producing an average of 12.89 lakh tonnes per year on an area of 7.59 lakh ha.
- ☆ Water is a crucial input for growing this crop. In India, it is predicted that by 2025, 33 per cent of the people of India would experience extreme scarcity. Low temperatures and damage from frost during the winter months are the most difficult times to raise high-value vegetables like chillies.
- ☆ The horticultural plant that is most vulnerable is pepper due to the high stomatal conductance, wide range of transpiring leaf surface, and drought stress as well as having a brief root system.
- ☆ So, the chilli crop greater levels of water need a good and accurate amount of water, yield and standard. The exact distribution of water and nutrients to the crop in this direction is greatly facilitated by drip irrigation.
- ☆ The drip irrigation has shown its advantages to other traditional irrigation techniques, particularly in fruit and vegetable crops, due to the exact and direct application of water in the root area. When the crop was irrigated using drip irrigation, the yield increased by 60 per cent for the chilli crop.

Cucumber

- ☆ The family Cucurbitaceae includes a significant and expansive collection of vegetables, including the cucumber.
- ☆ The role of cucumber in food and commerce is immense. The young fruits are used raw in salads.

- The cucumber plant is an annual, coarse prostrate plant with stiff hairs or spines on the leaves and stems. A section of each plant's root zone receives the daily amount of water it needs through drip irrigation, which also maintains a high soil matric potential in the rhizosphere to lessen plant water stress.
- Crop yield is always a crucial economic and effective indicator to take into account while developing a crop.
- With cucumbers, the amount of fruits per plant has a significant role in determining productivity. The different drip irrigation level produced the maximum fruits.

REFERENCES

Hakkim, V. M. A. and Jisha, A. R. C. (2014). Effect of Drip Irrigation Levels on Yield of Salad Cucumber under Naturally Ventilated Polyhouse. *IOSR Journal of Engineering*, 4 (04): 18-21.

Jha B. K., Mali, S. S., Naik, S. K. and Sengupta, T. (2017). Yield, water productivity and economics of vegetable production under drip and furrow irrigation in Eastern Plateau and Hill Region of India. *International Journal of Agricultural Science and Research*. 7(03): 43-50.

Lamont, W. J., Harper, J. K., Jarrett, A. R., Orzolek, M. D., Crassweller, R. M, Demchak, K. and Greaser, G. L. (2001). Agricultural Alternatives: Irrigation for Fruit and Vegetable Production. University Park, Pa: Penn State Extension,

Maughan, T., Allen, N. and Drost, D. (2017). Drip irrigation for commercial vegetable and fruit production. Utah State University Extension.

Pandey, A. K., Singh, A. K., Kumar, A. and Singh, S. K. (2013). Effect of Drip Irrigation, Spacing and Nitrogen Fertigation on Productivity of Chilli (*Capsicum annuum* L.). *Environment and Ecology*, 31 (1): 139–142.

Rolbiecki, R., Rolbiecki, S., Wichrowska, D., Wojdy³a, T., Ptach, W. and Chmura, K. (2014). Effect of drip irrigation of summer squash cultivated on the light soil. Infrastructure and ecology of rural areas (pp. 1467-1476).

Singh, D. K., Rajput, T. B. S. (2007). Response of lateral placement depths of subsurface drip irrigation on okra (*Abelmoschus esculentus*). *International Journal of Plant Production*. 1: 73-84.

Chapter 26

Use of Hydrogel in Horticulture

***Shalini Singh*[1]*, *Vibhu Pandey*[1], *Khursheed Alam*[2] *and Amit Kumar*[1]**

[1]Ph.D. Research Scholar, Department of Fruit Science,

[2]Ph.D. Research Scholar, Department of Vegetable Science,

Sardar Vallabhbhai Patel University of Agriculture and Technology, Meerut – 250 110, Uttar Pradesh

****e-mail: shalinisingh1523@gmail.com***

What is Hydrogel?

The demand for plant growth medium with greater water and nutrient retention capacity is significant (Wu and Liu, 2007. Such hydration techniques are crucial for reducing post-production yield losses in retail outlets in addition to lowering the frequency of irrigation during crop production. Polymer hydrogels, which have the capacity to absorb huge amounts of water, are a frequently found product for guaranteeing appropriate hydration.

Crosslinked polymers known as hydrophilic gels, also referred to as "hydrogels," or super absorbents, have a 400-1500-fold increase in water absorption capacity over their dry weight. Hydrogel is a soil conditioner that can retain plant nutrients and moisture in the soil. Commercial brands of hydrogel include Stockosorb, Raindrop and Agrosorb. When the soil close to the roots of the plants begins to dry out, hydrogel supplies water and nutrients to the plants. High molecular weight, synthetic, water-absorbing monomers are hydrogel polymers. The unique monomer building component, amount of water absorbed per gram of sample, particle size and distribution, reaction to salinity, and price are all ways that polymers differ from one another.

Early hydrophilic polymers were mostly used in non-agricultural applications, most notably in infant diapers, but they have recently been used in a wide range of applications including oil recovery, food processing, water purification and dressing of wound (Peterson, 2002).

General Description of Polymers

Polymers can be produced from a variety of monomers and are typically high-molecular-weight compounds. The materials that are produced can either be hydrophylic or hydrophobic (Tess and Poehlein, 1985). However, only hydrophylic-type polymers are ideal.

The crosslinked acrylamide: sodium polyacrylates, swellable starch, potassium polyacrylates, starch: acrylate copolymers, and acrylonitrile make up the chemical makeup of these hydrophylic polymers. In addition to serving as a physical barrier to the discharge of water from the gel, cross linking in polymers does seem to help with the storage of plant-available water (Johnson and Veltkamp, 1985).

Polymers can absorb distilled water at extraordinarily high rates (up to 1000 times their weight), but due to the salinity of most water sources, hydration rarely goes above 400 to 500 times their weight ($g\ g^{-1}$) under field conditions. The amount of hydration provided by the polymer reduces as the ion concentration in the water rises (Evans and Bowman, 1990). Within a single polymer and between different types of polymers, particle size can range from 5 m to 2 mm. The majority of polymers produced for the horticulture sector are designed to meet the following requirements: improve soil pore size and its number, enhance transplant survival rates, boost germination rates, and lessen or offset the impact of soil compaction on plant growth.

Classes of Hydrophilic Polymers

Hydrophilic polymers can be divided into three categories; however, each class's products can be chemically modified to have a different set of properties (Mikkelsen, 1994).

1. Starch-based polysaccarides, such as those generated from grains like corn and wheat, are naturally occurring polymers. In the food sector, natural polymers are most frequently utilised as thickening agents.
2. Cellulose is mixed chemically with petrochemicals to create semi-synthetic polymers. A polyethylene polymer mixed with sawdust was one of the first hydrogels especially created for horticulture (Erazo, 1987).
3. Petrochemicals are used to create synthetic polymers, and one of the most well-known polymers that are chemically bonded to prevent aqueous dissolution is polyacrylamide (PAM). Crosslinked polyacrylamide hydrogels are most frequently employed in

horticulture, while linear chain polyacrylamides are used for water purification, sealing of canals and erosion prevention (Peterson, D. 2002).

Various Groups of Polymers

1. Vinyl alcohol-acrylic acid co-polymers
2. Starch-polyacrylo-nitrile graft polymers (polyvinyl alcohols)
3. Cross-linked polyacrylamides with acrylamide sodium co-polymers[8]

Mode of Action for Hydrogel Polymer

When the hydrogel is combined with the soil, it hydrates to form an associated amorphous mass that resembles gelatin. This mass aids in long-term absorption and desorption, which slows the flow of water from the soil. The hydrogel particles are also thought of as "miniature water reservoirs" in the soil, from which water is withdrawn under root direction via an osmotic pressure difference (Woodhouse and Johnson, 1991).

As water is supplied to the plant, the hydrogel's volume respectably decreases, which frees up more room in the soil's pores for air and water infiltration, storage and root growth (Azzam, 1980).

As a result, the hydrogel polymer provided an effective property transfer as a basis for the delayed release of water and fertilizer dissolved in the soil. The hydrogel's water conservation offers a cushioned environment that is useful in short-term droughts and reduces losses during the institution phase.

Ability in water consumption and dry matter production are favorable crop reactions to hydrogel. Once polymers are added to soil, they preserve enormous amounts of water and nutrients, retaining up to 100 times their original weight and conserving about 95 per cent of the water that is kept available for plant assimilation and released as needed by the plant (Johnson, 1984).

As a result, plant growth was augmented with limited water supply, but in rainfall regions, a decrease in plant growth was observed.

Uses of Hydrogels in Horticulture

1. In Greenhouse

Many of the early suggestions were created for various greenhouse crops, and it was anticipated that the use of polymers would boost the water retention capacity of soil and hydroponic combinations used in the cultivation of flower and nursery crops in greenhouses (Bearce and McCollum, 1977; Foster and Keever, 1990; Gegring and Lewis, 1980; Letey *et al.*, 1992; Want, 1989). In addition to reducing the overall amount of water required to develop their crops to maturity, growers were looking for techniques that would lengthen the time between waterings of their crops. Sands or media with little organic matter seemed to have the best crop response to the incorporation of polymers in soil and/or soilless media.

2. In Vegetables

The nationwide drought of 1991 substantially advanced the use of polymers in the cultivation of vegetables. The kind of soil and amount of organic matter in the soil affect how much more water can be held by soils when polymers are added. The applications sought to reduce fertilizer applications and water consumption by 30 per cent to 50 per cent.

There has also been some evidence that polymers would stimulate root growth, leading to better plant growth. The crop responses to polymers are observed to be the eradication of plant growth cessation owing to drought stress, enhanced nutrient uptake, and improved crop yields. According to the recommendations based on academic and industrial research, dry polymer treatments made both broadcast and banded in the row result in a considerable decrease in moisture stress and higher yields compared to crops grown without the use of polymers. The price and overall quantity of polymer needed are the two application methods main differences.

3. In Fruit Crops

Findings from many studies revealed that orchards under stress continued to grow actively and had lower tree mortality in freshly formed orchards. Several injection devices have been developed in the western US to the interest in injecting hydrated polymers in orchards. The application equipment has been made to apply dry polymer at four or more separate locations near the drip line of trees in addition to applying polymers in the hydrated condition. According to early assessments from equipment developers, the equipment and the positioning of the polymer in the drip-line area produced positive results. The ideal polymer rate for orchard trees has been calculated to be 8 ounces of polymer in 80 gallon of water. However, factors such as tree size, species, location, soil composition and rainfall volumes and distribution patterns should be taken into account when determining the rate of polymer application at every location.

4. In Turf

Many researchers have found that there was a considerable reaction in the growth and look of grass following polymer applications because turf has shallow root systems in many species (as compared to most other horticultural crops) and continuous traffic that happens on a daily basis (Nus *et al.*, 1991; Washington and Van Hooze, 1991; Vlack, 1990).

Grass appears to be particularly susceptible to high levels of polymer (>3 lb 1000^{-1} ft^{-2} for established turf and >160 lb $acre^{-1}$ broadcast for newly seeded turf) introduced into the soil, out of all the other plant species tested with polymer trials (MacPhail *et al.*, 1980). Incorporating polymer into soccer grounds has boosted water retention, reduced hardness due to dry weather, maintained the health of the ground cover after games and may reduce sports injuries. However, it seems that injecting hydrated polymers into established turf is more effective and successful than applying dry polymers. Injectors have been created by a number of people or businesses to apply hydrated polymers to turf.

Factors Affecting the Efficiency of Hydrogels

The physical, chemical, and environmental characteristics of hydrogels, such as the dissolved salts in the surrounding water solution, all have an impact on how efficiently they can absorb substances.

1. Chemical and Physical Composition

Many hydrogel goods are only referred to as "polymers" or "poly acrylamide," despite the fact that the precise chemical makeup of these items is a trade secret. Most are made of inorganic materials; however, Zeba® is different because it is made of starch. Some hydrogel formulations additionally include bio stimulants, slow-release fertilizer or mycorrhizal fungus spores.

The physical differences in the particle sizes of the numerous hydrogel products have a big impact on how they should be utilized. For root dips, finer-textured materials work best, whilst coarser grades are preferable for incorporating dirt. For specific uses, such as soil integration, root dips, hydroseeding, seed coatings, or even to keep cut Christmas trees moist, Soil Moist® is available in a variety of compositions (JRM Chemical Inc).

2. Dissolved Salts

In comparison to monovalent ions like ammonium nitrogen (NH4+) and potassium (K+), divalent ions like calcium (Ca++) and magnesium (Mg++) are more restraining (Mikkelsen, 1994). In one study, commercially available hydrogels were soaked in distilled water, moderately saline tap water (electrical conductivity of 1.45 mS/cm), and a weak fertilizer solution, among other solutions (Wang and Gregg, 1990). The saturated hydrogels were allowed to drain after soaking in order to calculate their absorptive capacity. The weight of water retained in the distilled water treatment, in which the hydrogels varied greatly, reflects the best absorption. Viterra retained the most water compared to the other hydrogels, with Agrosoke absorbing and holding onto significantly less.

Application Equipments

Recently, a number of businesses began marketing polymer injection machinery across the country. The Aqua-Life Tree Injector was created by Gene Seifert (Condor Industries Inc., Ogden, Utah), and it can inject either dry or hydrated polymers. Before the polymer being injected into the soil through the injector, the unit's air compressor cracks the soil close to the tree stem. For the planting of trees and shrubs, the injectors have been tested in parks, orchards, street tree plantings and on the sides of roadways.

Three separate polymer injectors are marketed by Injection Aeration Systems Inc. (IAS) of Cerritos, California: for grass, for orchards and vineyards, and lastly for home yards and landscaping. Each of these IAS injectors are made to use a Venturi chamber and high pressure water injection (3,000 psi) to inject dry polymer. The dry polymer is drawn into the water stream by a vacuum created by the Venturi chamber, which draws it out of the polymer hopper. Depending on the particular crop application, the injectors can deposit the polymer 4 to 20 inches (10 to 50 cm) beneath the soil surface. The soil will be aerated and the polymer will be injected by both polymer injectors. Even more advantageous to the plant than applying the polymer itself can be soil aeration.

Application of Hydrogel in Agriculture

In dry and semi-arid areas, hydrogel polymers are essential structural components for fostering an environment favorable to plant growth. They can also be used as retention ingredients in a variety of ways, such as the following: [21]

1. Seed coatings or chemicals that help seeds germinate.
2. Roots of seedlings are dipped before establishing.
3. Preventing chemicals that help plants develop.
4. Slow-release coating of insecticides and herbicides for protection.
5. Biocides and herbicides made of polymers.
6. Water-insoluble polymers
7. Soil rehabilitating polymers.
8. Specifically, hydrogel collects water and soluble fertilizer before releasing them at the right time for plants.

Effect of the Hydrogel in Retaining the Water

In arid and semiarid regions with limited access to supplemental irrigation sources and salinity conditions that negatively impact crops' slow growth and productivity, hydrogel polymer has been used to improve the water reservoirs near the root zone and increase the field capacity of various soils. Additionally, hydrogel polymer has been used to increase both the quantity of water for plants and the duration of its availability (Montesano *et al.*, 2015). Additionally, prior research show that hydrogel polymer has strong water-retention, water-uptake, and water-use efficiency properties that improve plant performance and promote development by reducing water stress in plants (Belen-Hinojosa *et al.*, 2014).

The combination of the fertilizer and the polymer appears to be the mechanism through which fertilizer leaching occurs. Polymers are also being studied as potential carriers for protective agents like pesticides and herbicides (El-Hady *et al.*, 1981). When irrigation water is scarce in dry and semi-dry areas, the usage of hydrogels is especially beneficial (Bakass *et al.*, 2002).

Effect of Hydrogel on Soil Properties

In arid and semi-arid areas, soil moisture is seen as a barrier to agricultural development. Since the 1950s, polymers have been recognized as soil conditioners.

Through a number of ways, agricultural hydrogels can alter the various soil qualities, which include:

1. Use the soil's ability to retain water.
2. Improving the permeability of the soil.
3. Enhancing water absorption in various types of soil.
4. Improve the effectiveness of water utilization.
5. Extend irrigation periods to allow for the longer time needed to attain a permanent wilting threshold.
6. Reducing water runoff and soil erosion.
7. Use soil infiltration and penetration.
8. Reduce the propensity for soil compaction.
9. Ensuring better soil drainage.
10. Encourage crop yield in conditions such as low irrigation.

Effect of Hydrogel on Water Saving

Various soil conditioners, including Stockosorb, Raindrop, and Agrosorb, can hold onto water and plant nutrients. When the soil close to the roots of the plants begins to dry out, stockosorb provides water and nutrients to the plants. Planting on well-drained, sandy, gravely soils without adding additional irrigation or fertilizer gradually reduces crop output.

CONCLUSION

The use of hydrogel in dry and semi-arid areas enhances soil characteristics, increases the soil's capacity to hold water, improves soil water retention, boosts irrigation efficiency, accelerates the growth of different crops, and increases agricultural water productivity. Additionally, it creates a favorable environment for greater root growth in well-drained soils, which eventually boosts production. Hydrogels can be employed as absorbent in environment conservation in the agricultural sector as moisture retention, soil conditioners and nutrient carriers, as per their chemical and physical structures.

Since they have been employed in agriculture for more than 40 years, hydrophilic polymers have a wide range of applications, both in the nursery and on the out-planting site. Hydrogels absorb and hold onto water that would otherwise be lost via evaporation or leaching when added to growing media or soil at the nursery or on the out-planting site. It has also been demonstrated that hydrogels can stop cationic nutrients from leaching. Hydrogels, when used as a root dip, can shield roots from desiccation and improve root-to-soil contact following out planting. Trials in nurseries and fields have yielded conflicting results. It is impossible to make generalizations regarding whether to utilize hydrogels or not due to the great variance across products and environmental factors. Growers or planters considering the usage of hydrogels should carry out small-scale tests in their own settings, just like they would with any other product. Simply providing additional attention to maintain roots moist for plants receiving root dips improves out planting performance.

REFERENCES

Abobatta, W. (2018). Impact of hydrogel polymer in agricultural sector. Advances in *Agriculture* and *Environmental Science*. 1(2): 59-64.

Azzam, R. A. (1980). Agricultural polymers. Polyacrylamide preparation, application and prospects in soil conditioning. *Communications in Soil Science and Plant Analysis*. 11(8): 767-834.

Bakass, M., Mokhlisse, A. and Lallemant, M. (2002). Absorption and desorption of liquid water by a superabsorbent polymer: effect of polymer in the drying of the soil and the quality of certain plants. *Journal of Applied Polymer Science*. 83(2): 234-243.

Bearce, B. C. and McCollum, R. W. (1977). A comparison of peat-lite and non-composted hardwood-bark mixes for use in pot and bedding-plant production and the effects of a new hydrogel soil amendment on their performance. Flor. Rev. 10: 21-24.

Belen-Hinojosa, M., Carreira, J. A. and Garcia-Ruiz, R. (2014). Soil moisture pretreatment effects on enzyme activities as indicators of heavy metal-contaminated and reclaimed soils. *Soil Biology and Biochemistry*. 36(10): 559- 1568.

El-Hady, O. A., Tayel, M. Y. and Lofty, A. A. (1981). Super gel as a soil conditioner: its effect on plant growth, enzymes activity, water use efficiency and nutrient uptake. *Acta Hort*. 119: 257-265.

Erazo, F. (1987). Superabsorbent Hydrogels and Their Benefits in Forestry Applications. In: Landis TD, comp. Proceedings of the Intermountain Forest Nursery Association. Oklahoma City, Gen. Tech. Rep. RM-151. Fort Collins, C0: U.S. Department of Agriculture, Forest Service, Rocky Mountain Forest and Range Experiment Station: 14-17.

Evans, I. and Bowman, D. C. (1990). The effectiveness of hydrogels in container plant production is reduced by fertilizer salts. Foliage Dig. 3: 3-5.

Foster, W. J. and Keever, G. J. (1990). Water absorption of hydrophylic polymers (hydrogels) reduced by media amendments. *Journal of Environmental Horticulture*. 8(3): 113-114.

Gegring, J. M. and Lewis, A. J. (1980). Effect of hydrogel on wilting and moisture stress of bedding plants. *Journal of the American Society for Horticultural Science*. 105(4): 511-513.

Johnson, M. S. (1984). Effect of soluble salts on water absorption by gel forming soil conditioners. Journal *of Science and Food Agriculture*. 35(10): 1063-1066.

Johnson, M. S. and Veltkamp, C. J. (1985). Structure and functioning of water-storing agricultural polyacrylamides. *Journal of the Science of Food and Agriculture*. 36: 789-793.

Letey, J., Clark, P. R. and Amrhein, C. (1992). Water-sorbing polymers do not conserve water. *California Agriculture*. 46(3): 9-10.

MacPhail, J. M., Emens, J. L., Harney, P. M. and Tsujita, M. J. (1980). Effect of Viterra hydrogel on germination, seedling growth and sod establishment of Kentucky bluegrass. *Canadian Journal of Plant Science*. 60: 665-668.

Mikkelsen, R. L. (1994). Using hydrophilic polymers to control nutrient release. *Fertilizer Research*. 38: 53-59.

Montesano, F. F., Parente, A., Santamaria, P., Sannino, A. and Serio, F. (2015). Biodegradable Superabsorbent Hydrogel Increases Water Retention Properties of Growing Media and Plant Growth. Agriculture and Agricultural Science Procedia. 4: 451-458

Nus J., Boaz, M. and Wofford, D. Jr. (1991). Water-absorbing polymers show promise in turf. Golf and Sports Turf, 4.

Peterson, D. (2002). Hydrophilic polymers - effects and uses in the landscape. Restoration and Reclamation Review. Vol 7S. 16 p.

Tess, R. W. and Poehlein G. W. (1985). Applied polymer science. ACS Symp. Ser. 285. American Chemical Society.

Vlack, T. R. (1990). Creeping bentgrass responses to water absorbing polymers in simulated golf greens. The Grass Roots- Wisc. GCSA Publ., 3.

Wang, Y. T. and Gregg, L. L. (1990). Hydrophilic polymers – their response to soil amendments and effect on properties of a soilless potting mix. *Journal of the American Society for Horticultural Science* 115(6): 943-948.

Want, (1989). Medium and hydrogel affect production and wilting of tropical ornamental plants. *Hort. Science*. 24(6): 941-944.

Washington, D. C. and Van Hooze, G. (1991). Pain relief: Polymers save water and reduce injuries. Turf South; 4: 36-39.

Woodhouse, J. M. and Johnson, M. S. (1991). The effect of gel-forming polymers on seed germination and establishment. *Journal of Arid Environments*. 20: 375-380.

Wu, L. and M. Liu. (2007). Slow-release potassium silicate fertilizer with the function of superabsorbent and water retention. *Industrial* and *Engineering Chemistry Research*. 46: 6494–6500.

Chapter 27

Utility of Drone Technology in Horticulture

Anju Jayachandran

Ph.D. Research Scholar, Department of Fruit Science, College of Agriculture, Kerala Agricultural University, Thrissur, Kerala

**e-mail: anju26.jayachandran@gmail.com*

INTRODUCTION

Farmers today must deal with a wide range of issues. Global issues include climate change, soil quality, weed and pest incidence, population expansion, urbanisation, and deteriorating environmental conditions. The use of UAVs or drones, in agriculture, can greatly improve risk and damage assessments and transform how we plan for and respond to disasters that have an impact on the livelihood of vulnerable farmers and the nation's food security (FAO, 2018). The Indian government has made several announcements recently stating that it is focusing on boosting drones in the agricultural sector. According to the Indian federal budget for 2022–2023, the government is eager to employ Kisan drones to strengthen the nation's agricultural industry. Drones are frequently being used in farms in order to assist the farmers as a part of precision agriculture. UAVs and other high-tech tools are already being used by modern farmers for agricultural monitoring and forecasting. Drones can gather information on topics, including crop production, livestock health, soil quality, nutrient assessments, weather and rainfall patterns. This data is then used to create a more precise map of any problems that may already exist as well as solutions built on solid evidence.

Figure 27.1: Drone.

Definition and History of Drones

A drone is an unmanned aircraft. It was earlier known by the name unmanned aerial vehicles (UAVs) or unmanned aircraft systems (UASs) and they weigh in the range of 2-20 kg. A drone is essentially a flying robot that may be controlled by a remote or fly on its own using software-controlled flight plans in embedded systems in conjunction with onboard sensors and GPS. The idea of drones was first introduced in 1849 when Austria assaulted Venice with unmanned balloons carrying explosives. The Hewitt-Sperry Automatic Airplane, commonly referred to as the "flying bomb," made its first flight on September 12, 1917, during World War I, introducing the idea of an unmanned aircraft. Numerous unreferenced reports claim that John Clervaux Chaytor used a hot air balloon with movable tethers to spread seeds over a submerged valley bottom on the family farm's "marshlands" in Wairau, New Zealand, in 1906. This is the first recorded use of aerial application in agriculture. Soon after, in 1921, the US Army Signal Corps research station in Ohio and the US Agriculture Department employed a plane for crop dusting (Johnson, 2002).

Basic Components of a Drone

Drones have a power source such as a battery or fuel that enables them to fly. In order to reduce weight and improve manoeuvrability while in flight, drone frames are often composed of lightweight composite materials. Drones need a controller, which an operator may remotely launch, navigate and land. They employ WiFi and radio waves to connect with the drone. The various components of an agricultural drone include

- ☆ Frames
- ☆ Controller systems
- ☆ Propulsion systems
- ☆ Camera systems
- ☆ Navigation systems
- ☆ Batteries
- ☆ Other components (wires, connectors, carry cases, sprayers and sprinklers)

Types of Drones

According to Watts *et al.* (2012), drones are classified into seven categories namely Micro Air Vehicles (MAV), Vertical take-off and landing (VTOL), Low Altitude-Short Endurance (LASE), Low Altitude-Short Endurance Close, Low Altitude low endurance, Medium Altitude and High-Altitude Long Endurance (HALE).

There are also different kinds of UAVs available and be divided into the following groups:

- ☆ Fixed wing
- ☆ Rotary wing
- ☆ Tethered vehicle
- ☆ Lighter-than-air (LTA)

Drone-Based Agricultural Projects

Applications of Drones

Any crop in any place may be monitored with drones. Utilizing drone technology can increase crop yields, reduce time, improve long-term performance, and make land management more sustainable.

1. Soil and Field Analysis

Agricultural drones can be used for field and soil analysis to plan fields more effectively. Drones using remote sensing cameras gather information from the ground using the electromagnetic spectrum to examine the field and soil. Different materials reflect different wavelength ranges, making it possible to differentiate between them. They can used to assess soil fertility, nutrients and moisture content as well as the terrain and soil conditions and soil erosion.

2. Monitoring the Crop

Crop surveillance is the monitoring of a crop's development from the time the seeds are sown until the crop is ready to be harvested. This includes scheduling the fertilizer application, checking for pest infestations and tracking the impact of weather conditions. The only way a farmer can guarantee a timely harvest, particularly when working with seasonal crops, is through crop surveillance. Any mistakes at this stage could lead to crop failure. Crop surveillance aids in planning for next cropping season. Drones can assist in efficient crop surveillance by using infrared cameras to scan the field. Farmers can then act on the information they receive in real-time to improve the health of the plants in field.

Figure 27.2

3. Plant Health Monitoring

The Normalized Difference Vegetation Index (NDVI), a specialised imaging tool used by drones, uses precise colour data to evaluate the health of plants. This enables farmers to monitor their crop as they develop so that any issues can be quickly resolved to save the plants. Crop health is also monitored by drones equipped with "normal" cameras. To monitor crop growth, density, and coloration, many farmers already use satellite images. However, accessing satellite data is expensive and frequently less efficient than closer drone imaging. The fundamental idea of NDVI is based on the fact that leaves reflect a lot of near-infrared (NIR) light due to their spongy tissue. The spongy tissue collapses and the leaves reflect less NIR light but the same quantity of visible light when the plant is stressed or dehydrated. Combining these two signals can therefore aid in differentiating between healthy and diseased plants (Cruzan *et al.*, 2016; Yamamoto *et al.*, 2017; Martinelli *et al.*, 2015).

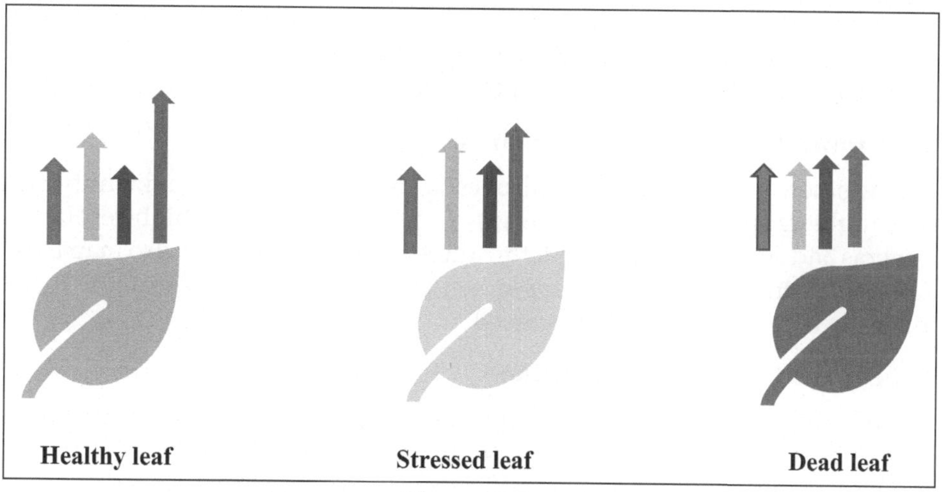

Figure 27.3

4. Irrigation Monitoring and Planning

Drones that have thermal cameras and remote sensing capabilities can aid with irrigation-related issues and can divide regions into different moisture regimes. This facilitates in precise irrigation planning.

Figure 27.4

5. Planting of Seeds

About 15 billion trees are removed each year for mining, urbanization and horizontal agricultural growth (Erin stone, 2017). Drone-seeders can be used to quickly afforest. These drones employ a pneumatic firing mechanism that shoots seed pods deeper into the soil even in hilly terrain and mangrove forests. Up to 40,000 seeds can be planted into the soil by two flying drones each day.

6. Drones for Detecting Deficiencies

Another important merit of drones is its ability to analyse, identity and survey the crops for any deficiencies. Their laser-equipped high-resolution cameras and sensors aid in completing these tasks fast. They can also map these deficiencies in real time, and subsequent crop-related decisions can be made using the data collected and processed.

7. Drones for Spray Application

When compared to conventional methods, agri-drones can spray chemicals on crops quickly since they have reservoirs that can be filled with fertilisers and insecticides. In order to boost efficiency and reduce chemical costs, drone sprayers offer very thin spray applications that can be targeted to specific areas. Therefore, drone technology has the potential to bring in a new era of precision farming.

Figure 27.5

8. Livestock Management

Farmers won large number of livestock. These livestock can be monitored and managed by drones through the high-resolution infrared cameras built into their sensors, which can quickly identify sick animals and take appropriate action.

9. Drones as Security

With the aid of a drone, monitoring the area may be done quickly instead of taking hours of walking. A daily overview of agricultural operations is provided by the drone camera to make sure everything is going according to plan. Additionally, they can be used to find objects, as well as injured or missing animals from grazing field.

10. Predict Weather Glitches

Weather can be a farmer's best friend and worst enemy. Weather conditions cannot be accurately predicted but with the use of drones, detection of upcoming weather patterns can be done quickly. Farmers can utilise this information to better prepare themselves. Planning the crop that would be ideal for the season and how to care for planted crops afterwards can be done with the help of early warning of storms or lack of rain with the help of drone technology (Erdeli *et al.*, 2017).

11. Geofencing

Drones equipped with infrared cameras can quickly identify people or animals. Therefore, drones can protect the fields from external animal damage, especially at night.

Benefits of a Drone

1. **Security:** Trained drone pilots operate the drones, so their misuse can be prevented
2. **High efficiency:** Drones operate instantly and can do tasks twice as quickly as humans.
3. **Saves water:** Drones use ultra-low volume (ULV) spraying technology, which uses less water than conventional spraying techniques.
4. **Less cost and easy maintenance:** Agri drones are reliable, cheap and require less maintenance. The main characteristics include a detachable container, a cost-effective frame, and precise pesticide spraying.

Limitations of Drones

1. **Connectivity problem:** Online availability is frequently poor in rural places. In such cases, a farmer will need to make an investment in internet connectivity, which might become an added expense.
2. **Dependent on weather:** Good weather is crucial for drone operations. Drone operation is not advised while it is raining or windy outside.
3. **Skill and knowledge:** Utilizing new technology is a positive improvement, but doing so requires great skill and adequate knowledge.
4. **Initial cost is high**
5. A traditional farmer is unable to analyse drone images. A farmer will need to acquire the abilities and knowledge of imaging software

to have some fundamental knowledge and abilities to operate an agriculture drone.

6. The threat to wildlife, especially birds, is one of the many serious adverse effects on the environment.
7. One must have permission from the government to utilise it.
8. The growing list of airspace restrictions could result in fines or other consequences due to the constantly evolving laws and regulations governing drone flight.
9. A drone survey's battery life is a drawback and this reduces the drone's flight time.

Regulation and Ethics of Drone

Countries all over the world are investing in the development of drones due to their great potential. They do, however, have some usage rules. The Remotely Piloted Aircraft System (RPAS) regulations were created by the Indian Directorate General of Civil Aviation (DGCA) to regulate aviation in India. The use of RPAS (drone/UAV) for agricultural applications is generally permitted by the DGCA, GOI regulations, with the exception of pesticide spraying until specifically authorised. The general guidelines for using drones are as follows:

1. It is crucial to stay away from densely populated regions that has plenty of crowd
2. Should only fly off in excellent weather and during the daytime
3. In sensitive places, such as government or military sites, within 5 kilometres of airports, or in areas where aircraft are operating, the use of drones or camera drones is prohibited
4. User of a drone must be a licenced drone pilot and be at least 18 years old
5. The drone must be equipped with a licence plate containing the name and contact details of the operator
6. Multiple UAVs cannot be controlled simultaneously by a single human
7. Basic drone regulations must be followed while flying a drone that weighs more than 250 g

Cost of Drone

Agricultural drones typically have fixed wings and may cost up to $25,000, depending on the capabilities and sensors needed to perform the task for which they are intended. Some drones are more expensive because imaging sensors, software, hardware, and other components are all included in the price (Pathak *et al.*, 2020).

Government Initiatives to Support Drone Usage

The Indian government has offered a number of incentives on drone purchases in an effort to lower the cost of drones for farmers and other stakeholders and to encourage drone use. The Indian Agriculture Ministry is offering a 50 per cent or INR 0.5 million maximum subsidy to SC-ST, small and marginal, women, and farmers in the north-eastern states to encourage the usage of Kisan Drones. Financial aid will be provided to other farmers up to 40 per cent, or a maximum of INR 0.4 million.

The Farm Machinery Training and Testing Institutes, Institutions of the Indian Council of Agricultural Research, Krishi Vigyan Kendra (KVK), and State Agricultural Universities (SAUs) are each given a 100 per cent discount on the purchase of the drone for use in farmer fields. Farmers Producers Organizations (FPOs) receive grants at a rate of 75 per cent for the procurement of drones for use in field demonstrations. Existing and new Custom Hiring Centres (CHCs) under Cooperative Society of Farmers, Farmers Producer Organizations (FPOs), and Rural Entrepreneurs offer financial support for drone purchases up to Rs. 0.4 million (40 percent of the total cost of the drone). Additionally, graduates in agriculture who establish CHCs are entitled for financial aid at 50 per cent of the cost of the drone, up to a maximum of Rs. 0.5 million. Therefore, with these incentives in place, agri-training and research organisations can use drones for free. Furthermore, implementing agencies are eligible for hectare-level subsidies if they hire drones for demonstrations, rather than purchasing drones (PIB, GoI).

In fact, as an outreach activity to showcase and promote drones, Bharat Drone Mahotsav (Festival) 2022, India's biggest drone festival was held on May 27 and 28, 2022 in New Delhi. Prime minister of India, Shri Narendra Modi inaugurated the ceremony. The Prime Minister's participation in several drone-related events in a short period of 6-7 months clearly demonstrates the importance that the government holds on the growth of the drone industry in India, particularly Kisan drones ('India can become global drone hub: PM Modi at country's biggest drone festival', Hindustan Times, May 27, 2022)

CONCLUSION

Drones for agriculture are growing like any other kind of farm equipment, and they appear to be a truly unique technology that will revolutionise Indian agriculture. The detection and management of pests and diseases, analysis of soil fertility, monitoring of animal health, and identification and control of weeds all employ the visual imageries produced by integrated cameras. Although there are many agri- and horticultural uses for drones, their adoption is being constrained by regulatory issues. Drones face some significant obstacles to being scaled up in order to find a better role in agriculture, including their costly initial investment, operational and maintenance expenses, shorter flight times, frequent battery replacement, and regulatory obligations. This technique can be widely used and has benefits including site-specific precise information on the spatial variation of various management strategies, which can raise productivity and improve the efficiency of input utilisation at the farm level.

REFERENCES

'What is kisan drone? Five things you should know', Livemint, (Oct 19, 2022). https: //www.livemint.com/news/india/what-is-kisan-drone-five-things-you-should-know11645248791806.html.

Application of Drone Technology in Crop Fertilization. https: //www.equinoxsdrones.com/blog/application-of-drone-technology-in-crop-fertilization#: ~: text=One per cent 20of per cent 20the per cent 20most per cent 20critical,them per cent 20with per cent 20the per cent 20needed per cent 20nutrients.

Application of Drones in Agriculture in India (2022). https: //tropogo.com/blogs/application-of-drones-in-agriculture-in-india.

Bharat Drone Mahotsav 2022: PM Modi to inaugurate India's biggest drone festival, https: //economictimes.indiatimes.com/industry/transportation/airlines-/-aviation/bharat-dronemahotsav-2022-pm-modi-to-inaugurate-indias-biggest-dronefestival/articleshow/91805895.cms.

Cruzan, M. B., Weinstein, B. G., Grasty, M. R., Kohrn, B. F., Hendrickson, E. C., Arrendondo, T. M. and Thompson, P. G. (2016). Small unmanned aerial vehicles (micro-UAVs, drones) in plant ecology. *Applications in Plant Sciences,* 4(9), 1600041.

Drone Technology In Agriculture. https: //www.croptracker.com/blog/drone-technology-in-agriculture.html.

Erdelj, M., Natalizio, E., Chowdhary, K. R. and Akyildiz, I. F. (2017). Help from the sky: Leveraging UAVs for disaster management. *IEEE Pervasive Computing,* 16(1): 24-32.

Erin Stone (2017). Drones spray tree seeds from the sky to fight deforestation. Available: https: //www.nationalgeographic.com/science/article/dronesplant-trees-deforestation environment. Accessed on 20-10-2022.

FAO, (2018). E-agriculture in action – Drones for Agriculture; Available: http: //www.fao.org/documents/card/en/c/I8494EN. Accessed on 20- 10-2022.

Johnson, A. A. (2002). McCook Field 1917–1927. Landfall Press, Chicago.

Martinelli, F., Scalenghe, R., Davino, S., Panno, S., Scuderi, G., Ruisi, P. and Davis, C. E. (2015). Advanced methods of plant disease detection. A review. *Agronomy for Sustainable Development,* 35(1): 1-25.

Pathak, H., Kumar, A. K., Mohapatra, S. D. and Gaikwad, B. B. (2020). Use of Drones in Agriculture: Potentials, Problems and Policy Needs. ICAR-NIASM 12(2), 12-23

Samvaad with Scindia (2022). Minister of Civil Aviation Interacted with farmers on benefits and challenges of Kisan Drones. https: //pib.gov.in/PressReleasePage.aspx?PRID=1826135#: ~: text=Drone per cent 20(Amendment) per cent 20Rules per cent 2C per cent 202022,of per cent 20a per cent 20drone per cent 20pilot per cent 20licence.

Watts, A. C., Ambrosia, V. G. and Hinkley, E. A. (2012). Unmanned aircraft systems in remote sensing and scientific research: Classification and considerations of use. *Remote sensing,* 4(6): 1671-1692.

Yamamoto, K., Togami, T. and Yamaguchi, N. (2017). Super-resolution of plant disease images for the acceleration of image-based phenotyping and vigour diagnosis in agriculture. *Sensors,* 17(11), 2557.

Chapter 28

Vertical Farming in Vegetable: A New Approach of Agriculture

Khursheed Alam[1], Manoj Kumar Singh[2], Shalini Singh[3], Vibhu Pandey[3] and Muzeev Ahmad[4]

[1]Ph.D. Scholar, [2]Associate Professor, Department of Vegetable Science,

[3]Ph.D. Scholar, Department of Fruit Science,

Sardar Vallabhbhai Patel University of Agriculture and Technology, Meerut – 250 110, Uttar Pradesh

[4]Assistant Professor, Department of Agriculture, Integral Institute of Agricultural Sciences and Technology, Lucknow – 226 026, Uttar Pradesh

**e-mail: alamk1996@gmail.com*

INTRODUCTION

Growing vegetables vertically is a cutting-edge trend in the development of safe ground technology. Due to the necessity to conserve energy, nutrient-rich soil, and water under the circumstances of an expanding global population and expanding metropolitan areas, it is relevant. The term "urban farms" refers to the practise of growing plants in bare urban spaces, such as rooftops, tunnels, and enclosed greenhouses. These technologies set themselves apart from conventional farming techniques because they require substantially less space and consume much less energy and water. A vertical vegetable garden is a simple solution to increase growing space, decrease insect and disease concerns, and enhance the appearance of decks and patios.

Vertical gardening in urban and semi-urban areas could be an efficient method to merge food production and consumption locally (Al-Kodmany, 2018).

Extreme weather conditions, biotic and abiotic causes, and diminishing land resources all pose a threat to crop productivity. Land and clean water are finite resources that are being used up every day. Predictions show that by 2050, food production will need to increase by 60 per cent. (Tilman *et al.*, 2002; Green *et al.*, 2005; Alexandratos and Bruinsma, 2012) to provide food for the estimated 9.5 billion people.

Vegetables grow up tall instead of across the ground in a vertical garden. For gardeners with limited area, vertical vegetable gardening is a viable option. Vertical vegetable gardening is quickly becoming one of the most popular ways to make the most of the space available in a greenhouse. Vertical gardening is exactly what it sounds like: growing plants in vertical rows rather than horizontal rows. In other words, the plants are piled one on top of the other in various arrangements. Vertical planting allows greenhouse horticulturists to cover the entire greenhouse with vegetation from the bottom to the top.

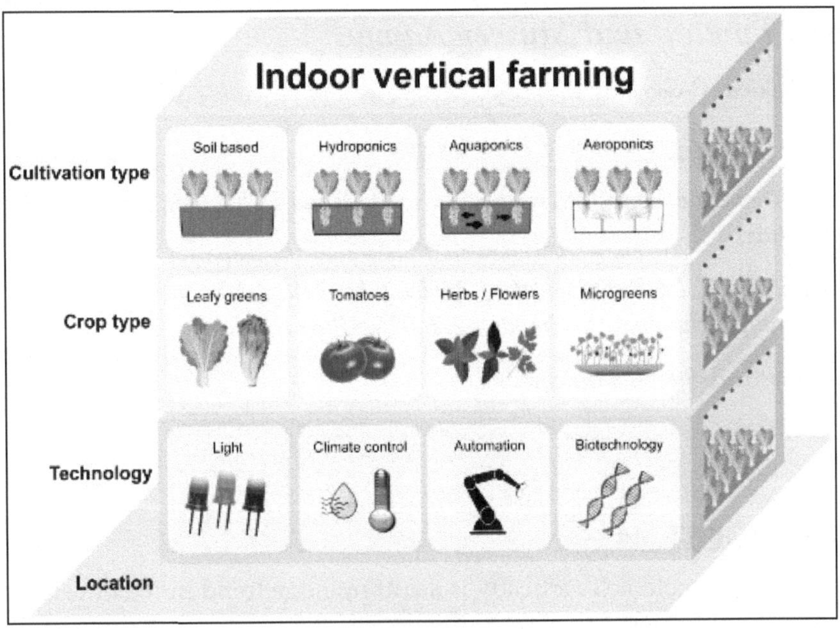

Figure 28.1

In 2015, the volume of the vertical vegetable market amounted to 1.15 billion dollars. By the end of 2016, a 25 per cent growth was observed, and an increase in the market volume to 6.0 billion dollars is expected by 2025 (GIR, 2014-2025). The main market segment (42 per cent) is occupied by the countries of the Asia-Pacific Region (China, India, Japan, Taiwan, Singapore, South Korea), in which the shortage of acreage and water for irrigation is already felt. Then the countries of Europe (30 per cent), namely United Kingdom, Germany, France, Belgium, the Netherlands follow, and then the countries of North America (21 per cent), namely Canada, the USA and Mexico follow. 7 per cent of the market is occupied by other countries (Report, 2019). It should be noted that the main volume of the market are green crops and strawberries. The main vegetable crops of protected ground (tomato, cucumber, sweet pepper) are represented to a small extent (Markets and Markets, 2019). The domestic market of vertical vegetable growing is in the process of formation.

Background History of Vertical Farming

The idea of vertical farming is not brand-new. Examples of it can be seen in Philon's Seven Wonders of the Ancient World, including the Hanging Gardens of Babylon, which were constructed circa 600 BC. Gilbert Ellis Bailey, who also wrote the book "Vertical Farming," first used the term "vertical farming" in 1915. Additionally, he contended that hydroponically farming in a controlled vertical environment would be advantageous for both the economy and the environment. William Frederick Gericke, a professor at the University of California, Berkeley, invented hydroponics in the early 1930s and is regarded as its creator. Ake Olsson, a Swedish ecological farmer, also suggested vertical farming as a way to grow fresh vegetables in urban areas in the 1980s. He is credited with creating a spiral-shaped rail system for plant growth (Corvalan *et al.*, 2005; Healy and Rosenberg, 2013; Thomaier *et al.*, 2015; Despommier, 2010, 2013, 2014). An American environmentalist and professor of public health named Dickson Despommier fervently championed the idea of vertical farming at the turn of the century. As "the mass cultivation of plant and animal life for commercial purposes in skyscrapers," he defined the vertical farm. As seen vines are planted on rooftops in India as a very attractive example of a vertical green wall.

Advantages Vertical Farming

The following are a few of the advantages of vertical gardening:

Increased Crop Production

Crops can be grown indoors all year round. Depending on the crop, all-season farming increases the productivity of the cultivated surface by a factor of 4 to 6. The factor could reach 30 with some crops, like strawberries. Crops wouldn't need to be transported between production and sale, reducing the chance of spoilage, infestation, and energy use compared to traditional farming. They would also be marketed in the same environment in which they are grown.

Protection From Weather-Related Problems

The productivity of vertical farms would be largely independent of weather and protected from extreme weather events because vertical plant farming offers a controlled environment. The nature of geological and meteorological events, such as monsoons, hailstorms, tornadoes, flooding, wildfires, and extreme droughts; or unfavourable temperatures or amounts of precipitation.

Environment Friendly

Significantly shortens the distance travelled, saving money, energy, and the planet's carbon imprint.

Growing Higher Quality Produce

No need for the use of toxic pesticides or herbicides; superior quality produce with increased nutritional content and a longer shelf life.

Conservation of Resources

With less demand for new farmland owing to population growth, vertical farming will conserve numerous natural resources and prevent deforestation and desertification brought on by agricultural encroachment on natural biomes.

Vertical Farming Flexibility

Adjustable from minor to extremely large food enterprises; able to grow over 80 different varieties of leafy greens, microgreens, and strawberries; works on non-arable areas; around major markets or urban centres.

Methods Used in Vertical Farming

The three vertical farming techniques used in India are hydroponics, aeroponics, and aquaponics. Here is a description of these three terms. Let's have a look, then.

Hydroponics

Growing plants without soil is referred to as hydroponics. In hydroponic systems, plant roots are immersed in liquid solutions containing trace elements including iron, chlorine, manganese, boron, zinc, copper, and molybdenum

Figure 28.2

as well as macronutrients like nitrogen, phosphorus, sulphur, potassium, calcium, and magnesium. In addition, soil replacements like gravel, sand, and sawdust are used to nurture the roots because they are inert (chemically inactive) mediums. Hydroponics has the capacity to boost yield per area while consuming less water.

Aeroponics

The National Aeronautical and Space Administration's (NASA's) quest to develop a productive method of growing plants in space served as the inspiration for the development of aeroponics.

In aeroponics, nutrient-rich liquid solution is misted into air-filled chambers where plants are suspended. It is the most environmentally friendly method of soilless growing because it requires no change of the growing medium and consumes up to 90 per cent less water than the most effective traditional hydroponic systems.

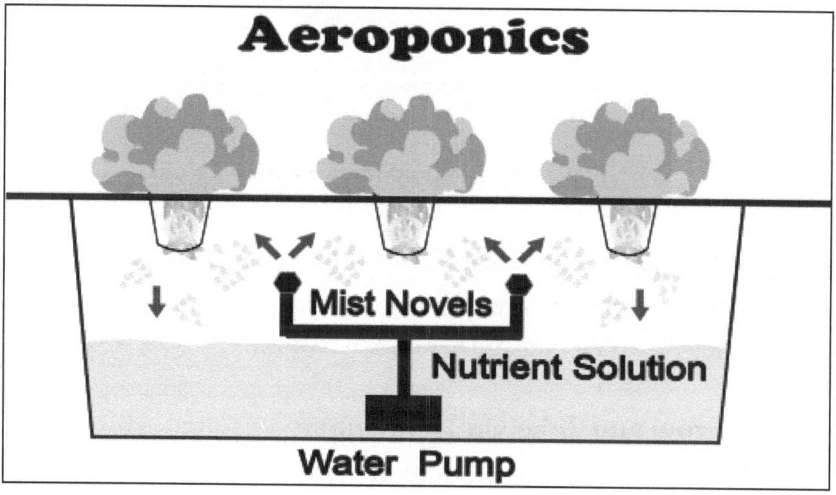

Figure 28.3: Aeroponics System.

Aquaponics

The words aquaculture, which describes the practise of raising fish, and hydroponics, which describes the process of growing plants without soil, were combined to form the term aquaponics. By combining the growth of terrestrial plants and the growth of aquatic organisms in a closed-loop system that resembles nature, aquaponics advances hydroponics.

A solid removal unit filters nutrient-rich effluent from the fish tanks before directing it to a bio-filter, where poisonous ammonia is changed into nutrient-rich nitrate. The plants clean the effluent, which is then recycled back into the fish tanks, while absorbing nutrients.

Additionally, to save energy at night, the water in the fish tanks absorbs heat and aids in maintaining the greenhouse's temperature by consuming the carbon dioxide produced by the fish.

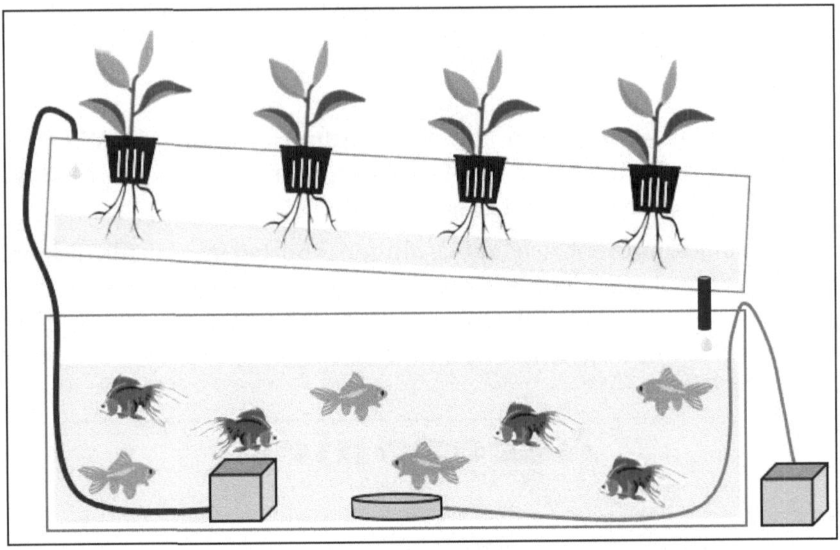

Figure 28.4: Aquaponics System.

Why Vertical Farming is Best for Herbs and Leafy Greens?

Leafy greens and herbs are grown more frequently by vertical farmers than other types of crops for a variety of reasons. Let's take a closer look at a few of them.

Simple to Grow and Take Up Less Space

Leafy greens are simple to produce in vertical systems because they take up little space and are ready for harvest in a few weeks, such as lettuce, kale, swiss chard, and spinach. Farmers typically stack lush greens on top of one another to optimise space.

Furthermore, these vegetables can regenerate their leaves, permitting a second and third harvest if carefully harvested and given continuous watering. This not only lowers supplemental expenses like recurrent seed purchases but also makes it possible for farmers to offer high-quality products all year long.

The Plant Can Be Sold in Large Parts

Traditionally, cash crops like tomatoes and eggplants are planted for their fruits. The plant's residual components, which can be rather large, are either disposed of or recycled after harvest. On the other hand, with leafy greens, vertical growers can sell a significant portion of the plant, preventing much of it from going to waste. This reduces additional expenses (particularly energy) and time-consuming procedures like composting or properly disposing of plants.

Excellent Variety and Consistent Demand

Leafy greens and herbs come in a variety of edible kinds and grow in comparable environments. There are at least four commercial lettuce varieties: romaine, butterhead, iceberg, and curly leaf. Additionally, there are many other varieties of basil, including the well-known Genovese basil and Greek dwarf basil.

All through the year, there is a consistent demand for leafy greens and herbs that are produced vertically. The Covid-19 epidemic caused a large surge in 2020. The epidemic hampered the conventional food supply chain systems, which increased demand for products cultivated close to home.

Given the increasing demand for items produced vertically, it is crucial for farmers to diversify their product offerings and be ready to cater to a wider clientele.

Plants that Could Potentially Grow in Vertical Farms

Numerous vegetable crops can be cultivated in vertical farms, including peppers, aubergines, cantaloupes, strawberries, mushrooms, and pumpkins. For vertical farmers, growing these crops presents a number of technical and financial problems. While some of these problems can be quickly fixed by modifying current vertical systems, others call for significant adjustments.

For instance, when grown in vertical systems, pumpkins and eggplants can develop heavy fruits that can be harmed. By including grow cages or support platforms in their vertical farms, growers can address this issue.

Pollination is a crucial concern as well. Indoor agriculture cannot provide insect pollination, which is necessary for vine crops to produce fruits. Vertical farmers are therefore forced to hand-pollinate these plants or use flower set sprays.

Vertical farming offers a strong and healthy foundation for the cultivation of medicinal plants in addition to the aforementioned crops. It is well known that these plants' secondary metabolism is enhanced by controlled growth circumstances and a consistent supply of nutrients. When grown in hydroponic systems, several medicinal plants, like echinacea and basil, responded particularly well to these conditions by growing quickly and accumulating significant biomass in a short period of time.

Without jeopardising or depleting their native populations, vertical farming promises to give pharmaceutical businesses a year-round supply of high-quality therapeutic plants.

The Balance between Biological and Economic Viability

Comparing biological and economic viability Economic and biological viability are two important aspects that can make or break the success of any vertical farm, as I said at the beginning of this essay.

The ability of plants to endure, develop, and reproduce under particular growing circumstances is referred to as biological viability. Plants grown indoors must adjust to significant shifts in photoperiod, light intensity, temperature, and–most significantly–the absence of soil. In other words, this is simply the technical capacity to raise plants in a specific setting. In our example, the inside setting.

Economic viability, on the other hand, relates to the projected advantages and financial success from carrying out a specific project. Vertical farming's economic viability is severely hampered by the significant up-front and ongoing costs.

We can see that not every crop that can be grown indoors is economically viable by keeping this distinction in mind. As a result, when choosing which crops to grow, farmers should exercise caution and always keep in mind the financial viability of their operations.

For instance, despite ongoing need and the ability to be cultivated indoors, commodities like rice and corn are incompatible with vertical farming. They initially take up a lot of room. In addition, a sizable portion of the plant is inedible, wasting most of the solar energy required to grow it.

Vertical Vegetable Gardening Models

Growing a Vertical Vegetable Garden Almost any vegetable that can be grown in a garden can also be produced in a container. Vegetable plants can be grown in almost any sort of container. As long as they provide proper drainage, old washtubs, wooden crates, gallon-sized coffee cans, and even five-gallon buckets can be used to produce crops.

Shelves

Although most vegetables can be grown in containers, shelves allow you to grow a variety of vegetables on each shelf, as high as you can reach or as much space as you have. You can place the vertical vegetable garden in such a way that all of the plants get enough sunshine at the same time. Although any style of shelving can be used, slatted shelving is the best. This will allow for better air circulation, and surplus water from the top shelves will drip down to the lower shelves between watering intervals. If shelves aren't your thing, tiers of containers can be used to create a vertical effect. Alternatively, vegetables can be planted in trellises or in hanging baskets.

Hanging Baskets

Hanging baskets can be hung from the balcony or on suitable hangers. Many vegetables, especially ones with trailing features, can be grown in hanging baskets. Peppers and cherry tomatoes, as well as trailing plants like the sweet potato vine, not only look great in hanging baskets, but they also flourish in them. Hanging baskets, on the other hand, are more prone to drying out, especially during hot times, so keep them watered on a daily basis.

Vertical Planters

Vertical planters are made-to-order or homemade planting structures designed to hold plants vertically. A plastic 55-gallon drum with holes cut into the side where the plants can be inserted serves as a vertical planter in the DIY version. The majority of vertical planters are made to be utilized with potting soil. These planters are naturally watered from the top, allowing gravity to distribute moisture evenly throughout the container. Strawberries and lettuce are two crops that can be grown successfully in vertical planters.

Trellises

Trellises can be used to keep trailing or vine crops in check. A fence can also be used as a trellis for vine vegetables like squash and cucumbers, as well as beans, peas, and tomatoes. Corn stalks are another good method to make unusual pole supports for beans and other climbing crops while taking advantage of vertical space. To keep vine-growing plants like pumpkins under control, use a stepladder as a makeshift trellis. The rungs of the ladder can be used to train the vines while the vegetables are placed on the ladder's steps for added support. This method works well with tomato plants.

How to Get Started with Vertical Vegetable Gardening

Choose a Location for the Vertical Garden

Choose a sunny site in the vertical garden because most veggies require at least six hours of sunlight. A sunny balcony may be an option if you reside in an apartment complex. If you have a yard, consider planting a sunny spot against your house's south-facing wall.

Select Vertical Structures

Plants require a vertical maintenance system in order to grow vertically. Trellises, tripods, pyramids, walls, fences, wire cages, and walls are all common support structures. While supporting crops, arches or arbores can act as focal points. Vertical Vegetable Gardening can be done on almost any building that points to the sky. Consider using bamboo poles, tree branches, ladders, sunflowers, or corn stalks as unique decorations.

Prepare the Soil

The soil feature is essential for your vertical garden, whether you put veggies in the ground or in containers. Use soil that is weed-free and rich with manure. Mix peat moss or perlite into the soil in pots to promote good drainage.

Choose Containers as Applicable

You may use practically any kind of container to produce veggies in containers. Make sure the container is large enough to hold the type of vegetable you're planting. Vegetables that are larger and heavier will require larger and more durable containers. Hanging baskets, urns, window boxes, washtubs, coffee cans, clay pots, and wooden crates are all excellent options. Drill small holes in the bottom of the container before filling it with soil if it does not have drainage holes.

Understand the Basics of Growing Vegetables in Vertical Vegetable Farming

Any kind of vegetable can be grown vertically as long as the container is large enough to sustain the plant and the vertical stacking is sturdy enough to support it. Find out what your garden's conditions are. Based on the light, wind, heat, and moisture in your garden, choose crops that will thrive. If you're growing vertical vegetables in containers, you'll need to water them on a daily basis.

Prefer Suitable Vegetables for Vertical Vegetable Farming

While most vegetables can be grown vertically, there are a few that are well-known for their vertical gardening success. Consider these vegetables if you're just getting started with vertical gardening.

Peas

Peas are a spring harvest crop. Train them up chicken wire or a fence by spacing them only two inches apart. Replace the vines with beans, squash, or any other vine crop once they've dried up.

Winter Squash

With the exception of the huge and heavy Hubbard and Banana squash, all winter squash can be grown vertically without fruit maintenance. Vertically grow winter squash and gourds. Winter squash and gourds have long vines that can reach ten feet (3 metres) in length for the squash and 25 feet (7.6 metres) in length for the gourds. You must give a solid, stable vertical support since the vines grow so long and become so heavy. Wrap cloth twine or pieces of pantyhose around each fruit and tie to the vertical supporting structure to provide heavy winter squash, such as butternut, added maintain. String should not be used to keep the squash in place since it may cut into the stems. Some garden centres provide materials designed specifically for sustaining huge veggies like winter squash and gourds, such as winter squash and gourds.

Beans

Pole beans are available all year (bush beans, mature all at once, and then stop).

Melons and Pumpkins

In a vertical garden, grow melons and pumpkins. Melons and pumpkins, like winter squash, have long vines that require extra care when grown vertically. Melons and pumpkins should be grown on robust, stable vertical supports. To provide needed support, wrap slings made of old sheets, rags, towels, or fabric scraps around the melons and pumpkins and fasten them to the fence or trellis.

Root Vegetables

Hydroponic systems can be used to effectively grow root vegetables like carrots and beets, but doing so indoors is far more difficult than doing so with leafy greens or herbs. They need a lot of space, nutrient-rich water, or a growing medium to grow normally. They also require up to 16 hours of light each day. Farmers' production costs are greatly increased by these demands, rendering them unviable economically at least right now.

Cucumbers

Cucumbers have a vertical growth pattern. Cucumbers are great for vertical gardens since they climb. They can be trained to climb nearly any structure, such as an A-frame, cage, or trellis, but it must be robust enough to support the vines' weight. Cucumbers may require some aid to adhere to the vertical structure, but once the tendrils coil around the support, they will climb vertically on their own. Cucumbers will hang vertically from the plant, making harvesting them simple.

Potatoes

Plant multiple varieties of potatoes in your pots in the early spring, before it's time to plant other veggies, and you'll have bumper yields. In the loose soil that has been prepared, they will grow quickly.

Tomatoes

In a vertical garden, tomatoes are grown. Tomatoes need a strong vertical structure because they are hefty plants. Tie the branches to the structure using soft thread or cotton cloth at regular intervals as the tomato plants expand. Any vertical structure, including strings, will support indeterminate or vining tomatoes that keep ascending long from the vine's tip. Tomatoes are typically planted in cages or on poles, with the plants growing vertically.

Summer Squash

When grown vertically, summer squash produces fruit throughout the length of the vine. Summer squash should be tied in one-foot lengths.

Disadvantages of Vertical Farming

- ✰ The biggest problem is the initial cost of setting up the vertical farming system. It covers the price of automated racking and stacking systems, climate control systems, and software for remote control.
- ✰ Crop pollination may be a problem with vertical farming systems because there are no insects there.
- ✰ High energy costs since artificial lights are used exclusively for plant growth.
- ✰ The primary municipal water supply could become contaminated by extra nutrients utilised in vertical farming.
- ✰ With vertical farming, a lot of trash, plant waste, *etc.*, can be produced around the buildings.
- ✰ At first, a skilled labour won't be available; they'll need to be trained.

CONCLUSION

The goal of intensive vertical farming is to get the maximum output out of a limited amount of area. A dense garden uses up as little space as possible. Intensive gardening is not simply for those with a small garden space; rather, it emphasizes efforts to produce an optimum plant environment, resulting in higher harvests. A vertical vegetable farming is simple to design and construct. Vertical vegetable farming has a number of advantages, including saving space, making harvesting easier, improving air circulation, keeping veggies off the ground, and producing higher yields.

REFERENCES

Alexandratos, N. and Bruinsma, J. (2012). World agriculture towards 2030/2050: the 2012 revision. ESA working paper no. 12-03. FAO, Rome.

Al-Kodmany, K. (2018). The vertical farm: a review of developments and implications for the vertical city. Buildings 8(2): 24. https: //doi. org/10.3390/buildings8020024.

Corvalan, C., Hales, S. and McMichael, A. J. (2005). Ecosystems and human well-being: health synthesis. World Health Organization, Geneva.

Despommier, D. (2010). The vertical farm: feeding the world in the 21st century. Thomas Dunne Books, New York

Despommier, D. (2013). Farming up the city: the rise of urban vertical farms. *Trends Biotechnol.* 31: 388–389

Despommier, D. (2014). Encyclopedia of food and agricultural ethics (vertical farms in horticulture). Springer, Dordrecht.

Global Industry Report, 2014-2025 2017, Report ID: IVR 1-68038-797-1.

Green, R., Cornell, S., Scharlemann, J. and Balmford, A. (2005). Farming and the fate of wild nature. Science 307(5709): 550–555. https: //doi.org/10.1126/science.1106049.

Healy RG, Rosenberg JS (2013). Land use and the states. Routledge, New York

Markets and markets (2019). http: //www.marketsandmarkets.com/Marker-Reports/verticalfarming-market-221795343.html (Accessed 03 04 2019).

Thomaier, S., Specht, K., Henckel, D., Dierich, A., Siebert, R., Freisinger, U. B. and Sawicka, M. (2015). Farming in and on urban buildings: present practice and specific novelties of zero-acreage farming (ZFarming). *Renew Agric Food Syst*. 30: 43–54.

Tilman, D., Cassman, K., Matson, P., Naylor, R. and Polasky, S. (2002). Agricultural sustainability and intensive production practices. Nature 418(6898): 671–677. https: //doi.org/10.1038/nature01014.

Vertical farming market analysis report by offering (lighting, climate control), by growing mechanism (hydroponics, aquaponics), by structure, by fruits, vegetables and herbs, and segment forecasts, (2019). Available at: http: //www.grandviewresearch.com/industry-analysis/vertical-farming-market (Accessed 03 04 2019).

Index

D

E

F

T

U

V

W

Z

9 789359 191980

Printed at Repro India Ltd.